EXPLORING THE SECRETS OF THE AURORA

ASTROPHYSICS AND SPACE SCIENCE LIBRARY

VOLUME 346

EXPLORING THE SECRETS OF THE AURORA

Second Edition

SYUN-ICHI AKASOFU
International Arctic
Research Center, USA

Syun-Ichi Akasofu
Director
International Arctic Research Center
PO Box 757340
930 Koyukuk Drive
Fairbanks, Alaska 99775
USA
sakasofu@iarc.uaf.edu

Cover illustration: Credit: Jan Curtis

Library of Congress Control Number: 2006932468

ISBN 978-0-387-45094-0 ISBN 978-0-387-45097-1 (eBook)

Printed on acid-free paper.

9 8 7 6 5 4 3 2 1

springer.com

Preface to the Second Edition

In the first edition, I described the evolution of magnetospheric physics from my own personal point of view (I used the word "evolution" rather than "development" because, unlike development, evolution can go right or wrong). However, I was not interested in simply relating a history of magnetospheric physics. I wanted to emphasize the fact that we still have many long-standing, unsolved problems from the early days and to suggest that there is a possibility that the present paradigm may not be headed in the right direction. We have to recognize that several fundamental problems remain; what is needed is not just improvements to traditional theories or a mopping-up of residual problems. I am convinced that new thinking is needed to solve long-standing unsolved problems. What I suggest in this book may be far from the correct way to understand them, but I hope that my suggestions will serve the readers to recognize that many unsolved problems exist and that different ways of thinking are needed to solve them. It is for this very reason that I emphasized in the first edition that a successful scientist must be able to conceive new ideas that do not fit established theories. Obviously, new ideas will encounter confrontations; the scientific community does not want to see the framework it has already built collapsed by new ideas; it is bound to reject the new ideas. In that first edition, I tried to describe how one might overcome such difficulties based on my experience.

In the Epilogue of the first edition, I reached the conclusion that a different way of thinking, paradigm change, namely the conception of a new idea, is an act of synthesis based on a set of facts, including other researchers' viewpoints and theories. In this age of infinite specialization, the importance of synthesis should be emphasized more now than ever. This is because we have to deal with systems, and a study of systems must be based on either integration or synthesis. This is why I emphasize the importance of synthesis in the second edition.

Unfortunately, synthesis is foreign to most scientists, because they are trained to focus on a narrow subject of ever-increasing specialization. Nevertheless, synthesis could produce an important epoch-making advance in science. In 1981, I was honored to be chosen as one of the 1000 most quoted scientists in all fields of science. Looking back on my contributions in science, my work that has endured the longest and has been quoted the most has been synthesis work, rather than my topical work. Thus, I believe that synthesis is an important task in science, because it can often lead to a paradigm change.

My dream of a grand synthesis is to bring solar physics, interplanetary physics, magnetospheric physics, and upper atmospheric physics together in terms of space weather research. Although undoubtedly incomplete, I tried to synthesize the results from the four disciplines in a new Chapter 4 and also in Chapter 8 in the second edition. It is my hope that these provide some idea of the synthesis process.

As mentioned in the Preface in the first edition, this book is intended to be neither a textbook nor the standard monograph on solar-terrestrial physics. In this book, I wanted to emphasize three points:

(1) The first is that readers should appreciate how much effort is required for new findings to become common knowledge. The contents of one chapter in this book may be condensed into one sentence or one page in a standard textbook. On the other hand, this book describes how science actually evolves. This cannot be learned in a textbook.
(2) The second is that readers will learn that a major advance in a scientific field requires a new synthesis of observed facts. The second edition emphasizes this point in the new Chapters 4 and 8.
(3) The third point is that there are many long-standing, unsolved problems in advancing solar-terrestrial physics that are waiting for new ways of thinking by young researchers. When readers find my own views strongly expressed on the present paradigms, they might infer that the problems are not yet solved, although the standard textbooks may imply they are understood as basic knowledge. My views are only intended to initiate new ways of thinking.

I am also hoping that this book would have a narrative feel rather than a textbook "drowned-by-equations" feel, and that it will be read like a history book. Nevertheless, I hoped it would still provide younger researchers an accurate portrayal of the background required to pursue magnetospheric physics and solar physics.

For all these reasons, as those also mentioned in the Preface in the first edition, this book should not be considered as one of the standard textbooks, monographs, or reference books. It is for this reason that references are not given in this book. The year that follows quoted authors refers to the period when their works were written. A list of some of the standard monographs is provided at the end of the book.

I would like to thank many colleagues who are mentioned in this book and others for their advice and discussion throughout my research career. In a way, this book is a joint product with my colleagues, in particular Yasha Feldstein, Ghee Fry, Kazuyuki Hakamada, Tony Lui, Ching Meng, Takao Saito, and Lee Snyder. We have worked together for three decades or more. I would like to also thank Mrs. Kimberly Hayes for her most dedicated work in completing both the first and second editions of this book.

Preface to the First Edition

My purpose in writing this book is to describe my own experiences, from my graduate student days in the 1950s to the present (2001), when I came upon phenomena or facts that did not support the prevailing ideas and theories, or even contradicted them. In some instances, the encounters began with nothing more than the naïve questions I posed as a graduate student to my professors regarding a well-established fact. Others were the result of questions my graduate students asked me. Essentially, this is an account of my personal encounters with some of the ideas and theories that once prevailed but were later eliminated in the history of auroral science.

I believe that young researcher's success as scientists depends on how they deal with new phenomena or facts that do not fit established theories. One cannot be a researcher unless he/she can address such a problem, because such an encounter is the very first step for new progress. Some may put the discordant facts on the shelf or sweep them under the rug, so to speak. Others may try hard to shoehorn new facts into prevailing ideas by modifying or improving them. Yet others may try to establish a new idea, scheme, or theory by adapting their findings, and those of others, by abandoning the prevailing dogma or interpretation of the phenomena or facts. It has been my experience that it is the people in this last group who produce epoch-making progress in science.

The choice of what to do when facing this situation is not easy and depends on many factors. First of all, researchers have to know where they stand at that point in the history of their scientific discipline. It is therefore crucial to have a deep historical knowledge of the background of a prevailing idea or the established interpretation of a phenomenon. To choose a course of action without knowing the background would be like starting to run in the dark without a sense of direction or of the surroundings. Unfortunately, I see too many young scientists doing just that, particularly those who believe that technological advance is everything. Often, a mentor provides the history, not necessarily in a classroom setting, but through daily interactions. I was fortunate to have a very good mentor, Sydney Chapman, who guided me during my early days.

It is also my hope for this book that young researchers will learn that even a simple, one-line statement in a standard textbook, such as "The aurora lies along an oval-shaped belt." endured a decade of struggle before acceptance by the scientific community. My point here is that it is important to learn how to proceed during the period of controversy and struggle, which requires skills not taught in a textbook. However, it is not the intent of this book to provide a general methodology, even if one existed, on how to overcome such problems. I show several examples, right or wrong. The creative approach taken by individual researchers is crucial at this point. In science, we may eventually reach the same or a similar conclusion, but the creative approach taken depends greatly on the individual, as the history of science proves. Science is a human endeavor and is not a dry subject at all.

It is obvious, first of all, that new ideas or theories in science should explain more observational facts than the old ones did. However, that an idea is great (or better) does not guarantee its immediate acceptance by the scientific community. Scientific accuracy is a necessary condition for acceptance, but is not in itself a sufficient condition for it. The readers of this book will see examples, not a methodology, of how such situations were dealt with in the history of auroral science by researchers who made significant advances in understanding auroral phenomena. The most serious problem in a scientific discipline occurs when a given idea or theory dominates utterly. The longer a particular prevailing idea dominates, the more damage it does, retarding progress as researchers, young and old, begin to feel that there is nothing major left to be done.

Looking back at the history of auroral science, one can find that our pioneers had dreams. Our generation also has dreams. Some of the recent advances have made their and our dreams a reality. In order to make this book a little more than just my own ramblings, I have added several highlights concerning those advances in some of the chapters.

Despite the considerable progress in the disciplines of solar-terrestrial physics, a number of long-standing fundamental problems have remained unsolved for many decades. It is my belief that some of these problems remain unsolved because no doubt has been cast on the guiding concept behind the prevailing ideas, not because we presently lack the technology to solve them. In order to stimulate new or different ways of thinking, I have decided to provide some unconventional ideas, although they will certainly be criticized or ignored by those who believe that they are on the right track and that their difficulties are only technical in nature. However, it must be noted that all the materials used here were at least accepted and published in standard scientific journals; many of my unsuccessful geophysical research projects will be described elsewhere.

Space physics must evolve. The future of space physics depends on the creativity of the young generation with a wide range of interests in other fields of science. With a solid background in space physics and at least one other field, the young generation should be able to create a new field of science. I have suggested the exploration for life on planets of distant stars by searching for oxygen emissions in their aurora. That is just an example, and there may be many new unexpected fields of science.

Obviously, this book is not a textbook, or an autobiography, or a treatise of facts and theories for a particular prevailing idea or two. It is a sort of reflection on my research endeavor during the last 40 years or so. Since I have an instinctive tendency to avoid prevailing ideas and theories, I am perhaps not a normal scientist, but I hope nevertheless that this book will be useful, particularly for graduate students and young scientists, especially in helping them think beyond the box of accepted wisdom.

I thank my senior and junior colleagues in many countries, and my former graduate students, who participated in my research activities and helped guide me. Without their close interaction over my research career, I would not have written this book. Those who are not mentioned in the main text are acknowledged in the figure captions. I have also worked with many other close colleagues who are not mentioned in this book, but could not mention them in order to focus on the subject areas specifically dealt with in this book.

Note: At the end of the book, further readings are listed for those who are interested in the history of auroral science, but this is not a reference list. The names of authors with the year in parentheses may look like citations, but instead they indicate the year their papers were published. I have used the full first name of those authors with whom I had at least some acquaintance. For all the rest, I have given only their initials.

Table of Contents

Prologue

The story in this book had a fascinating beginning that can best be described by R.C. Carrington (1860) with his own words:

While engaged in the forenoon of Thursday, September 1, in taking my customary observations of the forms and positions of the solar spots, an appearance was witnessed, which I believe to be exceedingly rare – two patches of intensely bright and white light broke out . . .

Simultaneous with this first sighting of what is now called a white-light solar flare (a most intense type of solar activity), the terrestrial magnetic field record made at the Kew Magnetic Observatory in Greenwich, England, showed a distinct magnetic variation.[1] About 16 hours after this remarkable event, a great geomagnetic storm began and a brilliant auroral display appeared over northern Europe and many other places. Carrington suspected that the geomagnetic storm was related to what he had observed on the Sun, but hesitated to assert the connection. The footnote in Carrington's report to a meeting of the Royal Astronomical Society reads:

While the contemporary occurrence may deserve noting he would not have it supposed that he even leans towards hastily connecting them. "One swallow does not make a summer."

It is in this way that solar-terrestrial physics was born. Lord Kelvin (1892) took up Carrington's extremely modest suggestion of the solar-terrestrial connection during the Anniversary Meeting of the Royal Society of London, England, in 1892.

Kelvin, then the president of the society, attempted to explain the observed geomagnetic variations in terms of the solar magnetic changes observed at a distance of 200 solar radii and found that the expected changes of the dipole moment of the Sun were too large to be reasonable. Thus, he concluded:

[1] This magnetic change is a result of *augmentation* of the ionospheric current by an enhanced conductivity of the Earth's ionosphere (Sqa), which is caused by the flare's radiations.

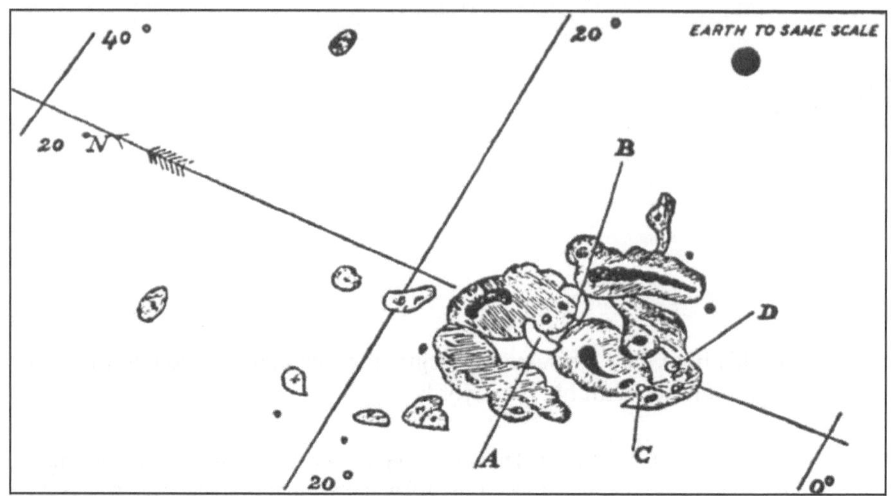

R.C. Carrington's sketch of a sunspot group. He was the first to witness a solar flare (A, B, C, D).
Source: Carrington, R.C., *Mon. Not. Roy. Astronom. Soc.,* **20**, 1860

Lord Kelvin (1824–1907)
Source: Terr. Magn. Geoelect

...Guided by Maxwell's "electro-magnetic theory of light," and the adulatory theory of propagation of magnetic force which it includes, we might hope to perfectly overcome a fifty years' outstanding difficulty in the way of believing the Sun to be the direct cause of magnetic storms in the Earth, though hitherto every effort in this direction has been disappointing. It seems as if we may also be forced to conclude that the supposed connection between magnetic storms and Sunspots is unreal, and that the seeming agreement between the periods has been mere coincidence.

E.W. Maunder (1851–1928)
Source: Courtesy of R.H. Eather

His difficulty is understandable. Without the concept of a medium (which now is known as solar plasma flow) that carries the effects of solar disturbances out into interplanetary space, it is not possible for the Sun to cause the magnetic changes recorded on the Earth.

E.W. Maunder (1905) made a new approach to this problem by noting that geomagnetic disturbances generally reoccur every 27 days, the so-called 27-day recurrence tendency. After an extensive study of magnetic and solar records, he concluded:

First: The origin of our magnetic disturbances lies in the Sun: not any body or bodies affecting both. This is clear from the manner in which those disturbances mark out the solar rotation period...

Second: The areas of the Sun giving rise to our magnetic disturbances are definite and restricted areas... not due to a general action or influence diffuse over the whole solar surface.

Third: The areas of the Sun, wherein the magnetically active areas are situated, rotate with the speed of the chief spot-bearing zones, viz., latitudes 0° to 30°.

Ninth: ...though Sunspots and magnetic disturbances are intimately connected, large Sunspots will often be observed when no disturbances are experienced, whilst sometimes disturbances will be experienced when no spots with which they can be associated are visible...

The first statement was the most definitive in history in suggesting that the sun is responsible for geomagnetic disturbances. The other remarks are also remarkably accurate in spite of the very limited amount of data available to Maunder at that time. The spot-free region he referred to is what we now call a coronal hole. In his concluding remark, Maunder noted:

> That, therefore, which Lord Kelvin spoke of twelve years ago as "the fifty years outstanding difficulty" is now rendered clear...

A. Schuster (1905) immediately criticized Maunder's conclusion by an argument similar to that presented by Kelvin:

> ...I cannot, therefore, agree with his somewhat boastful claim that he has rendered clear what Lord Kelvin has called a "fifty years' outstanding difficulty." He has, no doubt, added a new fact and made an important contribution to the subject. He has given a renewed interest to it and brought out the urgent importance of further investigation, but the mystery is left more mysterious than ever. The facts have become harder to understand and more difficult to explain.

In the history of solar-terrestrial physics, as in any other field of science, such controversies among experimenters, observers, and theorists have been a common occurrence. However, through such controversies, their efforts have been interwoven, resulting eventually in a better understanding of natural phenomena.

After such exciting beginnings, the concept of the Earth's electromagnetic environments has evolved dramatically (Figure I). K. Birkeland viewed the interaction between the solar gas and the Earth's magnetic field in terms of motions of solitary charged particles in a dipole field. He set up an elaborate discharge chamber to study the trajectories of electrons around what he called a *terrella*. Stimulated by Birkeland, C. Störmer began his lifelong study of trajectories of charged particles in a dipole field. His life work was summarized in his book, *The Polar Aurora* (1955). In their studies, both Birkeland and Störmer assumed that the Earth's magnetic field was unaffected by the advancing solar gas.

In order for this particular field of science to make substantial progress, however, we had to wait for Sydney Chapman and Vincenzo Ferraro (1931) to introduce the concept of confinement of the Earth's magnetic field in a cavity carved in the solar gas flow. Chapman and Ferraro considered the solar gas to be consisting of an equal number of positive and negative particles (plasma in present terminology) and attempted to understand the behavior of the plasma flow as it approached a dipole field. They inferred that the solar plasma flow forms a comet-like structure around the Earth, extending in the anti-solar direction and confining the Earth and its magnetic field in it. Chapman and Julius Bartels summarized the development of the field in their classic treatise *Geomagnetism* in 1940.

The discipline of geomagnetism evolved into magnetospheric physics after the International Geophysical Year (IGY), the historic geoscience enterprise in 1957–1958, namely during the beginning of the space age. Tomy Gold (1959) coined the term *magnetosphere* by defining it as "*the region above the ionosphere in which the magnetic field of the Earth has a dominant control over the motions of gas and fast charged particles.*"

The Earth's electromagnetic environment is continuously monitored by recording changes of the Earth's magnetic field. The record shows from time to time characteristic changes $\Delta \boldsymbol{B}$ of the Earth's magnetic field. At a low-latitude observatory, the magnetic variations begin with a step-like increase for a few hours, which is then followed by a decrease of a larger magnitude for a day or so. The upper diagram of Figure II shows magnetic records of the north-south

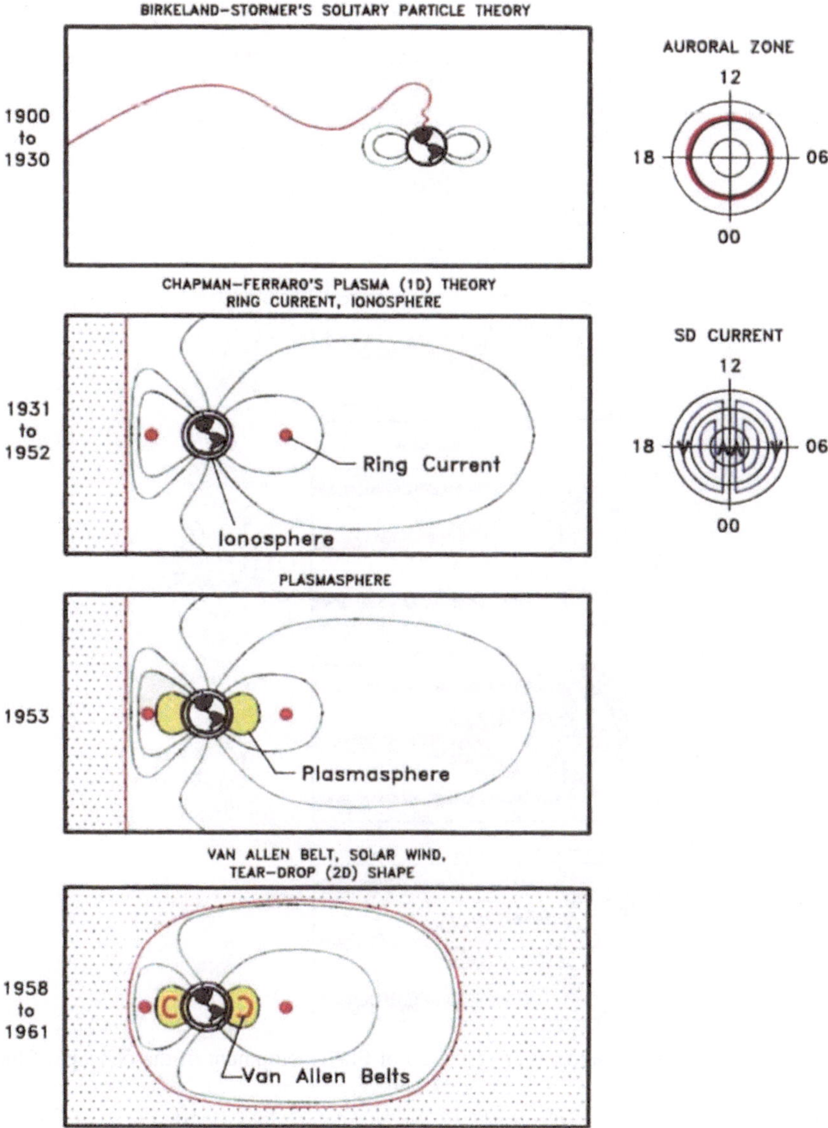

FIGURE I.

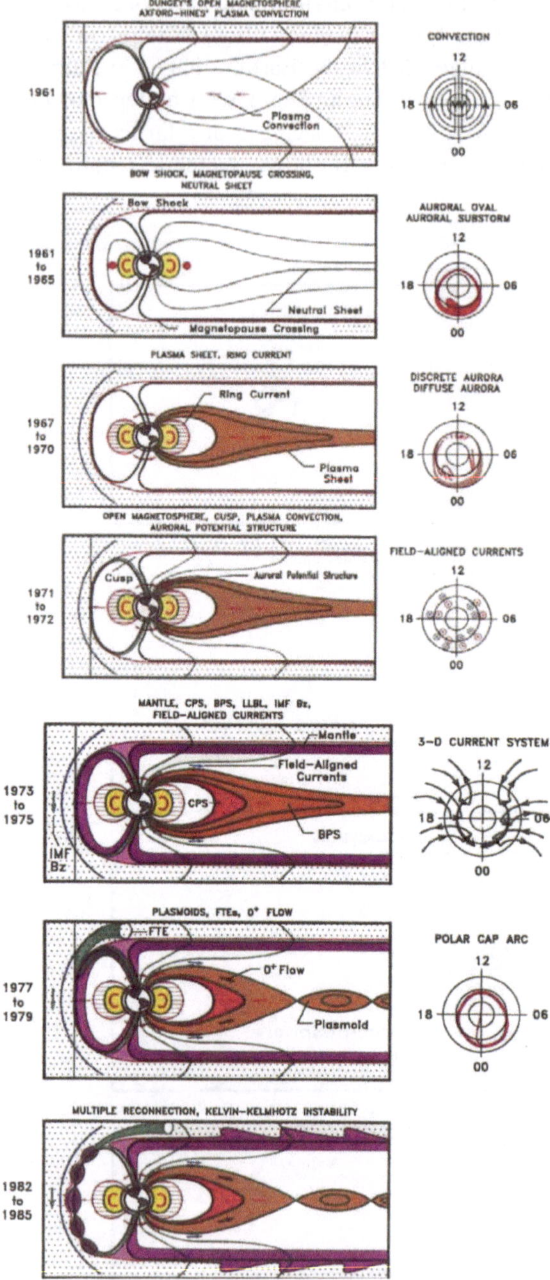

FIGURE I. (*Continued*) Schematic presentation of the development of the concept of the magnetosphere.
Source: Akasofu, S.-I., *Magnetospheric Substorms*, ed. by J.R. Kan, et al., AGU Monograph, 64, p. 3, Washington D.C., 1991

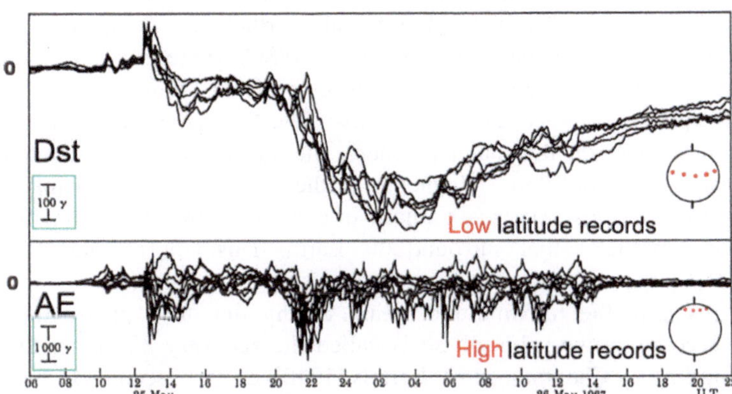

FIGURE II. Upper – Superimposed magnetic records (the north-south component) on May 25–26, 1967, from six low-latitude observatories separated widely in longitude. Lower – Superimposed magnetic records on the same dates from nine high-latitude observatories separated widely in longitude. Note the difference of the scale between the upper and lower diagrams.
Source: Akasofu, S.-I. and S. Chapman, *Solar-Terrestrial Physics*, p. 542, Oxford University Press, Oxford, 1972

component from several low-latitude stations widely separated in longitude; northward changes are recorded as positive changes, while southward changes are recorded as negative changes. The first increase and the subsequent larger decrease are observed at all stations, indicating that those changes occurred on a global scale. This phenomenon is called the *geomagnetic storm*. The development of the study of geomagnetic storms is one of the important subjects of this book. It may be noted that the term magnetic storm was coined by A. Von Humboldt in his treatise *Cosmos* (1871).

The geomagnetic storm field $\Delta \boldsymbol{B}$ is produced by various electric current systems that develop around the Earth when solar disturbances reach the Earth. The field $\Delta \boldsymbol{B}$ is thus superposed on the Earth's main field $\boldsymbol{B}_o$, which does not change in days or months.

During a geomagnetism storm, at high-latitude observatories, fluctuations of a much greater magnitude than those seen in low latitudes, consisting of a number of simultaneous impulsive changes, can be observed. In the lower diagram of Figure II, magnetic records from a number of high-latitude stations are shown; note the difference of the scale for the low- and high-latitude records. Those impulsive changes are magnetic manifestations of what we now know as *magnetospheric substorms*. During a geomagnetic storm, a number of such intense impulsive disturbances occur.

Birkeland classified fluctuations of the Earth's magnetic field in terms of equatorial positive/negative and polar positive/negative changes. As far as I am aware, Chapman was the first who established the present concept of the geomagnetic storm. It consists of the *storm sudden commencement* (SSC), a step-function-like increase in the horizontal (north-south) component and the *main phase*, a larger decrease that follows the SSC. There is often a relatively steady period of a few hours after the SSC, which is followed by the main phase; this period is called the *initial phase*. The SSC is caused by the impact of the shock wave on the magnetosphere; the shock wave is generated by a solar plasma/magnetic cloud advancing in the solar wind after being ejected during solar activities. The main phase is caused by the formation of a belt of energetic particles that surround the Earth. This belt is called the *ring current belt*.

After reaching the maximum decrease during the main phase, the storm tends to recover slowly; this phase is called the *recovery phase*. In the book *Geomagnetism*, by Chapman and J. Bartels (1940), an early account of the development of a study of geomagnetic storms is outlined. Chapman told me that there was great difficulty in publishing it, as world tension was mounting before World War II.

K. Birkeland (1867–1917)
Source: Courtesy of University of Oslo

Sydney Chapman (1888–1970)
Source: Courtesy of Geophysical Institute

After World War II, in the 1950s and 1960s, there were several important developments in a study of the electromagnetic environment between the Sun and the Earth and beyond. First of all, until that time, interplanetary space was thought to be practically a vacuum, except for the streams suggested by Maunder and clouds ejected by solar flare activity. Thus, the Chapman–Ferraro cavity was thought to form only *occasionally*, as the solar plasma engulfed the Earth. Meanwhile, Ludwig Biermann (1951, 1953) suggested a continuous flow of solar plasma. Gene parker (1958) theorized that the Sun blows out plasma continuously with a supersonic speed from the whole solar surface under a certain temperature profile in the corona; he coined the term "*solar wind*." The subsequent detection of the solar wind by the Mariner 2 spacecraft in 1962 brought about a significant change in the concept of the magnetosphere. The magnetosphere is now considered a permanent feature of the Earth, so long as the solar wind blows, rather than forming only occasionally when the Earth is engulfed by intermittent solar plasma flows.

Second, an extensive tail of the magnetosphere was first revealed by the IMP-1 satellite, reported by Norman Ness and his colleagues (1965), as had been suggested by Jack Piddington (1960) . It was found later by space probes on their way to outer planets that the magnetotail extends to a distance of about 1000 Earth radii and perhaps farther.

Third, it was found that the Earth is surrounded by an extensive atmosphere of ionized gases. Based on the study of atmospherics (radio emissions generated by thunderstorm lightning), L.R.O. Storey (1953) found that atmospherics can propagate approximately along the geomagnetic field lines from one hemisphere to the other. The propagation requires an extensive ionized atmosphere to a distance of several Earth radii. This ionized atmosphere has been named the *plasmasphere*. The ionosphere feeds the ionized gases to the plasmasphere.

The Space Age and space research by rockets and satellites were initiated by James Van Allen. In his effort to explore the origin of auroral electrons and cosmic rays, his first attempt was to study auroral electrons near Greenland by *rockoons*, a combination of a rocket and a balloon. It is worth noting that the space age was initiated by the curiosity of scientists like him, who were pursuing the causes of auroral and geomagnetic phenomena. It was his pursuit of auroral electrons by satellites, which led him to the discovery of the Van Allen radiation belts. Subsequently, the ground-based discipline of geomagnetism, together with satellite-based studies, developed into magnetospheric physics. In theoretical space research, Hannes Alfvén stimulated my generation most by introducing many creative concepts, including the concept of the guiding center, magneto-hydrodynamics (MHD), Alfvén waves, dusty plasmas, and many others.

In 1968, Sam Bame and his colleagues at Los Alamos National Laboratories discovered the most extensive region of plasma, called the *plasma sheet*, in the tail region of the magnetosphere. Thus, the magnetosphere has been found to be not an empty cavity, but to consist of several plasma domains. In the 1970s, solar wind-like plasmas were found well inside the boundary of the magnetosphere, and the region occupied by such plasmas is called the *plasma mantle*. The plasma in the plasma mantle flows in the anti-solar direction with a speed of about 100 km/sec, appreciably less than that of the solar wind. Certainly, the plasma in the plasma mantle is of solar wind origin. This finding indicates that the magnetospheric boundary does not exclude completely the solar wind from the magnetosphere, as Chapman and Ferraro originally envisioned in their theory.

Hannes Alfvén (1908–1995)
Source: Courtesy of H. Alfvén

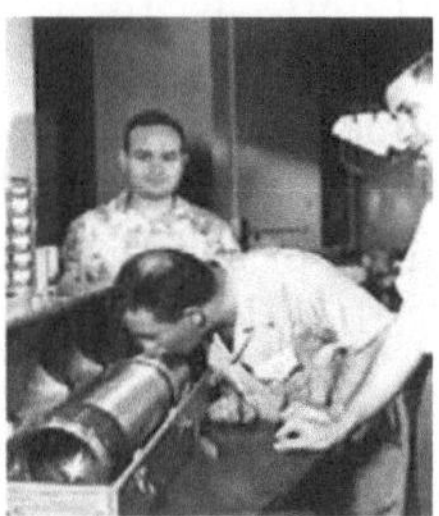

James Van Allen with Carl McIlwain (left) and George Ludwig (right), giving a farewell kiss to their detector to be carried by one of the first U.S. satellites.
Source: Courtesy of J.A. Van Allen

This was also a major change of the concept of the magnetosphere. Another unexpected finding by D.C. Hamilton and his colleagues in 1988 was that oxygen ions (O^+) of ionospheric origin, instead of solar wind protons, become the dominant ions in the ring current belt during intense geomagnetic storms.

Jim Dungey (1961) made the most drastic addition to, or more appropriately the most fundamental revision of, Chapman–Ferraro's original theory. He suggested that the magnetic field lines carried by the solar wind are connected with some of the geomagnetic field lines across the boundary of the magnetosphere. Such a magnetosphere is said to be *open*, while the Chapman–Ferraro model is called a *closed* magnetosphere. The difference between the two theories is that Dungey considered magnetized solar wind plasma, while Chapman and Ferraro considered it a diamagnetic plasma.

Dungey envisaged that the connection process, called *reconnection*, takes place on the dayside magnetopause and that the connected field lines are then transported in the anti-solar direction by the solar wind, resulting in the magnetotail. Subsequently, the field lines are reconnected there and then transported back to the dayside magnetosphere. Dungey's view was that this transport process may occur intermittently and that magnetospheric disturbances, such as magnetospheric substorms, are a manifestation of such a transient process.

In this book, we consider that this interaction between the magnetized solar wind and the magnetosphere constitutes a dynamo that converts a small fraction of the kinetic energy of the solar wind into electrical energy. Magnetospheric disturbances are various manifestations of the power generated by the *solar wind-magnetosphere dynamo*. Chapter 1 describes efforts toward this understanding based on my own experience.

The aurora can then be understood as the only visible manifestation of electrical discharge processes that are powered by the dynamo. Its output power is usually one million megawatts or more. The discharge takes place in an oval-shaped belt, called the *auroral oval*, in the polar upper atmosphere. On the basis of this finding, it could be expected that a magnetized planet with an atmosphere, such as Jupiter, Saturn, Uranus, and Neptune, would have a similar auroral oval, while a non-magnetized planet, such as Venus and Mars, would have no auroral oval. Indeed, the Hubble Space Telescope Project succeeded in imaging the auroral ovals of Jupiter and Saturn (see Figure 2.29), while the Venus and Mars orbiters could not image any indication of the auroral oval.

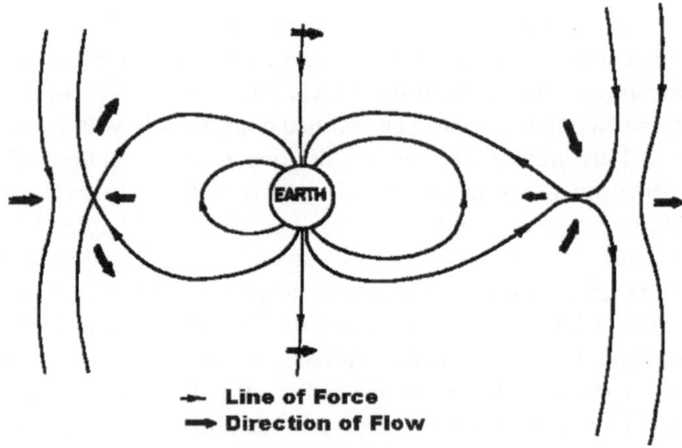

Dungey's open magnetosphere.
Source: Dungey, J.W., *Phys. Rev. Lett.*, **6**, 47, 1961

As mentioned earlier, geomagnetic disturbance fields $\Delta \boldsymbol{B}$ are the magnetic fields produced by the discharge currents generated by an enhanced solar wind-magnetosphere dynamo power. *Thus, auroral activity and geomagnetic disturbances are only different manifestations of an enhanced dynamo power.* Obviously, the two subjects cannot be discussed separately in understanding magnetospheric disturbances. One purpose of studying auroral phenomena and geomagnetic disturbances is, among other things, to infer the configuration of the discharge current system in the magnetosphere and the dynamo process that feeds the current. Chapters 2 and 3 describe our efforts in this endeavor during the early days of the space age.

A typical geomagnetic disturbance field $\Delta \boldsymbol{B}$ undergoes a specific sequence of changes, as shown in Figure II. We now understand that a geomagnetic storm is the magnetic manifestation of what we call a *magnetospheric storm* that results from a large increase of the dynamo's power. Similarly, an auroral storm is its visible manifestation.

It also has been found that the magnetosphere has a specific response to an increased power of the solar wind-magnetosphere dynamo for a few hours. The results of these responses are called *magnetospheric substorms*, their manifestations being the *polar magnetic* and *auroral substorms*. It may be noted here that MHD-based magnetospheric physicists are interested in interpreting magnetospheric disturbances in terms of magnetic flux transfer, not in terms of the dynamo. A magnetospheric storm results from a frequent occurrence of intense magnetospheric substorms. That is to say, *the substorms are the basic elements of a storm*. This is because substorms are responsible for feeding oxygen ions (O^+) from the ionosphere into the ring current belt. During recent years, considerable progress has been made in studying the distribution of the injected oxygen ions, by using an imaging method, called the energetic neutral atom (ENA) imaging. In Chapter 4, we synthesize selected important facts in an attempt to understand processes associated with magnetospheric substorms.

Planetary magnetism is an important subject for all geophysicists, solar physicists, and astrophysicists. It has been a great surprise that the dipole fields of both Uranus and Neptune appear to be inclined considerably with respect to their rotation axis and are greatly off-centered. So long as the generation of planetary magnetism relies on the planetary rotation, it is difficult to explain why the magnetic axis is inclined greatly from the rotation axis. Chapter 5 provides a nontraditional interpretation of planetary magnetic fields. In this attempt, it is assumed that the photosphere of the Sun corresponds to the core surface of the magnetized planets and a spherical surface of 3.5 solar radii, called the *source surface*, corresponds to the planetary surface, where the field is more or less dipolar. Thus, a study of the relationship between the magnetic fields of the photosphere and the source surface might provide a new way of interpreting the observed planetary magnetic field.

The solar wind stretches the dipolar field lines on the source surface all the way to the outer boundary of the heliosphere, where the solar wind interacts with interstellar gas. As the Sun rotates with a period of about 25 days, the stretched field develops a spiral structure. The heliospheric current sheet is formed as the extension of the magnetic equator of the Sun. The current sheet divides the stretched dipolar field lines (the interplanetary magnetic field lines) into two regimes, away and toward the Sun, in terms of the orientation. The axis of the dipolar field on the source surface is inclined with respect to the rotation axis. As a result, the heliospheric current sheet develops a wavy structure as the Sun rotates.

As the solar wind and its magnetic field are continuously changing, the power of the solar wind-magnetosphere dynamo varies as a result. In particular, after a few days of intense solar activities (including solar flares), coronal mass ejections, an intensified solar wind, together with its shock wave, reaches the

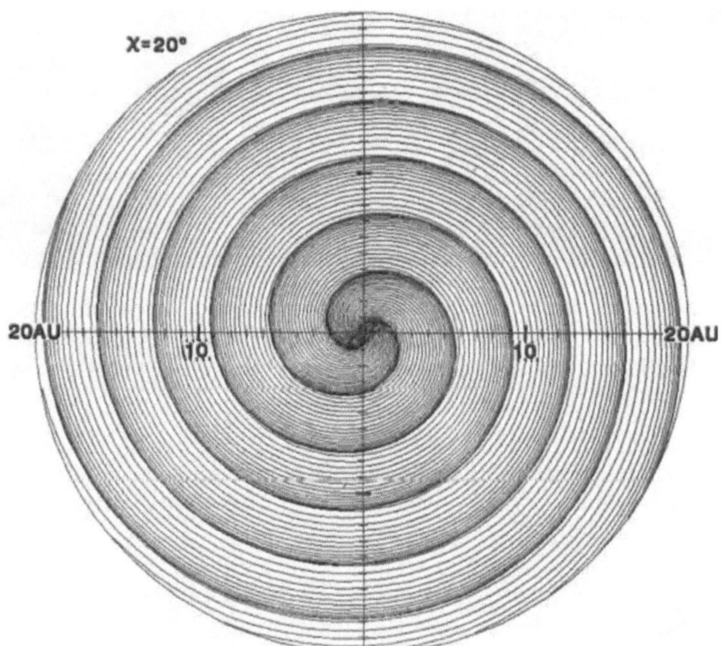

The spiral structure of the interplanetary magnetic field lines within a distance of 20 AU; the Sun is located at the center.
Source: Hakamada, K. and S.-I. Akasofu, *Space Sci. Rev.*, **31**, 3, 1982

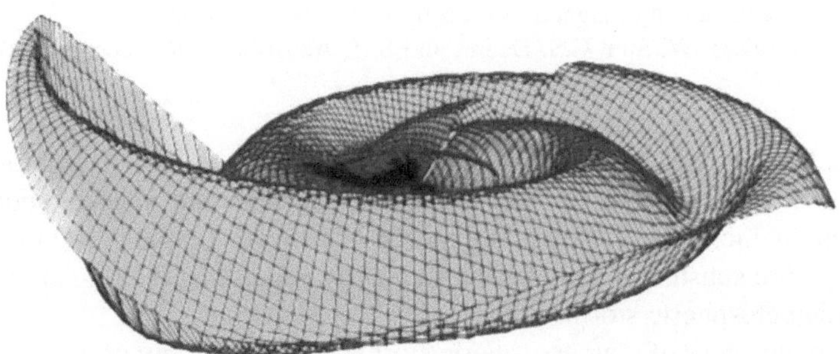

Heliospheric current sheet that extends from the Sun for Carrington Rotation 1654. The Earth's orbit is also shown.
Source: Akasofu, S.-I. and C.D. Fry, *J. Geophys. Res.*, **91**, 13, 679, 1986

magnetosphere. The shock wave compresses the magnetosphere. As a result, the Alfvén waves are generated at the front of the magnetosphere and propagate into the magnetosphere, causing the SSC. The intensified solar wind in the form of plasmoids or magnetic flux ropes are thought to generate the shock wave

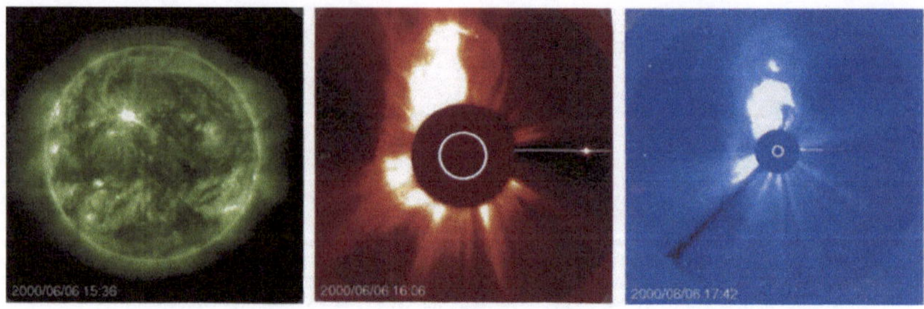

Solar activity (solar flare) and a coronal phenomenon called "coronal mass ejection (CME)", observed by the Solar and Heliospheric Observatory (SOHO).
Source: SOHO/(Instrument) consortium. SOHO is a project of international cooperation between ESA and NASA

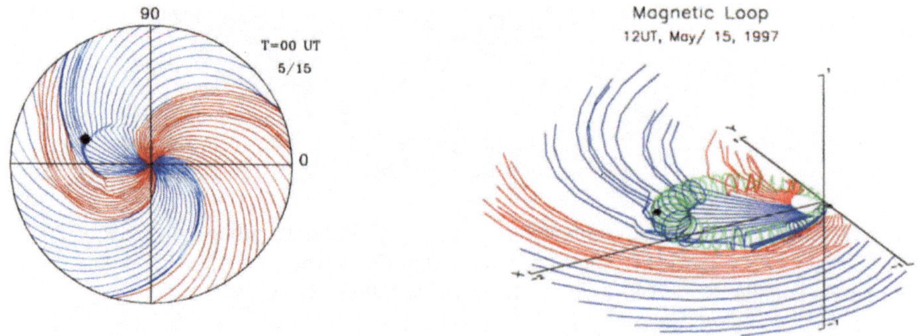

Left: The propagating of a shock wave during the May 1977 event. Right: An advancing magnetic flux rope with a helical structure.
Source: Saito, Takao, W. Sun, C.S. Deehr, and S.-I. Akasofu, *JGR*, Accepted, 2006

in the background solar wind and arrive at the front of the magnetosphere. When the magnetic field in the plasma cloud has an intense southward component, it increases the dynamo power, causing a frequent occurrence of magnetospheric substorms and thus subsequently generating the ring current belt and the magnetospheric storm.

The ultimate cause of magnetospheric storms is thus a variety of transient solar activities. In spite of more than a half-century of intense research, however, the causes of sunspots, solar flares still remain as long-standing unsolved problems. The nature of coronal mass ejections is still in debate. Most solar physicists consider that solar activities are directly related to hypothetical thin magnetic flux tubes beneath the photosphere, their uplift by buoyancy and magnetic reconnection among them after their uplift. In Chapter 7, it will be pointed out that magnetic flux tubes are nothing but a hypothesis, perhaps an unworkable one. It will also be pointed out that a dynamo process in the solar photosphere must generate the source of energy for solar activities, since solar activities are basically electromagnetic phenomena. In this book, I offer a non-traditional idea about the sunspot formation and causes of solar activity.

In this short review on the progress of solar-terrestrial physics, one can see that investigators of the four disciplines (solar physics, interplanetary physics, magnetospheric physics, and aeronomy) have made considerable progress in the twentieth century after Carrington's finding. However, for these disciplines to progress further, in particular in terms of space weather research, it is important for solar physicists, interplanetary physicists, magnetospheric physicists, and upper atmospheric physicists to work together. There are many missing links among the four disciplines that will only be noticed if one attempts to synthesize space weather research. Chapter 8 is devoted to the integration of the four fields.

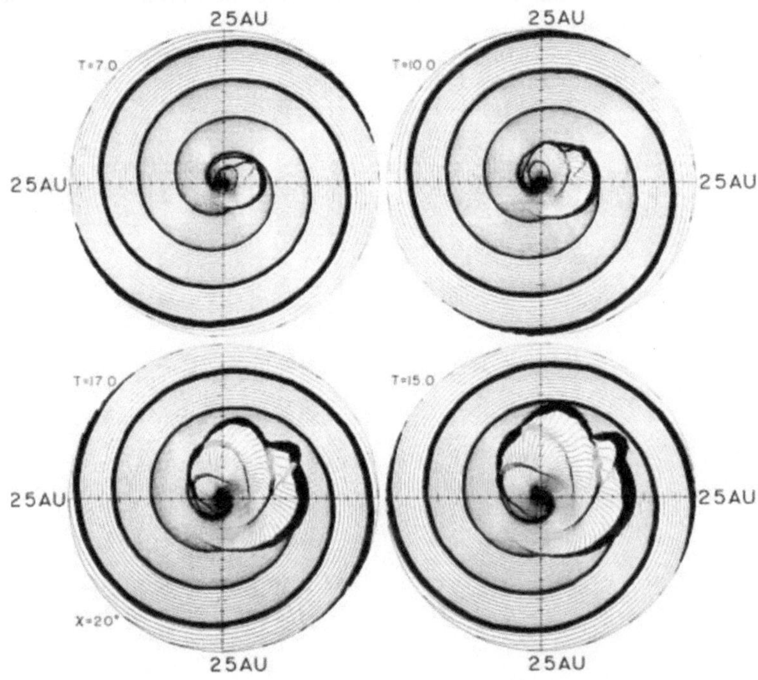

Interplanetary disturbances caused by a series of solar flares.
Source: Akasofu, S.-I., K. Hakamada, and C.D. Fry, *Planet. Space. Sci.*, **31**, 1435, 1983

The disturbed solar wind caused by various solar activities advances into the interplanetary structure, well beyond the distance of the Earth. It is quite exciting that both Jupiter and Saturn appear to show auroral activities. It is hoped that what we learn about the Earth's magnetosphere will be useful in studying their magnetosphere and vice versa. Shock waves also form a barrier for cosmic ray particles that enter from the outer boundary of the heliosphere, causing a reduction of the cosmic ray intensity in the heliosphere. This phenomenon was discovered by Scott Forbush and is called the *Forbush effect.* It is likely that the so-called "11-year cycle variations" of cosmic ray intensity result from an accumulated effect of the shock waves. Chapter 9 describes the magnetic field structure of the heliosphere and how it is disturbed by solar activities.

It is hoped that the readers of this book will find a number of long-standing unsolved problems in the four disciplines and that my non-traditional ideas stimulate better ideas than mine. I believe that many of the difficulties the present generation is facing are not due to the lack of basic knowledge and technical problems (for example, the capability of a supercomputer), but to our inability to recognize fundamental flaws in the presently prevailing concepts, namely paradigms. The Epilogue is devoted to discussing this issue. The new generation of scientists are encouraged to challenge the present paradigms and advance our understanding of electromagnetic phenomena around the Earth, in interplanetary space, and the heliosphere.

PLATE 1. An active auroral curtain near the zenith. This form is sometimes referred to as the corona-type display. Photographed by Jan Curtis.
Source: Photographed by Jan Curtis

PLATE 2. The curtain-like form of the aurora. The upper part of the curtain is rich in the dark red emissions from atomic oxygen. Photographed by Jan Curtis.
Source: Photographed by Jan Curtis

PLATE 3. An active auroral display called the westward traveling surge. The upper part of this particular aurora shows the dark red emission (660 mm) from atomic oxygen. Photographed by Jan Curtis.

Source: Photographed by Jan Curtis

PLATE 4. A typical red aurora. Photographed by Jan Curtis.
Source: Photographed by Jan Curtis

PLATE 5. An all-sky photograph of the aurora: photographed by J. Yokota.
Source: Photographed by J. Yokota

1 Search for the Unknown Quantity in the Solar Wind

1.1. Solar Corpuscular Streams

One of the most remarkable things about E.W. Maunder that was mentioned in the prologue, is that most of his conclusions have remained valid. In his sixth conclusion, Maunder (1905) stated:

> ...such a relation can only be explained by supposing that the Earth has encountered, time after time, a definitive stream. A stream which, continually supplied from one and the same area of the Sun's surface, appears to us, at our distance, to be rotating with the same speed as the area from which it rises.

During the first ten years of the twentieth century, Maunder's idea was gradually accepted. As indicated by the statements by Lord Kelvin and A. Schuster quoted in the prologue, the acceptance did not result from an elaboration and extension of the theoretical framework, but from intuitive associations between solar flares and geomagnetic disturbances, between geomagnetic disturbances and the aurora, and also between the aurora and the glow in a cathode ray tube. Recognizing the close association of geomagnetic storms and auroras is one of the important contributions toward the acceptance of the nature of the solar-terrestrial relationship Kelvin doubted. In his book *Cosmos*, A. Humboldt (1871) described his own finding:

> The mysterious course of the magnetic needle is equally affected by time and space, by the Sun's course, and by changes of place on the Earth's surface.... It is affected instantly, but only transiently, by the distant northern light as it shoots from the pole, flashing in beams of coloured light across the heavens.

This statement was based on his own incredibly strenuous observations. In a biography of Humboldt, L. Kellner (1963) noted:

> Humboldt had rented a small cottage in the garden of a rich brandy distiller on the outskirts of Berlin where he set up his instruments.... Here, Humboldt carried out more than six thousand observations, from May 1806 until June 1807. Glued to his post, he spent, at one time, seven days and nights, in succession, at his instruments, taking half-hour

readings... In December, he was lucky enough to observe a display of the aurora and simultaneously a violent perturbation of the magnetic needle.

Meanwhile, many physicists in the second half of the nineteenth century were convinced that the aurora was an electrical discharge phenomenon based on some similarity between the auroral phenomena and phenomena observed in high-voltage vacuum discharge tubes. Electrical discharge in a vacuum tube was one of the hottest topics among physicists in those days. Their studies eventually led to the discovery of electrons by J.J. Thomson in 1897. In the classic book on gaseous discharges *Conduction of Electricity through Gases*, by J.J. and G.P. Thompson (1903) it was noted:

We may, thus, regard the Sun, and probably any luminous star, as a source of negatively electrified particles that stream through the solar and stellar systems. Now, when electrons moving at a high speed pass through a gas they make it luminous; thus, when the electrons from the Sun meet the upper regions of the Earth's atmosphere they will produce luminous effects. Arrhenius has shown that we can explain, in a satisfactory manner, many of the periodic variations in the Aurora Borealis, if we assume that it is caused by electrons from the Sun passing through the upper regions of the Earth's atmosphere.

Recognizing the possible association between the aurora and an electron beam, K. Birkeland became one of the first proponents of what was once called the *corpuscular school*, the students of which proposed that the aurora and geomagnetic storms were caused by an electron beam ejected from the Sun. Birkeland (1908) stated:

It has gradually come to be recognized that aurora and magnetic perturbations should be regarded as rather moderate manifestations – at present, the only ones there are for us to observe – of an unknown cosmic agent of solar origin, and quite different from light, heat, or gravitation. It has long been supposed that this unknown aspect was in some way or other of an electrical nature.

Birkeland's observational, analytical, laboratory, and theoretical activities have been well documented in his three-inch-thick book, *On the Cause of Geomagnetic Storms and the Origin of Terrestrial Magnetism.* I am fortunate to have a copy of this book; it was given to me by the Committee of the Birkeland Symposium on Aurora and Magnetic Storms, which was held at Sandefjord, Norway, on September 18–22, 1967.

Birkeland's *terrella experiment* is displayed at the Auroral Observatory, University of Tromsø (Figures 1.1a and 1.1b). Stimulated by Birkeland, C. Störmer began his lifelong research on the aurora (Figures 1.2a and 1.2b). He computed trajectories of single electrons in a dipole field by devising a special integration method of the equation of motion. He also made extensive observations of the aurora and he summarized his research in his book *The Polar Aurora* (1955). He dedicated the book to his wife: *To my wife Ada who never ceased to encourage me to work hard till this book was safely finished.*

FIGURE 1.1a. K. Birkeland and Olav Devik experimenting on the terrella in a large vacuum box.
Source: Courtesy of University of Oslo

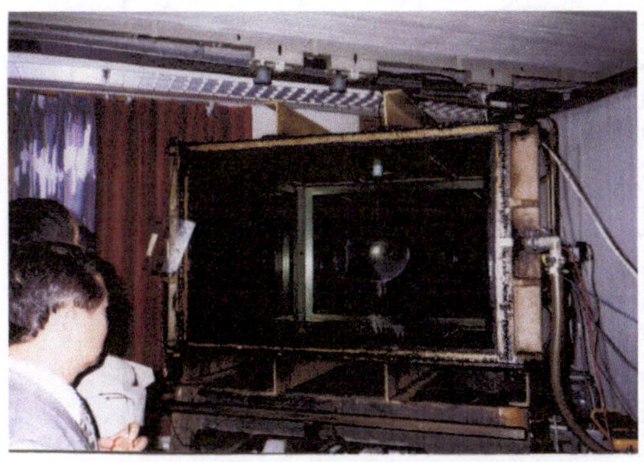

FIGURE 1.1b. Terrella experiment at the University of Tromsø.
Source: Courtesy of N. Fukushima

FIGURE 1.2a. C. Störmer (1874–1957).
Source: Courtesy of University of Oslo

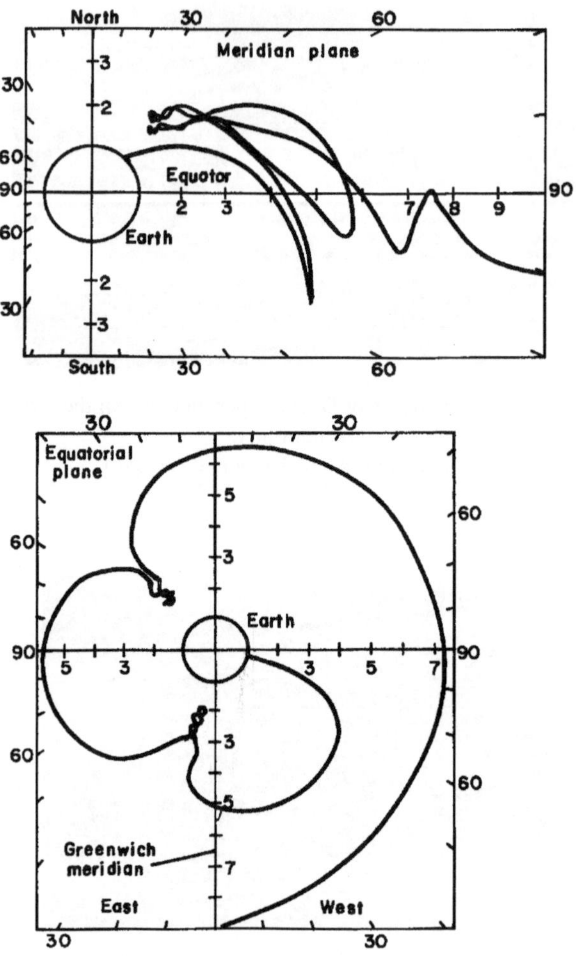

FIGURE 1.2b. An example of the trajectory of an electron in a dipole field computed by C. Störmer.
Source: McCracken, K.G., U.R. Rao, and M.A. Shea, *M.I.T. Tech. Rep.*, **77**, 2, 1962

Sydney Chapman (1918) also considered theoretically how a beam of either positive or negative charged particles could produce a global motion of the upper atmosphere as a cause of geomagnetic storms. The title of his paper was *An Outline of a Theory of Magnetic Storms*. He noted later (1967):

> I certainly misnamed this paper in calling it "An outline of a theory of magnetic storm." The observational part was useful, the theory was quite phony ...

Eventually, A. Schuster (1911), Chapman's professor, became aware of the possibility of the existence of beams of charged particles and stated:

This verdict (Lord Kelvin's argument) was generally accepted until recently, when the theory of a direct solar action has been revived in a form, which is assumed to be free from the objection raised. The magnetic actions being supposed to be due to a swarm of electrified corpuscles ejected by the Sun.

Similarly, F.A. Lindemann (1919) stated:

There seems to be no doubt that terrestrial magnetic storms are connected in some way with solar disturbances.

On the other hand, both Schuster and Lindemann noted that something is not quite right in this idea and criticized Chapman's theory based on a beam of positive or negative charged particles. Schuster (1911) stated:

... We must conclude that a swarm of electrons packed with sufficient density to cause a magnetic effect would soon get dissipated laterally into space until its magnetic action becomes negligible.

Lindemann (1919) examined the ionization rate of a hydrogen gas cloud by developing an equation similar to the Saha equation, which provides the ionization rate of a gas for a given temperature. Based on it, he concluded that the solar gas is highly ionized. Thus, he pointed out the importance of the electrostatic force between positive ions and electrons, which can prevent the repulsion and the subsequent diversion of charged particles of one sign. This suggestion became the basis for Chapman to launch a fresh approach to the problem of the interaction between solar corpuscles and the Earth's dipole field.

Chapman and Ferraro (1931) developed the first theory of magnetosphere formation by considering the solar corpuscles as a gas consisting of positive ions and electrons in equal number. Such a gas is now called *plasma.* The term plasma had been introduced by I. Langmuir (A.H. Rosenfeld, 1966) for a gas consisting of an equal number of positive and negative ions, but they were not aware of Langmuir's work.

During the first half of the twentieth century, interplanetary space was considered practically a vacuum, and it was believed solar corpuscular streams or clouds were emitted from time to time from two kinds of source regions. The first, discovered by Maunder, is the spot-free region that emits a stream of plasma for many months. Julius Bartels later named this region the M Region. It was said that "M" signifies *magnetically active* or *mysterious*, because it causes magnetic disturbances in spite of the fact that it is a region with no spots. This region is now called the *coronal hole*. The second one is an active region around a large sunspot group. In modern terms, an intense transient activity in the solar atmosphere ejects an isolated magnetic cloud (detached from the Sun), or produces an expanding magnetic flux rope (rooted on the Sun) during an intense solar flare or a coronal mass ejection (CME).

It may be noted that the solar flare-geomagnetic storm relationship was doubted by some even in the 1950s. This was because flares near the limb of the solar disk do not necessarily cause major magnetic disturbances; in such a case, flare disturbances are directed 90° away from the Sun–Earth line. H.W. Newton clarified this point. I mentioned his work on this issue in one of my early papers. Newton's wife found my paper and read it to him (he was blind by then). He was very pleased, and sent me a copy of his book *The Face of the Sun* (Penguin Books; 1958). It is a delightful book that I recommend to today's investigators. Every sunspot cycle produces a new generation of solar-terrestrial scientists who are most welcome to the discipline. However, many do not learn what earlier sunspot generation researchers learned. As a result, many phenomena are rediscovered every sunspot cycle. Fault predictions of a great auroral display after a major solar activity are still issued from time to time without taking into account Newton's results.

1.2. The Chapman–Ferraro Theory

Chapman took up a theoretical study of the interaction between Maunder's stream and the Earth's dipole field. Chapman and his graduate student Vincenzo Ferraro recognized that Maunder's stream should be treated as what we today call *plasma*, not a swarm of protons (or electrons). Here is Chapman's account of the birth of their theory (Chapman, 1968):

> In my first year, there were only two honors students in mathematics.... The other student was an Anglo-Italian, Vincenzo Antonino Ferraro whose father had emigrated from Italy and was in the hotel business; he was manager or head of one of the big high-class restaurants in London.... His son was born in England, but he was very much influenced by his family; I think the mother was a dominant figure, too. He went to a good English school in London, and then came to the Imperial College, and did very well in the degree examination and went on to do research.
>
> By that time, the Ph.D. had become established in England. So I was expected to provide him with a problem and guide him in it. Now, Larmor would never have given to me the suggestion of tackling the central problem in kinetic theory, which was the one that I myself discovered and sought to attack and solve. I felt I was doing a dangerous thing in giving Ferraro the problem I chose and asking him to work on the causation of magnetic storms. We tried to work out deductively what would be the consequence of the impact upon the Earth of a stream of what is now called plasma, neutral ionized gas. This had been suggested by Lindemann when he criticized my first phony theory on magnetic storms. He had not only destroyed my theory, he had proposed the constructive alternative suggestion, that the influence from the Sun must not be as I had supposed (like Birkeland and Störmer and others); namely, a stream of gas formed of charges of one sign only. He said it must consist of charges of opposite signs in practically equal numbers, so that it could hold together.
>
> Lindemann never tried to develop what would be the consequences on the Earth of the impact of such a stream of gas. I made an attempt at that while I was Professor at Manchester in 1919–1924, but, unfortunately, I started at the wrong end. I tried to find out what would be the steady state, as if the stream had been going on forever. It didn't work out; so I was still wanting to find out what would happen. I proposed

this subject to Ferraro. We played about with this problem, often being quite at a loss to know what would happen and how to approach the problem. But, finally, there did come the breakthrough of realizing that the stream would be a good conductor. Looking into Maxwell's great work on electricity and magnetism, and using at first a very crude model, namely, the approach of a conducting metal sheet towards Earth, we considered what currents would be induced in this sheet by the approach to Earth's magnetic field, and would add their own field in the space around the Earth. There would be a repulsion between the sheet currents and the field, which would tend to slow down the sheet. The sheet in this model was rigid. If you think of it as a gas instead of a rigid sheet, the current having at first a plane front, the current would be induced in its surface, but owing to the unequal repulsion in different parts, as we realized, the sheet would be subject to weaker forces. So, we inferred that a cavity would be formed around the Earth, enclosing the magnetic field – which would be confined in this cavity. This was the first published note in "Nature" in 1930; Ferraro came to the College in 1924, so this was in his sixth year. The work was later published in full, over a period of time, in 1931–32, in the journal that is now called the "Journal of Geophysical Research." For a long time people didn't believe in it or they were very dubious about it, didn't read it, or took no notice of it. Not long before it was actually demonstrated by satellites that this is what does happen, it came to be at last considered, and on the whole, accepted by a number of people. But, one of my American friends, Hulburt, who was an excellent director of the Naval Research Laboratory, developed an alternative theory of Ferraro's and mine, in one of his publications, as a matter now only of historical interest. I criticized Hulbert's theory, and I think it is quite dead now, as dead as my first theory of magnetic storms. However, Hulburt and I are very good friends.

In his article in *The Earth and Science* edited by A.H. Cook and T.F. Gastell for the occasion of Chapman's eightieth birthday, Vincenzo Ferraro wrote:

In 1927, I became one of Chapman's first research students at Imperial College; this was the beginning of a long and fruitful collaboration which afforded me much pleasure and in which, Chapman once told me, he much enjoyed, as indeed he must have enjoyed his collaboration with other people. Chapman and I undertook afresh the problem of the approach of the neutral ionized stream in the Earth's magnetic field and during the years 1927–33 we developed a new theory of magnetic storms. Only the theory of the first phase was then fully developed. We found that during its advance in the Earth's magnetic field electric currents are induced in the surface of the solar corpuscular stream. The surface currents shield the interior of the stream from the Earth's magnetic field so that particles in the stream can describe a rectilinear path up to the point where they enter the surface (current) layer of the streams. The action of the Earth's magnetic field on the surface currents repels the surface of the stream, the retardation being greatest over the part of the surface nearest the Earth. A cavity is thereby formed in the surface of stream, which deepens until equilibrium is reached between the kinetic and magnetic pressures. The geomagnetic field is thereby compressed by the solar cavity, the resulting increase in the horizontal force at the Earth's surface being identified as the increase in the horizontal force during the first phase of a magnetic storm.

In the first of a series of their papers on this subject, Chapman and Ferraro (1931) obtained a formula, which is basically similar to the Debye length (p. 94 in their paper), and showed that protons and electrons in the stream are strongly coupled in motion as they flow around the Earth's dipole field. Thus, they showed that the solar gas must be treated as *plasma* not as a cloud of solitary particles.

The basis of the Chapman–Ferraro theory is to regard the solar corpuscular stream as diamagnetic superconducting plasma. Thus, the gas cannot penetrate into the Earth's magnetic field, as strong shielding currents flow on the front surface of the advancing stream. This current is called the *Chapman–Ferraro current* (Figures 1.3a–1.3d). As a result, the Earth and its magnetic field are completely confined or compressed in a cavity. In this way, Chapman and Ferraro

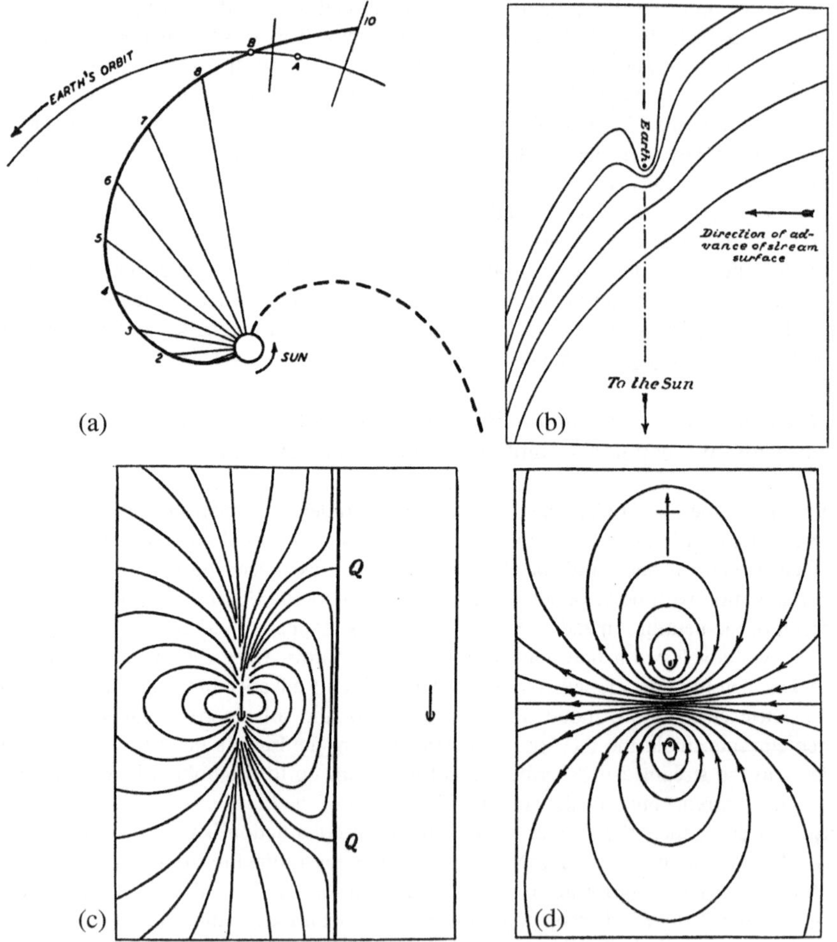

FIGURE 1.3. (a, b) The geometry of a stream from the Sun; (c, d) The compressed Earth's magnetic field and the shielding current (the Chapman-Ferraro current) on the advancing front of the stream.
Source: (a, b) Chapman, S. and J. Bartels, *Geomagnetism*, Oxford University Press, 1940. (c, d) Chapman, S., and V.C.A. Ferraro, *Terr. Magn.*, **36**, 77, 1931

explained successfully the storm sudden commencement (SSC) as a result of the impact of the solar plasma on the Earth's dipole field. In modern terms, the SSC is caused by the impact of a shock wave on the magnetosphere. The shock wave is generated when a high-speed solar plasma cloud or magnetic flux rope advances into the background slow-speed solar wind. It may also be noted that Chapman and Ferraro correctly envisioned the geometry of what we now call a high-speed stream from the coronal hole.

However, it was unfortunate that Chapman was convinced by this success that a theory of the main phase of geomagnetic storms (namely, the formation of the ring current belt) should be built upon the Chapman–Ferraro theory, namely, the interaction between *diamagnetic* plasma and the Earth's magnetic field. As is now known, an efficient energy transfer of as much as $10^{19} \sim 10^{20}$ erg/sec from the solar wind to the magnetosphere cannot be achieved by a diamagnetic plasma, because the diamagnetic plasma stream tends to flow around the magnetosphere without introducing much energy into it. In fact, Chapman agonized over this problem for almost 30 years (after the Chapman–Ferraro theory was established), because he was unable to find the energy transfer mechanism from the solar wind into the magnetosphere.

Actually, Chapman was considerably encouraged when he learned that the Explorer 12 satellite crossed the magnetopause in 1962, and demonstrated that the magnetic field just inside the magnetopause is close to twice that of the dipole field value, as expected from the Chapman–Ferraro theory. This satellite observation event is described in detail by Larry Cahill (1997). I recall that Chapman mentioned that it is rather rare that a theory becomes confirmed more than 30 years after its inception. He was all the more convinced that a theory explaining the main phase must be an extension of the Chapman–Ferraro theory. This was an exciting period during the early days of space exploration. I personally witnessed the occasion when Explorer 10 crossed the magnetopause. Jim Heppner, Norman Ness, and Joe Cain, at the Goddard Space Flight Center, were trying to understand this newly observed phenomenon (Heppner et al., 1963).

1.3. The Solar Wind

The view that interplanetary space is a vacuum into which the Sun intermittently emitted corpuscular streams was changed radically by Ludwig Biermann (1951, 1953) who proposed, on the basis of his study of comet tails, that the Sun continuously blows its atmosphere out in all directions at supersonic speed. At that time, it had generally been accepted that the solar radiation pressure was responsible for the tendency of the comet tail to trail in the antisolar direction. However, pointing out the fact that the magnitude of acceleration of the ionized component of cometary tails is of the order of $\sim 10^2$–10^3 times the radiation pressure, he concluded:

It is, thus, found that the acceleration of the CO^+ and N_2^+ formations observed in the tails are easily explained in terms of the friction between the solar and the cometary ions.

In 1957–58, Chapman was interested in zodiacal light and tried to explain it in terms of an extended *static* solar corona, but told me that he could not publish his paper in a regular scientific journal. Later, Gene Parker pointed out that Chapman's solution had a problem in that the pressure of his corona is finite at infinity.

In order to explain Biermann's conclusion, Parker (1958) examined thermal conditions of the solar corona that could lead to a supersonic flow at the Earth's distance and found that a supersonic flow can occur if the temperature in the solar corona decreases less rapidly than $1/r$, where r denotes the solar radius. This conclusion requires that the corona must be heated over a very extensive height range. In suggesting such a possibility, he coined the term *solar wind.*

The first observations of the solar wind by the Mariner 2 space probe were considered to be the confirmation of Parker's theory (Marcia Neugebauer and C.W. Snyder, 1962). During the period between 1960 and 1980, his idea of the generation of the solar wind was considerably rearticulated and elaborated by a number of solar physicists. However, there is so far no conclusive theory on both the high temperature of the corona and the acceleration process of the solar wind. It is puzzling that a strong solar wind tends to flow out from a dark region in the solar corona, what we now call a coronal hole, although it is known to be a magnetically open region; the magnetic field lines originating in a coronal hole extend into interplanetary space rather than forming closed loops in the corona (Chapter 5). The source process of the solar wind is still one of the long-standing unsolved problems of space physics.

1.4. Interplanetary Shock Waves

Tomy Gold (1955) intuitively associated the sudden rise of the horizontal component of the geomagnetic field at SSC with the impact of a shock wave that propagates in interplanetary gas. During the symposium titled *Gas Dynamics of Cosmic Clouds*, held at Cambridge, England, in 1953, he stated:

I should like to discuss, in connection with the subject of shock waves, some of the magnetic disturbances on the Earth that are caused by solar outbursts. The initial magnetic disturbance at "Sudden Commencement" of a magnetic storm can be accounted for, very roughly, by an increase of pressure of the tenuous gas around the Earth. This increase of pressure may perhaps be described as the effect of a wave sent out by the Sun through the tenuous medium. This description would then correspond to that of a stream of particles, while in the presence of a medium the correct description may lie anywhere between an acoustic wave, a supersonic shock wave, or an unimpeded corpuscular stream. The observations of magnetic storms may, hence, give us a fairly direct proof of the existence of shock waves in the interplanetary medium.

However, H.W. Liepmann objected to Gold's suggestion (1955). Liepmann argued:

I would like to ask whether the picture of a shock wave really is applicable. The mean free path in the residual gas between the Sun and the Earth appears to be 4 or 5 times the solar radius... In order to get agreement with Gold's values, the mean free path would have to be considerably shorter, i.e., by a ratio of about 100, or else the mechanism of interaction of the wave with the field of the Earth has to explain the very sudden rise observed ...

Gold refuted this objection by stating:

In considering the interaction between the stream and the residual gas, one must not restrict oneself to the collision cross section of neutral particles, but one has to consider the much stronger electromagnetic interactions that may occur between the two ionized gases.

Gold's view has since been elucidated in great detail by many researchers. Some solar wind ions are accelerated by colliding with both the advancing shock wave and the shock wave that forms at the front of the magnetosphere.

1.5. The Modern Interpretation of the Chapman–Ferraro Theory

Chapman and Ferraro theorized, first, the SSC and the initial phase in terms of encounter by the Earth and its magnetic field with a discrete solar corpuscular stream. However, since the Sun is known to expel the solar wind continuously, the Earth and its magnetic field are always confined in a comet-like cavity (Figure 1.4). As mentioned earlier, this cavity is called the magnetosphere. Since Chapman and Ferraro predicted the cavity, their theory of the SSC and initial phase was later regarded as the first theory of the formation of the magnetosphere. The SSC and initial phase of a geomagnetic storm can now be explained as a simple extension of their theory; it is caused by an enhanced solar wind pressure associated with the compression of the magnetosphere by the shock wave (Figure 1.5), not by an impact of isolated solar gas clouds or streams. Although these findings may sound trivial from the present understanding of the magnetosphere, each finding was an epoch-making advance in magnetospheric physics in its early days of development.

1.6. The Main Phase of Geomagnetic Storms and the Ring Current

Chapman and Ferraro thought that the morning side of the boundary of the magnetosphere (the magnetopause) is slightly positively charged, while the evening side of the magnetopause is negatively charged.

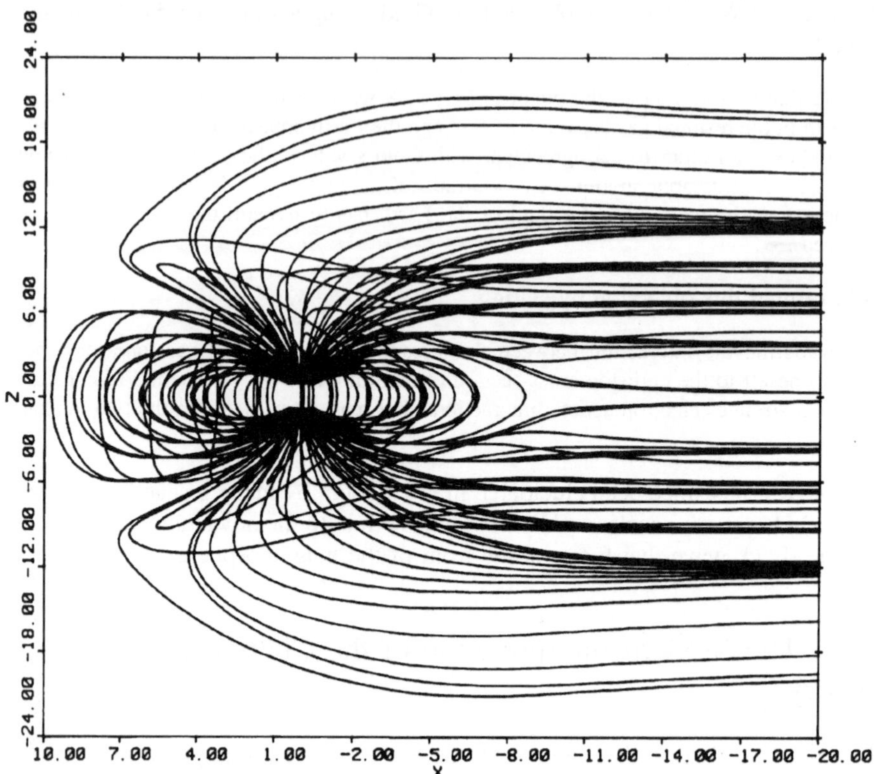

FIGURE 1.4. The Chapman-Ferraro theory predicted that the magnetic field configuration in the magnetosphere is completely confined in a comet-shaped cavity. *Source*: Roederer, M., see also S.-I. Akasofu, and D.N. Covey, *Planet. Space Sci.*, **28**, 757, 1980

They suggested that a dawn-dusk electric field, thus established, brings some of the particles into the cavity. Several theorists have independently considered this idea a few times during the last 30 years. Chapman and Ferraro thought that the particles thus brought into the magnetosphere form a toroidal westward-directed ring-like current around the Earth, explaining the large decrease in the horizontal component of the Earth's field during the main phase.

Unfortunately, as mentioned earlier, after his initial success in 1931, Chapman did not make much progress in explaining the development of the main phase and was still struggling with the problem when I joined him in Alaska in 1958; this problem is still a major issue today (Chapter 4).

I never took the opportunity to ask Chapman why he decided to come to Alaska after retiring from Oxford University. One possibility was that Störmer criticized his theory during the 1950 London, Ontario, conference – the first

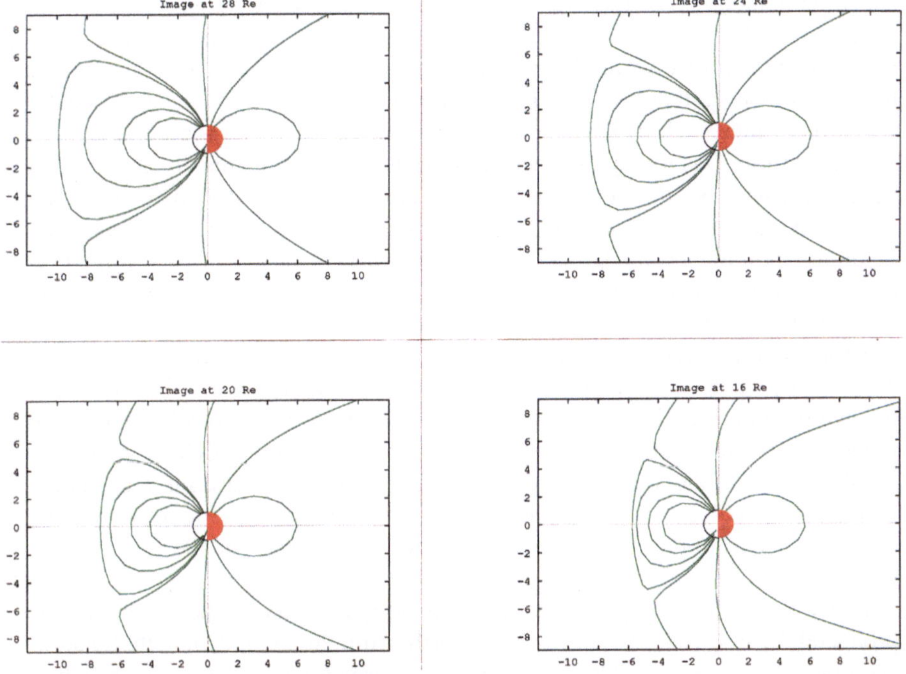

FIGURE 1.5. The arrival of an interplanetary shock wave compresses the magnetosphere (the noon-midnight cross-section).
Source: Akasofu, S.-I.

major conference on the aurora after WWII – saying that the Chapman–Ferraro theory cannot explain any specific aspect of the aurora. Störmer claimed that his theory, on the other hand, could explain details of many aspects of the aurora, including the auroral zone and the curtain-like structure of the aurora. Although the title of Chapman's paper was *Theories of the Aurora Polaris*, all he could do was to describe the Chapman–Ferraro theory; he could not find the mechanism by which the solar wind transfers its energy to the magnetosphere in causing the aurora. In Alaska, Chapman could continue to search for that mechanism.

When I was a graduate student at Tohoku University, Japan, an organization called *The Ionospheric Research Committee* was dedicated to research on solar-terrestrial physics. Top-level researchers in Japan attended its meetings and their discussions were stimulating for the young scientists. I was told in their meetings that a good understanding of the Chapman–Ferraro papers was a prerequisite to studying geomagnetic storms. Thus, I began that study and found the papers difficult to grasp, leaving me with a number of questions. I learned that Chapman worked at the Geophysical Institute, University of Alaska, and wrote to him in the spring of 1958. I included in my letter the questions I had, but did not expect a quick response. Chapman was the greatest authority on geomagnetism, and a contemporary of the great British astronomer A. Eddington, and I was simply a student who had just started learning geomagnetism. Thus, it was a great surprise to receive his letter in a matter of a few weeks. In his response, he said in essence that he could not answer some of my questions and asked if I would be interested in studying those questions under his guidance.

His response was totally unexpected. I wrote to him immediately, saying how delighted I was, but that it was beyond my dream to study abroad. To my great surprise again, I received a check from Chris Elvey, director of the Geophysical Institute, soon afterward. By then, however, I had been asked to be a member of the Japanese solar eclipse expedition party to go to the South Pacific. Thus, it was not until December 13, 1958, that I arrived in Fairbanks, Alaska.

Soon after my arrival in Alaska, the Van Allen belts were discovered. It became the dawn of scientific space exploration. James Van Allen correctly pointed out that energetic particles execute the motions studied by Störmer in his pioneering work on motions of charged particles in a dipole field (Figure 1.6). At the time of the discovery of the Van Allen belts, several researchers, including Fred Singer, suggested longitudinal drift motion constituted a westerly electric current around the Earth, causing the main phase of the geomagnetic storms.

In my Ph.D. thesis work (1961), I examined quantitatively the magnetic field produced by trapped particles in the Earth's dipole field, although I had to postulate the growth of what I called the *storm time radiation belts*, which consist of protons of a few kilovolts. The computed magnetic field, produced

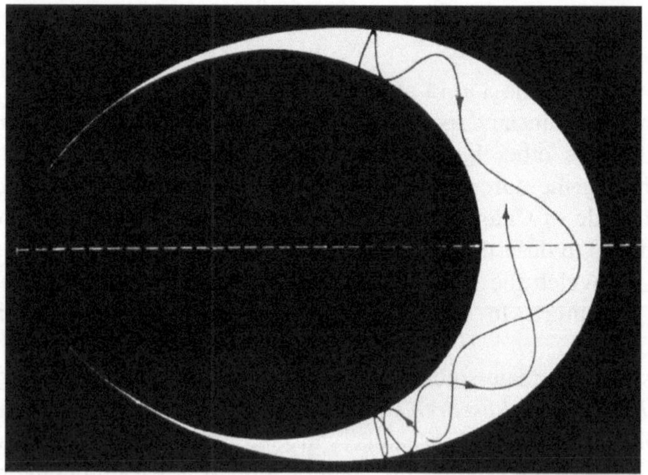

FIGURE 1.6. Störmer found a group of trajectories of electrons that are confined in a region isolated from infinity. An example of the trajectories is shown. James Van Allen thought that high-energy particles in the radiation have similar trajectories. *Source*: Störmer, C., *The Polar Aurora,* Oxford University Press, Oxford, 1955 (quoted by Van Allen, J.A., in Proceedings of the General Assembly, Berkeley, 1961, p. 99, Academic Press, London, 1962)

by the motions of protons in the belts, was found to point almost uniformly southward around the Earth, explaining the large depression of the field during the main phase (Akasofu and Chapman 1961); Figure 1.7. It should be noted that, unlike what most researchers believe today, the ring current field arises mainly from diamagnetism of the trapped particles in the dipole field, not the westward drift motion of positive ions. In fact, for an isotropic pitch angle distribution, the westward drift does not contribute to the current. In the inner half of the belt, the current flows eastward, while in the outer half, the current flows westward. The outer current is more intense than the inner current. It is for this reason that the ring current appears to flow westward. An IBM 7090 computer at the Goddard Space Flight Center was used for this computation; it was the fastest computer of its time. The computed results agreed with the ground-based observations and some of the earliest satellite observations of the magnetic field produced by the ring current reported by Paul Coleman and Larry Cahill. Further, the belt I postulated was surprisingly not too far from what Lou Frank, University of Iowa, detected later by a satellite.

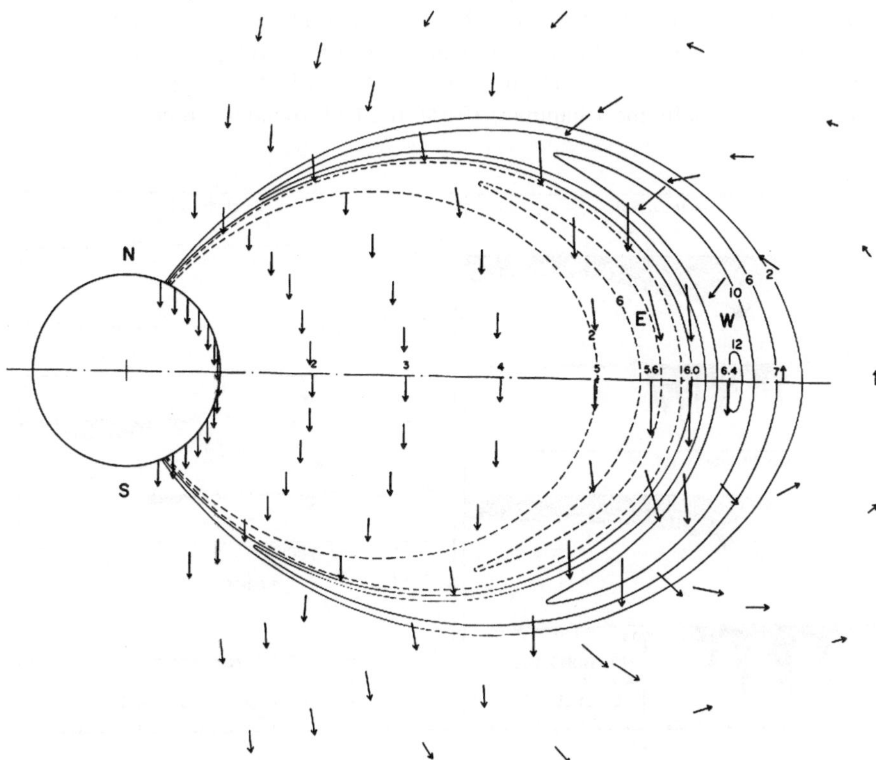

FIGURE 1.7. The distribution of electric currents produced by a belt of trapped particles in the Earth's dipole field (E, eastward, and W, westward) and the magnetic field produced by the currents.
Source: Akasofu, S.-I., J.C. Cain, and S. Chapman, *J. Geophys. Res.*, **66**, 4013, 1961

1.7. Variety of the Development of Geomagnetic Storms

Chapman, Van Allen, and I were quite happy about the computed results. Confirming that protons of a few kilo electron-volts in the trapping region can produce the desired westward current for the main phase decrease, we tried to find ways to bring solar wind protons deep into the trapping region across the magnetopause on the basis of the Chapman–Ferraro theory. However, after much struggle, we came to the realization that the Chapman–Ferraro theory actually tells us that a diamagnetic plasma tends to flow around the Earth, confining it into a cavity *without* transferring much energy into the magnetosphere. There is no way to bring solar wind protons to a distance of several Earth radii across the dayside magnetopause. However, the prevailing idea at that time was that only a tiny fraction of the solar wind energy was needed to cause the main phase, and that the problem should not be difficult to solve (Alex Dessler, and his colleagues, 1961). So our conclusion did not get any attention in the community.

After much thinking, I proposed to Chapman that I should examine the development of a number of *individual* geomagnetic storms, instead of a typical or an idealized storm, in an effort to study how the main phase actually develops. It immediately became evident that geomagnetic storms develop in a great variety of ways. In order to demonstrate the point, we published a paper that included Figure 1.8a (Akasofu and Chapman, 1963). If I had to choose three of the most

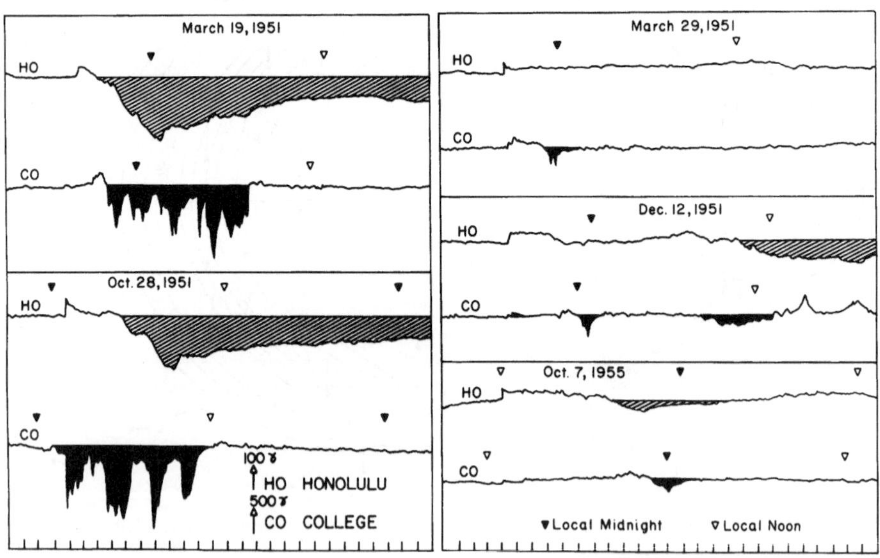

FIGURE 1.8a. Two types of geomagnetic storms that begin with SSC of similar magnitudes. Left: Storms with an intense main phase (HO: the Honolulu record) and polar magnetic substorms (CO: the College record); Right: Storms with weak or no appreciable main phase and substorms.
Source: Akasofu, S.-I. and S. Chapman, *J. Geophys. Res.*, **68**, 125, 1963

important figures published in my research career, this would be one of them, although I believe it would be impossible to publish such an unsophisticated figure in the *Journal of Geophysical Research* today. Figure 1.8b shows also the great variety of ways in which geomagnetic storms can develop. In the first storm the main phase did not develop, in spite of the fact that a strong solar wind blew around the magnetosphere for many hours after the SSC, as can be seen by the fact that the increased field level was maintained for many hours.

The third one is what has been thought to be a *typical* storm as Chapman conceptualized it. However, in many cases, the main phase can start to develop even before the SSC. In many other cases, a large main phase can develop without the SSC (see the last example in Figure 1.8b).

This type is known as *gradually commencing storms*, but was ignored because they are not *typical*. In fact, a major geomagnetic storm can develop even after what is known to be a *negative sudden impulse*, namely a sudden decrease of the solar wind pressure. In modern terms, such a sudden change is associated with the passage of an interplanetary discontinuity in the solar wind. An obvious conclusion from this study was that an increased solar wind pressure is neither a necessary nor sufficient condition for a geomagnetic storm to occur. However, such a conclusion was not acceptable to the scientific community at that time.

1.8. Unknown Quantity

After much discussion, Chapman and I reached an important realization: the Chapman–Ferraro theory implies that a diamagnetic plasma flow around the Earth can confine it in a cavity (the magnetosphere), but does not transfer the energy into it. We concluded (Akasofu and Chapman, 1963, p. 129):

> The variety of development of the storms seems to suggest some intrinsic differences between the solar streams far beyond what we would expect from a mere difference between their pressures. The nature of their intrinsic differences is at present unknown.

This conclusion annoyed and even outraged some prominent theorists, since it was so firmly believed that the Chapman–Ferraro theory was all that was needed and that a geomagnetic storm is a result of a stronger solar wind. Some even told Chapman that he was trying to destroy his own life work. Most researchers thought that there could not be any unknown quantity in physics, except for some elementary particles.

Nevertheless, we thought that the theory of the main phase must be built upon the Chapman–Ferraro theory, because it is successful in explaining the SSC. Thus, I once thought that the unknown quantity was neutral hydrogen atoms, which can penetrate across the dayside magnetopause without difficulty and can become energetic protons of the ring current belt after a charge exchange process (Akasofu, 1964). In Figure 1.9 the variety of the development of magnetic storms shown in Figure 1.8b was interpreted in terms of the degree

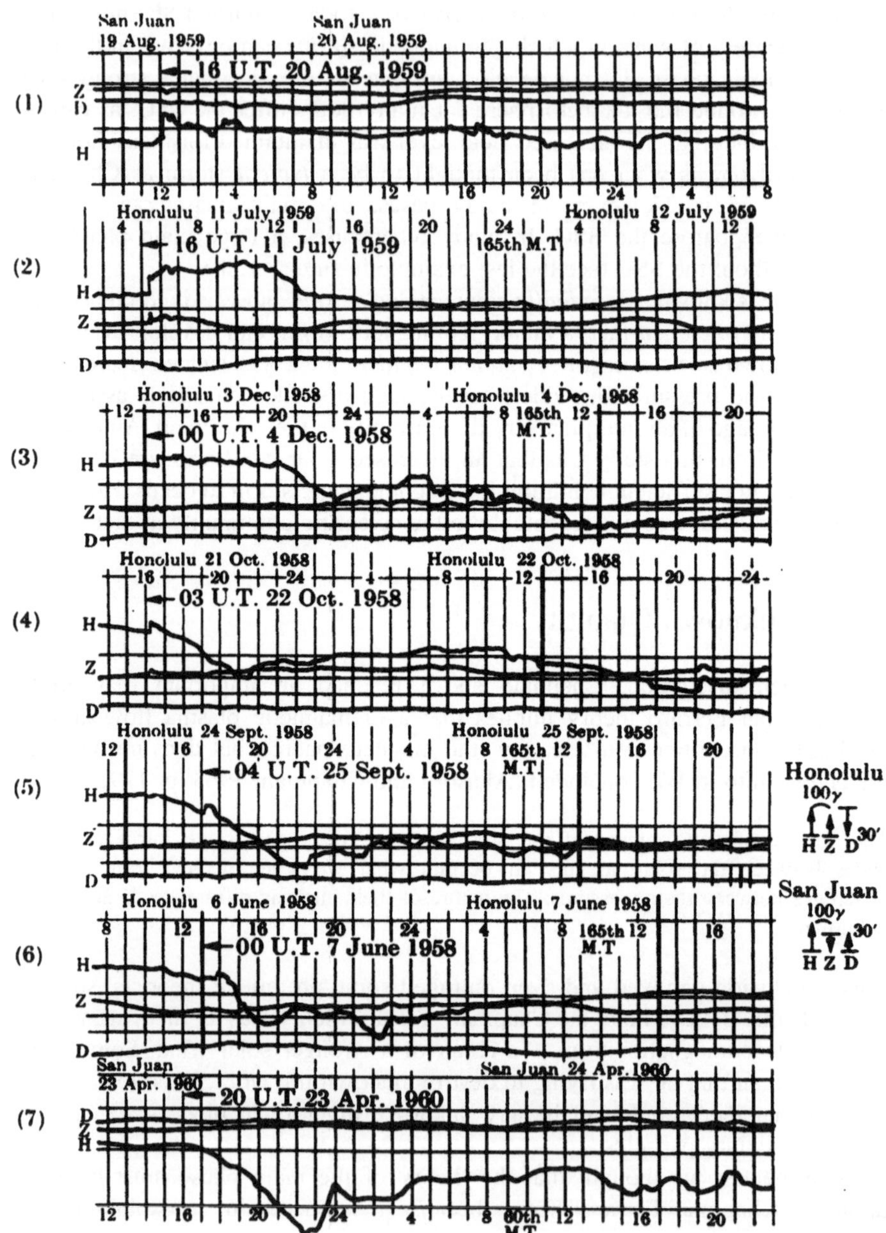

FIGURE 1.8b. A collection of magnetic records showing a great variety of ways in which geomagnetic storms develop.

Source: Akasofu, S.-I. and S. Chapman, *Solar-Terrestrial Physics*, Oxford University Press, Chapter 8, 1972

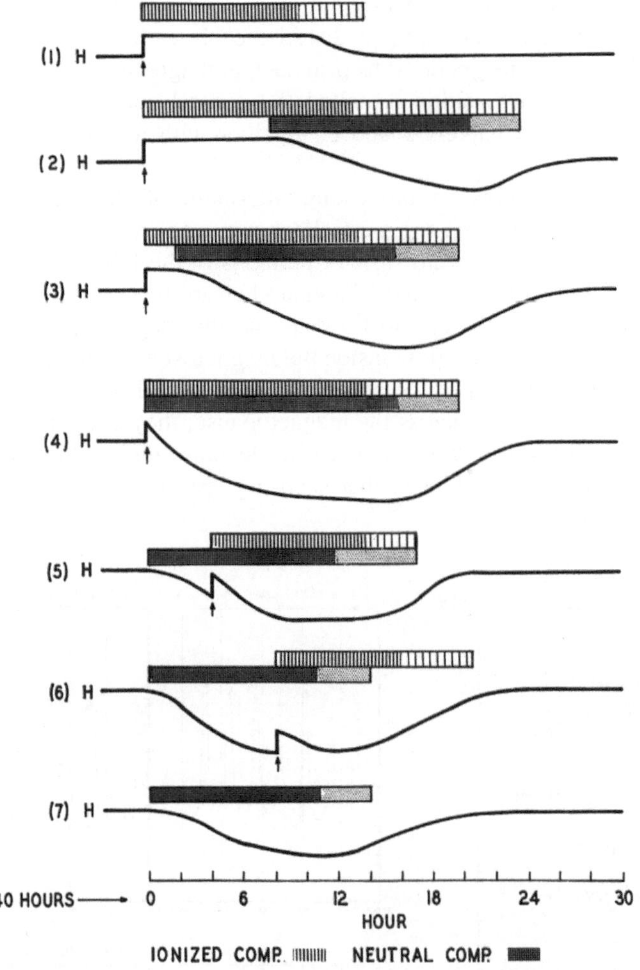

FIGURE 1.9. An attempt to explain the great variety of development of geomagnetic storms in terms of differences of the ionization rate of the solar plasma flow. *Source*: Akasofu, S.-I., Space Sci. Rev., **28**, 121, 1981

of the ionization of the solar wind. The first type is caused by fully ionized plasma, which corresponds to the case of the Chapman–Ferraro theory. On the other hand, the last type is produced by essentially un-ionized plasma atoms (which do not cause the compression (SSC) of the magnetosphere, but cause the main phase after penetrating into the magnetosphere and exchanging the charge).

As will be explained in the next section, this unknown quantity is now identified as the southward component ($-Bz$) of the interplanetary magnetic field (IMF), or more accurately, a specific combination of the solar wind speed V, the IMF magnitude B, and its polar angle θ. The magnetosphere responds to the southward component of the IMF lasting for a few hours in a very specific way. This mode of magnetospheric disturbance is called the *magnetospheric substorm*. The polar magnetic substorm is the magnetic manifestation of it, while its auroral manifestation is called the auroral substorm. These aspects will be discussed in Chapters 2 and 3, together with a synthesis/summary in Chapter 4.

Incidentally, there was an interesting aftermath of this neutral hydrogen story. When I was studying magnetotail phenomena in Los Alamos with Ed Hones, I found an anti-sunward flow of particles in the lobe of the magnetotail (Figure 1.10). At that time, as the Chapman–Ferraro theory indicated, the fully ionized solar wind was thought to flow around the magnetosphere, so that the entrained particles could not flow inside the magnetosphere and the magnetotail. One obvious interpretation of the observation was that this flow was composed of neutral hydrogen atoms across the magnetopause; they could become ionized by colliding with the detector. After a few sleepless nights, however, I found that a simple calculation showed that such a possibility was unlikely. On the

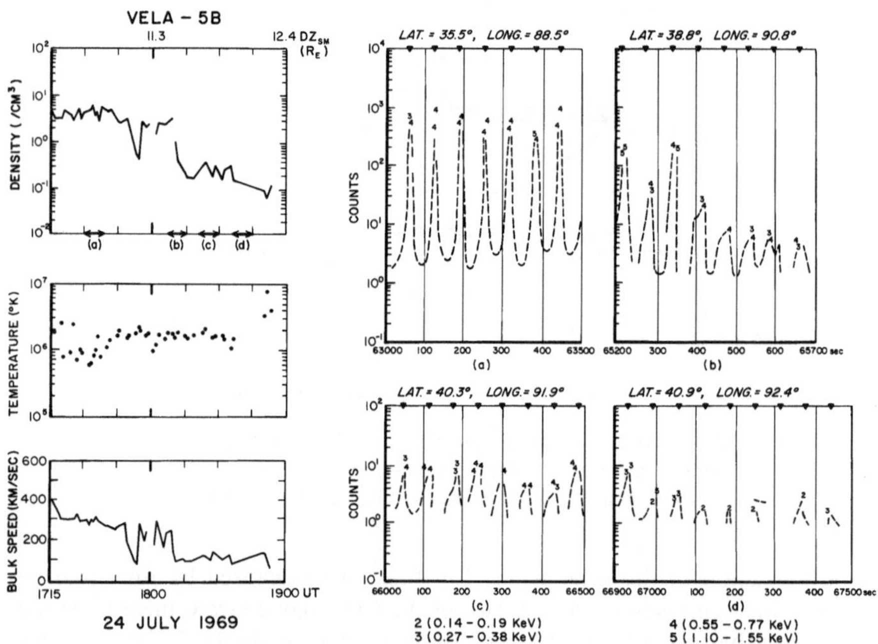

FIGURE 1.10. Example of observation of the plasma flow as a satellite (Vela-5B) enters from the solar wind regime into the magnetosphere. As the satellite spins, the unidirectional solar wind produces a large-intensity modulation in the detector (a). However, even after the satellite entered into the magnetosphere (b), a weak flow from the direction of the Sun is still observed (c and d).

Source: Akasofu, S.-I., E.W. Hones, Jr., S.J. Bame, J.R. Asbridge, and Lui, A.T.Y., *J. Geophys. Res.*, **78**, 7257, 1973

other hand, this finding led to the discovery of the mantle flow (1973). Speaking of neutral hydrogen atoms, I should point out a common misconception, which is that a solar prominence is fully ionized plasma. Actually, it is only partially ionized plasma, because the observed prominence emissions are from neutral hydrogen atoms (the Balmer alpha line). Further, it is clear from the observations of exploding prominences that neutral hydrogen atoms can escape from the Sun before they become ionized. Unfortunately, so far there has been no attempt to observe them in interplanetary space.[1] These neutral hydrogen stories cannot be mentioned in the standard monographs. However, this is an example of how science develops in actuality.

I became more convinced of the validity of our conclusion on the existence of the unknown quantity when I examined the intense magnetic storm illustrated in the Honolulu and College magnetic records shown in Figure 1.11. After the sudden commencement at about 13:40 165° Local Mean Time (LMT) on December 3, 1958, a strong solar wind blew for as long as 6 hours, but the main phase began to develop only after 20:10 165° LMT without an additional large enhancement of the solar wind pressure, which would be recorded in the horizontal component of magnetic records if it happened; intense auroral activities also began at the same time. Some unknown quantity in the solar wind must have arrived around the Earth at that moment to cause the main phase and the auroral activity. Thus, our research for the unknown quantity began.

1.9. The ε Function

A new understanding of the energy transfer process from the solar wind to the magnetosphere began when Don Fairfield (1967) found a close association between the so-called *southward turning* of the IMF vector and geomagnetic disturbances, or the southward component of the interplanetary magnetic field. Fairfield concluded that the IMF southward component can be identified with what Chapman and I thought to be the unknown quantity. This finding was based on Jim Dungey's suggestion of magnetic reconnection; Dungey credited this to F. Hoyle. He elaborated on his suggestion later and published a paper in *Physical Review Letter* in 1971. However, the significance of his open model in substorm processes was not well recognized by most magnetospheric physicists for almost ten years. In fact, in that period, many sketches of the magnetosphere model did not include the interplanetary magnetic field lines (see Figure I in the Prologue).

Figure 1.12 shows the development of the geomagnetic storm of February 15–16, 1967. One can see clearly the arrival of the interplanetary shock wave at about 23:50 UT on February 15, manifested by a step function-like increase of the field magnitude B, which nearly coincided with the SSC on the ground. However,

[1] At the time when this chapter was being prepared, I was not aware of the paper titled "Observations of neutral atoms from the solar wind," published by M.R. Collier, et al., in *J. of Geophys. Res.*, **106**, 24, 893, 2001.

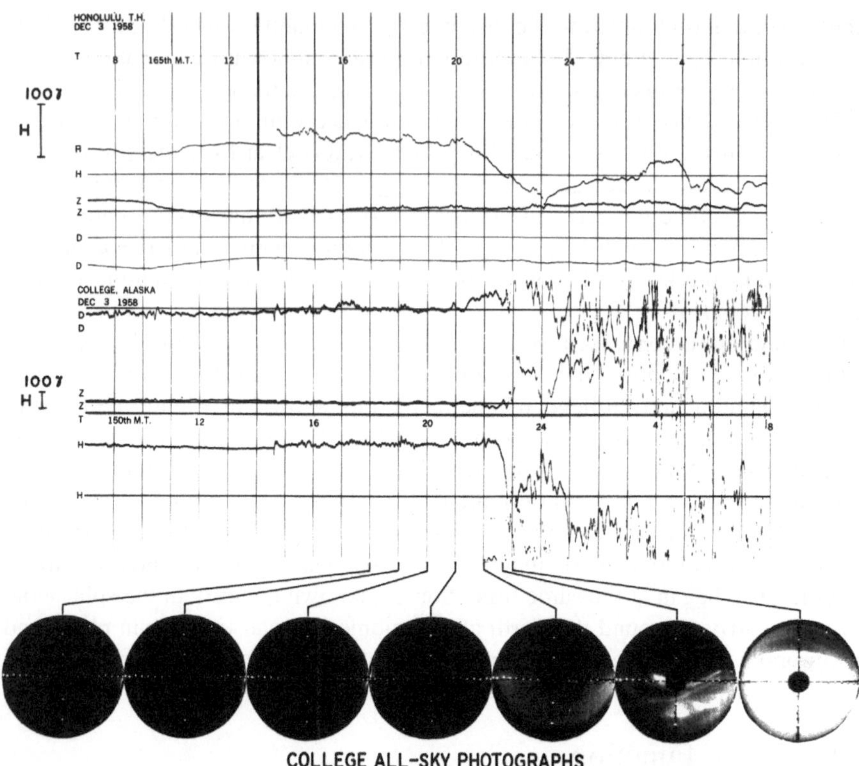

FIGURE 1.11. The development of the geomagnetic storm of December 4, 1958. From the top, the Honolulu magnetic record, the College magnetic record, and the College all-sky photograph on that day. Note that after the storm sudden commencement (SSC) at about 13:40 165° West Mean Time on December 3, the magnetosphere was exposed to a higher solar wind pressure for many hours. However, the main phase, high-latitude disturbances, and auroral activities developed 6 hours after the SSC. The nature of the solar wind must have changed at about 20:00, 165° LMT. Note that College (Fairbanks) was dark enough to observe the aurora during the intense flow.
Source: Akasofu, S.-I., *Planet. Space Sci.*, **12**, 801, 1964 (see also S.-I. Akasofu and S. Chapman, *Solar-Terrestrial Physics*, Chapter 8, Oxford University Press, Oxford, 1972)

the intense polar magnetic substorm activity (indicated by the auroral electrojet index (AE) and the associated development of the main phase (indicated by the Dst-ASY index) did not begin until about 09 UT on that day. One can see clearly that this time coincided with the arrival of the southward component (–*B*z) of the IMF, namely of the *unknown quantity* suggested by Fairfield.

However, in the 1960s and 1970s, most theorists in magnetospheric physics were preoccupied with the hypothesis of magnetic reconnection in the magnetotail as the energy supply process for the ring current. This was under the premise that magnetic energy was *gradually* and *continuously* accumulated in the magnetotail and that spontaneous magnetic reconnection suddenly converted

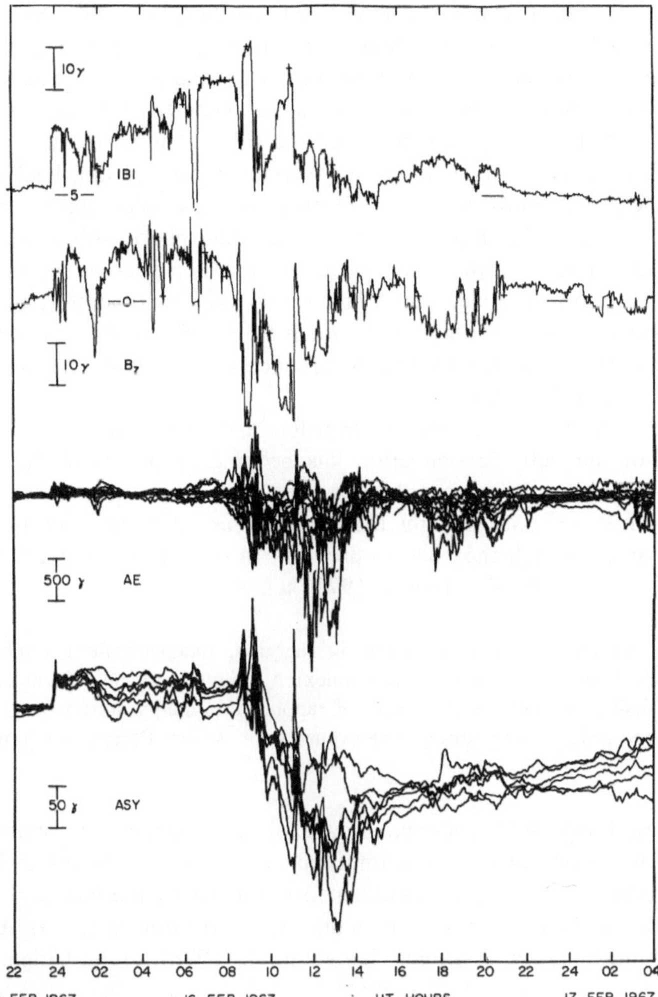

FIGURE 1.12. Superimposed magnetic records from both high and low latitudes, together with the interplanetary magnetic field magnitude *B* and the *Bz* component for the geomagnetic storm of February 16-17, 1967. The differences among the low-latitude observatories indicate the asymmetry of the main phase (ASY).
Source: Akasofu, S.-I., *Physics of Magnetospheric Substorms*, Chapter 5, D. Reidel Pub. Co., Holland, 1977

the magnetic energy thus accumulated into substorm energy. It was said that the magnetotail had more than enough energy for thirty substorms and that all we had to find was the process leading to magnetic reconnection. What Fairfield found was that each substorm requires a significant amount of input energy. However, Fairfield's paper and those that followed did not get the attention they deserved for many years.

As I mentioned earlier, there is little doubt that the energy for magnetospheric substorms is delivered from the Sun to the magnetosphere by the solar wind. Thus, in the last few decades, one of the most profound issues in magnetospheric physics, both theoretical and observational, has been to uncover the processes associated with the energy transfer from the solar wind to the magnetosphere and the subsequent transmission and conversion processes that lead to various magnetospheric substorm processes. Further, various polar upper atmospheric phenomena (such as the auroral substorm, the ionospheric substorm, the polar magnetic substorm, etc.) and also various disturbance phenomena in the inner magnetosphere and the magnetotail are mostly different manifestations of the magnetospheric substorm. Further, the magnetospheric substorm is perhaps the most basic type of magnetospheric disturbance *as a response to an increased energy input from the solar wind.*

In understanding these energy transfer and conversion processes, the hypothesis of magnetic reconnection has become such a powerful paradigm that reconnection has been considered to be the cause of most magnetospheric processes. Most theorists thought they had to base their theories upon it and many experimenters felt they had to prove it. In one of the standard references on this subject, Vytenis Vasyliunas (1975) stated:

> The process variously known as magnetic merging, magnetic field annihilation, or magnetic field line reconnection (or re-connexion), plays a crucial role in determining the most plausible, if not the only, way of tapping the energy stored in the magnetic field in order to produce large dissipative events, such as solar flares and magnetospheric substorms.

Indeed, from 1960–1980, understanding explosive magnetic reconnection was considered to be one of the most important theoretical problems to be solved in magnetospheric physics, as documented in reports by the National Academy of Sciences, the National Aeronautics and Space Administration (NASA), and various committees. For example, Colgate et al. (1978), in the Colgate Report, state:

> ...This magnetic reconnection may occur gradually or explosively. When it occurs explosively, it can lead to auroral substorms and solar flares ...

In the same report, magnetic reconnection is identified as the most important problem among six problems, which is described as:

> ... vital to further understanding of space plasmas ...

In fact, much of the past theoretical effort has been focused on finding mechanisms that make magnetic reconnection explosive in order to explain explosive phenomena, such as solar flares and substorms. At the same time, the resulting neutral line or the X-line has become a *magic* line. Many phenomena are blindly ascribed to unknown and unproved physical processes associated with the X-line.

For example, it was proposed without any definitive proof that auroral electrons were accelerated along the X-line, causing auroral arcs.

I avoided this particular paradigm and decided to go my own way. I must confess that this decision was not based on any rational thinking. It may be that I have an instinctive tendency to avoid a popular view.

In spite of this dominant trend of magnetic reconnection hypothesis until the 1970s, there was no serious attempt to examine observationally how the energy input rate $I(t)$ and the output rate $O(t)$ of the magnetosphere are related on a global scale, although such a study is crucial to examining whether explosive magnetic reconnection would be responsible for the magnetospheric substorms. For the purpose of this particular study, one may consider here three systems with very different relationships between $I(t)$ and $O(t)$. In the first system, time variations of the energy output rate $O(t)$ are almost identical to those of the energy input rate $I(t)$ (Figure 1.13a). In the second system, the energy is initially accumulated to a critical value, at which value it is suddenly unloaded. Therefore, in such a system (Figure 1.13b), $I(t)$ and $O(t)$ are expected to have different time variations. These two systems may be schematically represented by the so-called *pitcher model and the tippy bucket model*, respectively. In the pitcher model, $O(t)$ is more or less directly controlled by $I(t)$, and such a system may be called a directly driven system. On the other hand, in the tippy bucket model the amount of water in the bucket and the spring constant (equivalent to some magnetospheric threshold parameters) play important roles in controlling $O(t)$, and such a system may be called an unloading system.

If explosive magnetic reconnection is considered to be the primary process in generating substorm energy, we would expect that $O(t)$ will be significantly different from $I(t)$. This is because substorm energy would have to be accumulated in the magnetotail prior to its explosive conversion. There should be a delay, a time during which the energy is being accumulated -namely, the period between identifiable increases of $I(t)$ and $O(t)$; after substorm onset, $O(t)$ should increase sharply, regardless of how $I(t)$ varies, and O(t) at the peak time,

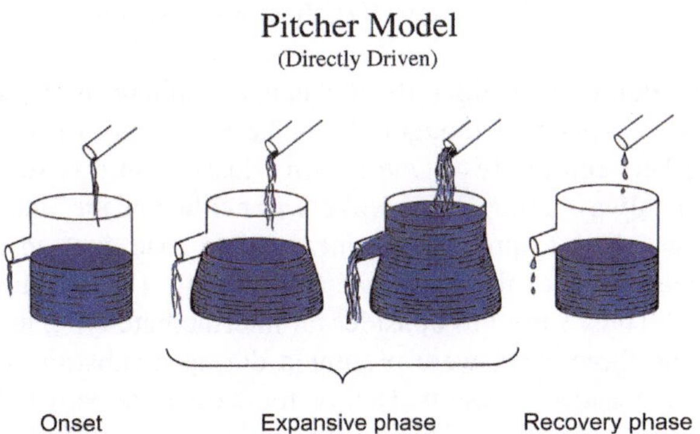

FIGURE 1.13a. The pitcher (directly driven) model.
Source: Akasofu, S.-I., *EOS*, **66**, 465, 1985

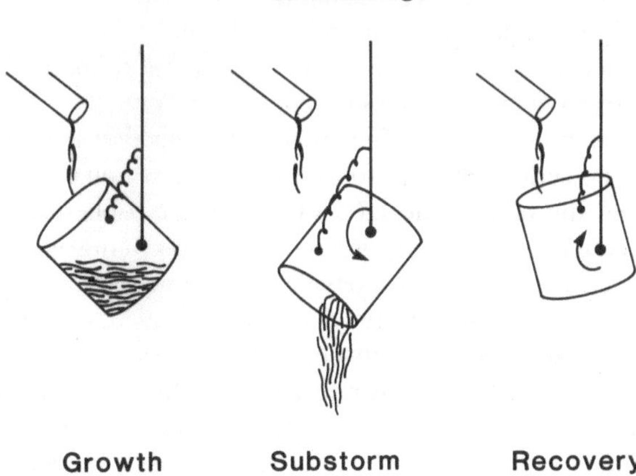

FIGURE 1.13b. The tippy bucket (unloading model).
Source: Akasofu, S.-I., *EOS*, **66**, 465, 1985

should be much greater than $I(t)$ at any time. Conversely, if $O(t)$ is found to be very similar to $I(t)$, there is little basis for hypothesizing explosive magnetic reconnection. It is for this reason that the relationship between $I(t)$ and $O(t)$ provides important information on the basic magnetospheric substorm process. In early energy transfer studies, investigators attempted to determine the correlation coefficient between a geomagnetic index (chosen from Kp, ΣKp, AE, Dst, etc.) and solar wind quantities (such as the solar wind speed V, the mass density mn, the southward component of the solar wind magnetic field $-Bz$, etc.). Among such correlation studies, the auroral electrojet index AE (which is a substorm index) is found to be highly correlated (the correlation coefficient being 0.7–0.8) with $-Bz$ or $V^2|-Bz|$. The high correlation coefficients have suggested that $I(t)$ and $O(t)$ are closely related. Unfortunately, however, neither $-Bz$ nor $V^2|-Bz|$ is $I(t)$; likewise, AE is not $O(t)$. It is not possible to compare apples and oranges.

When I was attempting to identify the magnetosphere as a pitcher-type or a tippy-bucket-type system, I thought about the possibility of a system that is an intermediate between the two. One lesson I learned in this study was that a natural system is always complex and likely is neither of the first two extreme cases. Thus, it was best to propose an intermediate case, instead of one of the two. If I had chosen one of them and had been wrong, I would have been criticized or ignored. Thus, I tried to consider an intermediate type, as illustrated in Figure 1.13c. The three cases were presented during a substorm conference in Los Alamos in 1978 and are illustrated a little more quantitatively in Figure 1.13d.

In order to examine the relationship between $I(t)$ and $O(t)$, Paul Perreault and I estimated the total output $U_T(t) = O(t)$, in units of power (erg/sec), on the basis of the two geomagnetic indices AE(t) and Dst(t) for a large number of geomagnetic

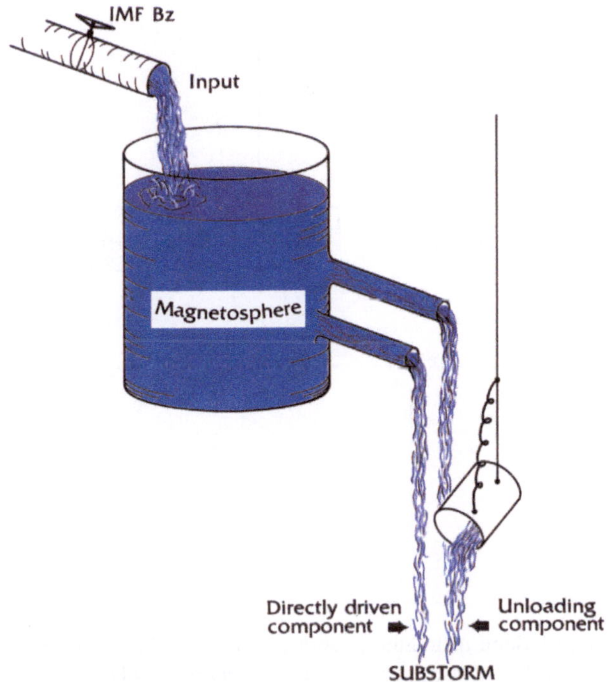

FIGURE 1.13c. Model combining both the pitcher and tippy bucket models.
Source: Akasofu, S-I., *EOS*, **70**, 529, 1989

storms, and then tried to find a combination of solar wind parameters that has the dimension of power and that resembles the output function in terms of time variations (Perreault and Akasofu 1978). The first input function we examined was the kinetic energy flux $((1/2)mnV^2)$. However, we found that there is no obvious relationship between this quantity and the output function. Actually, it was obvious even from the early study by Chapman and myself that an enhanced solar wind flow is not a necessary condition for the development of geomagnetic storms (Figure 1.14).

By then, it had been gradually confirmed by Roger Arnoldy (1971), Ching Meng et al. (1973), and many others that each substorm is associated with a specific change of the IMF *Bz* component, namely from a positive value to a negative value, as Fairfield observed first. Thus, obviously, the next simple combination of solar wind parameters that has the dimension of power (erg/sec) and that considers the *Bz* effect has the form of:

$$\varepsilon = VB^2 \sin^4(\theta/2) l_o^{\,2}$$

where V, B, and θ denote the solar wind speed, the IMF magnitude, and its polar angle; l_o is a constant ~ 7 Earth radii.

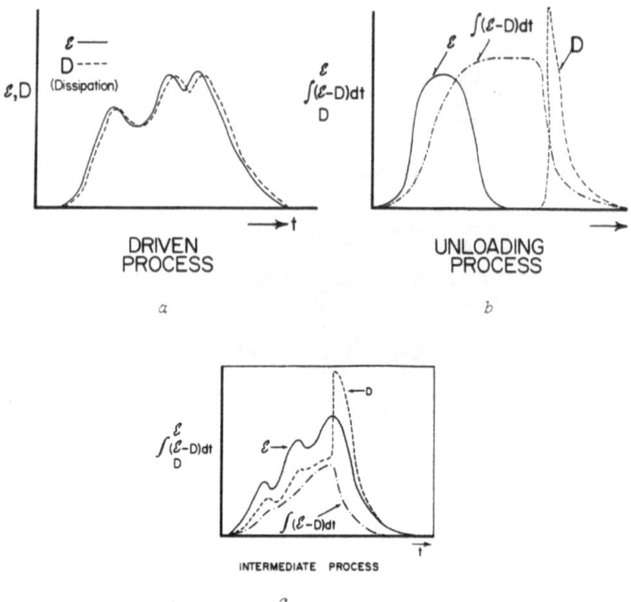

FIGURE 1.13d. Semi-quantitative representations of Figures 1.13a–1.13c. *Source*: Akasofu, S.-I., *Dynamics of the Magnetosphere*, p. 447, D. Reidel Pub. Co., Dordrecht, Holland, 1980

In this regard, an important development was that Mikhail Pudovkin and his colleague (1986) identified ε as the Poynting flux across the magnetopause. This is a theoretical confirmation that the ε function can be identified as the power generated by the solar wind-magnetosphere dynamo. More specifically, the magnetopause is where the solar wind-magnetosphere dynamo is located. M. Ali Alpar and J. Shaham (1989) applied the above formula of ε to pulsars.

By considering the range of variability of V, B, and θ in ε, θ is most crucial, then B, while effects of V are very small. We were surprised at how well the ε function reproduces the output function U_T. In Figure 1.14, we estimated the total energy dissipation rate U_T (the total output rate) from the AE and Dst indices (namely, pure magnetospheric quantities) and compared it with the kinetic energy flux (K) and ε (namely, pure solar wind quantities). One can easily recognize a close relationship between U_T (a magnetospheric quantify) and ε (a solar wind quantity), but not between U_T and K.

One of the most important conclusions derived from this study is that the magnetospheric substorm is the *element* of global magnetospheric disturbance. It is the response of the magnetosphere to a significant increase of the solar wind-magnetosphere coupling for a few hours or more. Most of what we call *auroral phenomena* are visible manifestations of electromagnetic energy dissipation processes of this particular global disturbance. Therefore, the solar

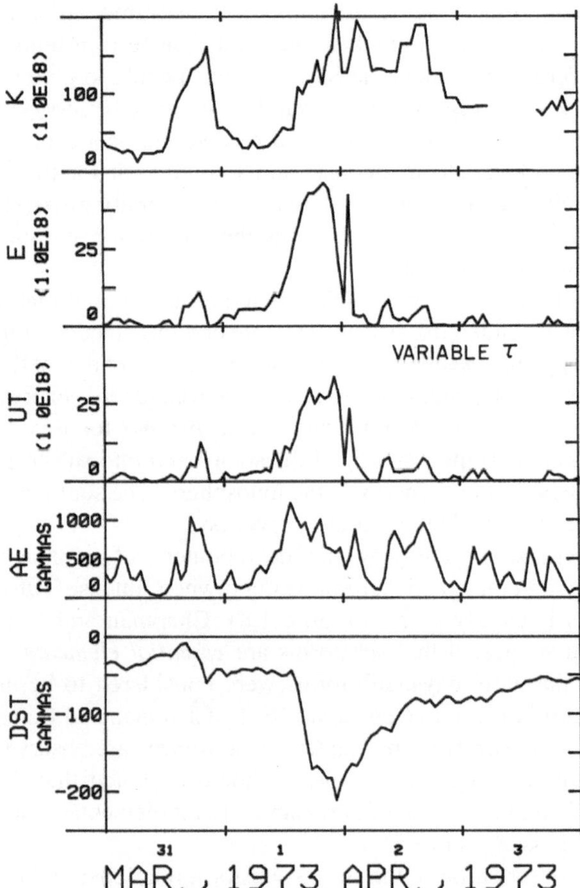

FIGURE 1.14. The kinetic energy flux (K), the ε (E) function, the total energy dissipation rate (UT) and the geomagnetic storm indices AE and Dst for the geomagnetic storm of March 31–April 3, 1973.
Source: Akasofu, S.-I., *Space Sci. Rev.*, **28**, 121, 1981

wind-magnetosphere coupling must constitute a dynamo that can supply the power for the dissipation process. Indeed, the ε parameter represents the power of this dynamo process for magnetospheric substorms. A typical magnetospheric substorm occurs when ε exceeds about 10^{18} erg/sec for a few hours. Soon after the publication of our results, Pat Reiff, et al. (1981) found the polar cusp potential is proportional to $\varepsilon^{1/2}$. This potential is approximately the voltage developed by the magnetospheric dynamo process; it is about 30–200 kilovolts (kV).

It so happened that the first libration point satellite, S3, was launched at that time, at the point of about 200 Earth radii distance, the gravitational pull of the Earth and that of the Sun are supposed to balance. It takes a little less than 60 minutes for the solar wind to reach from that point to the Earth. I was told that

I could receive the solar wind data from S3 on a real-time basis, free of charge, so long as my request was limited to two digits. Since ε in units of erg/sec is about 10^{18}–10^{20}erg/sec, I asked the S3 operations people to give me two digits, 2 and 8, if $\varepsilon = 2 \times 10^{18}$erg/sec, and 5 and 9 if $\varepsilon = 5 \times 10^{19}$erg/sec, and so on. This scheme worked well. Since we could receive the data every 5 minutes on a real-time basis, my graduate students and I could wait for the aurora on the roof of the Geophysical Institute building at the University of Alaska Fairbanks, when ε went up above 10^{18}erg/sec, assuring the occurrence of a substorm a little more than 30 minutes or so later.

In summarizing this section, I think it is important to note that the magnetosphere should be considered a system that converts the kinetic component of the solar wind energy into electromagnetic energy, since geomagnetic and auroral phenomena are various manifestations of electromagnetic energy dissipation processes. The magnetosphere must thus be a *dynamo* for this conversion. It transforms the kinetic (input) energy of the solar wind into substorm energy and eventually into heat (output) energy in the ionosphere. The southward component of the IMF facilitates this energy transfer process.

During the course of studying the development of geomagnetic storms, I realized that a geomagnetic storm occurs when intense substorms occur frequently. This is clearly seen in Figure 1.8a. Chapman and I concluded that this relationship suggested that substorms are *essential elements* of a geomagnetic storm. In the early days, substorms were considered to be unrelated to a geomagnetic storm. In fact, in *Geomagnetism* by Chapman and Bartels substorms were treated as *magnetic bays* in Chapter 10; substorms are observed as bay-like trace in mid-latitude magnetic records. It should be noted that the concept of substorms is different from that of Birkeland's polar elementary storms; see also Section 2.5 on the same subject.

Based on my observation of the storm–substorm relationship, I concluded that substorms are the cause of the ring current belt, injecting high-energy protons from the magnetotail into that belt. Carl McIlwain and his colleagues (1974) showed that both protons and electrons are injected into the ring current belt and drift around the Earth. Meanwhile, there was great surprise that oxygen ions (O^+) become the dominant ions in the ring current belt during an intense geomagnetic storm. Since the oxygen ions in the solar wind are highly ionized (O^{+7}), O^+ ions must be of ionospheric origin. Indeed, a recent observation shows that O^+ ions are ejected out from the ionosphere into the magnetotail at substorm onset. After reaching the magnetotail, these ions are injected into the ring current belt by a convective motion of plasma in the magnetotail. This aspect will be discussed in Section 4.5.

1.10. The Directly Driven and Unloading Components

The component of the output function that closely follows the ε function in time is now called the directly driven component, which is illustrated in Figure 1.13a. The rest is the unloading component, as illustrated in Figure 1.13b. The existence of the directly driven component had not been considered for many years, since the spontaneous reconnection paradigm was so powerful at that time. The directly driven component was *officially* recognized for the first time as late as 1987, in a joint paper by Gordon Rostoker and his colleagues (1987). That the magnetosphere must be driven first for substorms to occur and that substorms are not caused by a spontaneous process finally became clear.

In theorizing about the causes of the unloading component, it is crucial to know its characteristics, at least its time variations. It is rather surprising that proponents of magnetic reconnection have been theorizing substorm processes while ignoring characteristics of the time variations of the unloading component. I have learned that theorists tend to formulate their own problem in their own way and that they try to learn about only what they are interested in.

Observations are forgotten. In this particular case, they formulate a spontaneous and explosive reconnection problem, but are not concerned with the observed time variations. Further, they do not examine how it can possibly be stopped after it begins. I am not aware of even a single paper in this regard. If magnetic reconnection is so fundamental, each substorm should last until the entire tail is burned up.

However, we know that the magnetosphere has both components, as illustrated in Figures 1.13c and 1.13d. Now, the question is whether there is any method for separating the two components, so that we can learn about time variations of the unloading component. For this purpose, Wei Sun and his colleagues (1998) applied the Method of Natural Orthogonal Components (MNOC) to this difficult problem.

From this analysis, they found that the first natural component has a two-cell pattern, which is well known to be associated with the convection in the magnetosphere. It is enhanced during the growth and expansion phases of substorms and decays during the recovery phase of substorms. Further, it has a fair correlation with the ε function with a time lag of 20–25 minutes. Thus, this may be identified as the directly driven component (Figure 1.15).

The second natural component reveals itself as an impulsive enhancement of the westward electrojet, around midnight, between 65° and 70° latitude, during the expansion phase only. It is much less correlated with the ε parameter than is the first one. Thus, as a first approximation, we may identify it as the unloading component. Sun et al.'s analysis showed that the directly driven component tends to dominate over the unloading component, except for a brief period soon after substorm onset. This is the first clear determination of the time profile of the unloading component. Thus, knowing its characteristics and its time profile, it has now become possible to examine, for the first time, the physics of the unloading component and its cause. This problem will be further discussed in subsequent chapters.

The directly driven component is the one in which the energy derived from the solar wind is directly deposited in the magnetotail, the ionosphere, and elsewhere with a slight time delay. Thus, time variations of this component have approximately the same time variations as that of $\varepsilon(t)$. For this component, the equivalent current pattern in the polar ionosphere features two vortices; see the middle pattern in the upper part of Figure 1.15.

MATHEMATICAL SEPARATION OF DIRECTLY DRIVEN AND
UNLOADING COMPONENTS DURING SUBSTORMS

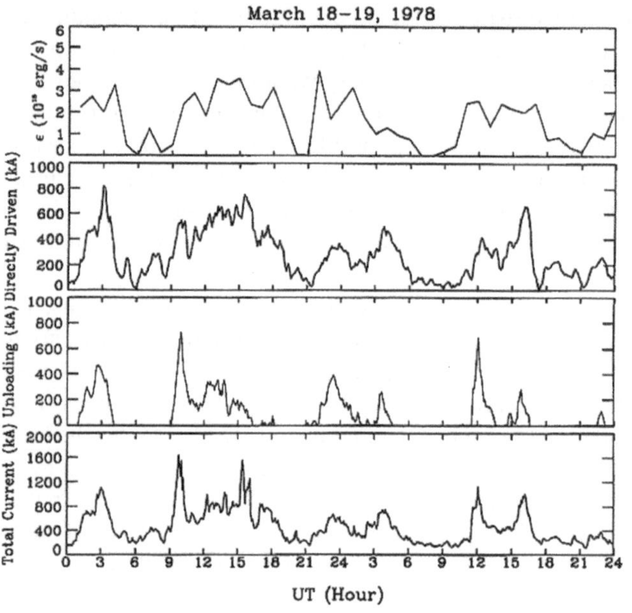

W. SUN, W.-Y. XU AND S.-I. AKASOFU [1998]

FIGURE 1.15. Separation of the directly driven component and the unloading component for the period of March 18–19, 1978.
Source: Sun, W., W.-Y. Xu, and S.-I. Akasofu, *J. Geophys. Res.*, **103**, 11,695, 1998

The unloading component must be caused by a magnetosphere–ionosphere (M–I) coupling process that is presently under intense debate (Chapter 4). The equivalent current pattern in the polar region associated with the unloading component has a single vortex involving a longitudinally confined westward electrojet centered around the midnight sector. Its time variations do not resemble the rate of energy derived from the solar wind (Figure 1.15). Magnetic reconnection and various instability processes beyond 10 Earth radii in the magnetotail have been proposed as the cause for the unloading component. Magnetic reconnection may occur as a result of the M–I coupling process, but it is unlikely to be the cause for substorm onset (Section 4.3). Simply put, the magnetotail (tail) cannot wag the ionosphere (dog). Many fascinating phenomena occur in the magnetotail, but we should not forget the very significant ionosphere.

1.11. The Open Magnetosphere

It appears that magnetospheric physicists did not consider seriously the open model until the beginning of the 1970s. They began to pay attention to the concept of an open magnetosphere when A. Vampola (1971) detected solar electrons of about 400 KeV uniformly over the entire polar region (Figure 1.16). According to Störmer's cut-off latitude calculation, these electrons could reach only very near the geomagnetic pole. There is no way to explain this phenomenon without considering the magnetosphere to be open. The only possible interpretation of this phenomenon is that these electrons reach the polar cap along the magnetospheric magnetic field lines, which are connected with the interplanetary magnetic field lines; these field lines are in turn connected to the Sun. Here, the polar cap is defined as the area where the open field lines originate. Further, the dayside and nightside boundaries of the area where the electrons had been detected coincided with those of the auroral oval (Chapter 2), indicating that the auroral oval delineates approximately the boundary of the polar cap.

I recall that during my visit to the University of Iowa in the early 1960s, my colleagues and I found a very strange phenomenon. Protons of energies well below Störmer's cut-off energies were sometimes observed deep in the so-called *forbidden region* (Figure 1.17). However, we had no idea how to explain this phenomenon, since the Chapman–Ferraro theory predicts that the equatorial boundary of the forbidden region is even higher when the Earth's dipole field is compressed by the solar wind. Now it may well be that this anomalous phenomenon is related to the fact that the magnetosphere is open.

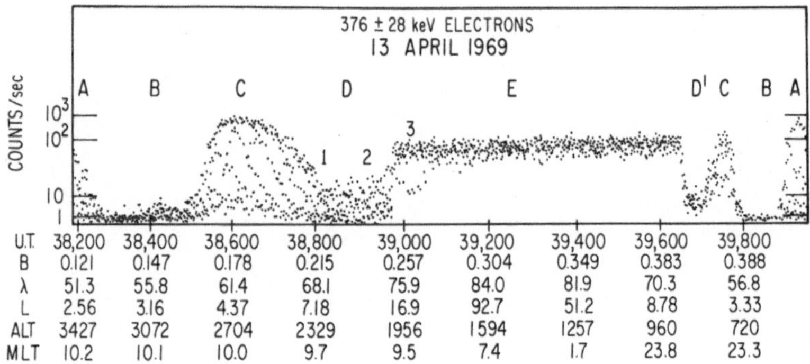

FIGURE 1.16. Observation of solar electrons as a satellite crossed the polar region from the late morning sector (9.5 MLT) to the late evening sector (23.8 MLT). Across the polar cap, the electron flux was uniform.
Source: Vampola, A.L., *J. Geophys. Res.*, **76**, 36, 1971

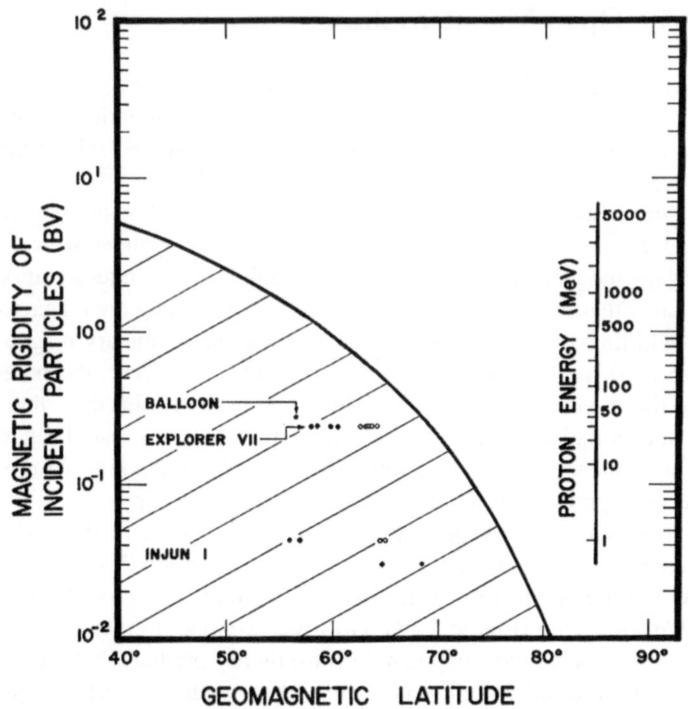

FIGURE 1.17. Störmer's *forbidden region* in magnetic latitude-magnetic rigidity diagram. Solar protons were observed deep in the forbidden region.
Source: Akasofu, S.-I., W.C. Lin, and J.A. Van Allen, *J. Geophys. Res.*, **68**, 5327, 1963

Chapman and Akasofu: Geophysical Institute, University of Alaska (1961).
Source: Akasofu, S.-I., Geophysical Institute, University of Alaska, 1961

Alfvén and Akasofu: Geophysical Institute, University of Alaska (1974).
Source: Akasofu, S.-I., Geophysical Institute, University of Alaska, 1974

Chapman and Akasofu: Geophysical Institute, University of Alaska (1964).
Source: Akasofu, S.-I., Geophysical Institute, University of Alaska, 1964

Alfvén and Akasofu in front of Chapman's bust at the Chapman Building of the University of Alaska Fairblanks (1974).
Source: Akasofu, S.-I., Geophysical Institute, University of Alaska, 1974

2
Confronting Paradigms: Aurora Research During the Early Space Age

2.1. My Earliest Association with the Aurora

My mother had a favorite song. It was a sentimental popular song, which she used to sing ever since she was a young girl. Its title is something like *A Drifter's Song* and starts with:

> I have to decide to go ahead or return home under the aurora. Russia is a big country, sunset in the western part, sunrise in the eastern part, a noon bell in the middle ...

My mother would sing this song to me, when I was five-years old. The only word I did not understand in the song was *aurora*, and I asked my mother about it. If I remember correctly, she told me that it was something she hadn't seen, but is a beautiful phenomenon in a far northern country. This was my first encounter with the word *aurora*.

I was born in a small town in the mountainous region of central Japan, only 10 miles from Mt. Asama, one of the most active volcanoes in Japan. One of my earliest childhood memories is a gigantic nighttime eruption, which I observed from my mother's back, while crying in fear. The elementary school I attended was small, but was very well equipped with scientific instruments. I recall I was, and still am, fascinated by lights from vacuum discharge tubes, which are closely related to the aurora, although I was obviously not capable then of associating the lights with the aurora.

The Department of Geophysics of Tohoku University, which I attended, was staffed with famous professors. Among them were Yoshio Kato (geomagnetism), Gi-Ichi Yamamoto (atmospheric sciences), Kokichi Honda (seismology), Hiroshi Kamiyama (ionospheric physics), among others. The department operated a magnetic observatory where I worked to earn wages. There, several magnetometers recorded magnetic changes. In the magnetometers, a light beam was deflected from a mirror attached to a magnet and produced a spot on a photographic paper wrapped on a rotating cylinder in a dark room. I was greatly attracted by movements of the light spot and learned that the movements were caused by the aurora, an electrical discharge phenomenon, in Siberia and Alaska. It was fascinating to imagine how a distant phenomenon like the aurora could cause delicate movements of the spot. It was during this time, my student days, when I associated the memory of my mother's song with what I was learning.

However, as mentioned in Chapter 1, it was Chapman–Ferraro's paper that brought me to Alaska.

2.2. The Auroral Zone to the Auroral Oval

E. Loomis (1860) was the first to assemble the first extensive collection of auroral appearances over the Earth and found that the aurora tends to appear most frequently along a fairly narrow belt centered around a point at the northwestern tip of Greenland, not at the geographic pole (Figures 2.1 and 2.2). H. Fritz (1873), using much more data covering the period from 503 B.C. to A.D. 1872, confirmed Loomis' findings and constructed his well-known map of isochasms the lines of equal average annual frequency of auroral visibility expressed by "M" nights per year. The maximum frequency of auroral visibility thus defined was found to lie approximately along Loomis' belt. This auroral belt has been called the auroral *zone*. The centerline of the aurora zone coincides well with a geomagnetic latitude (gm lat.) of 67°. The width of the auroral zone is about 5°–6° of latitude. Thus, on a geomagnetic longitude–latitude map centered around the geomagnetic pole (located near the northwestern tip of Greenland), the auroral zone is a circumpolar belt (Figure 2.3). Harry Vestine (1944) refined Fritz's isochasm map with the aid of additional data covering more than a century, including the two International Polar Years.

FIGURE 2.1. E. Loomis (1811–1889). *Source*: Courtesy of Yale University

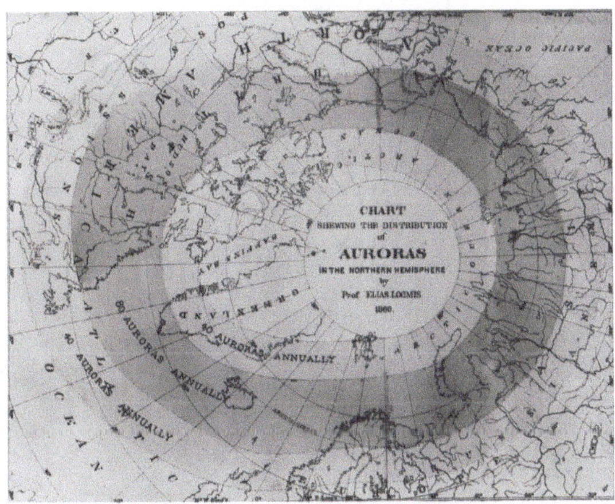

FIGURE 2.2. The auroral zone determined by Loomis.
Source: Loomis, E., *Amer. J. Sci. and Arts*, **30**, 89, 1860

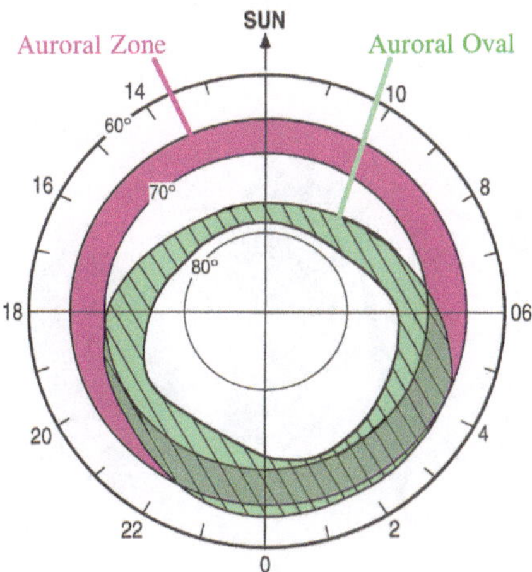

FIGURE 2.3. The auroral zone (red), the auroral oval (green) on the geomagnetic latitude-magnetic local time coordinate system.
Source: Akasofu, S.-I., *Aurora and Airglow*, ed. by B.M. McCormack, p. 267, Reinhold Pub. Co., New York, 1967

Figure 2.4 shows an auroral sketch made by N. Carlheim-Gyllensköld at Cape Thordsen in Svalbard during the First Polar Year (1882). This was one of the first scientific recordings of the aurora. A photographic method was introduced in auroral physics at the beginning of the twentieth century (Figure 2.5). A number of auroral expedition parties were dispatched to Greenland, Siberia, Canada, and many other countries during the Second Polar Year (1932). The isochasm map

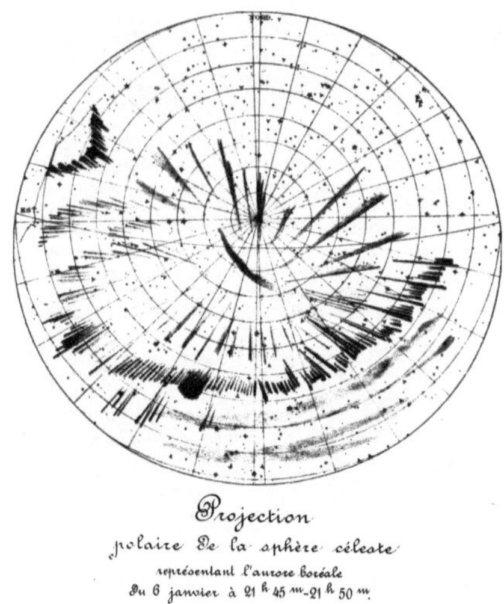

FIGURE 2.4. Sketch of the aurora by N. Carlheim-Gyllensköld at Svalbard (1907).
Source: Carlheim-Gyllensköld, N., *Observations faites au Cap Thordsen*, Spitzberg, vol. II, Stockholm, 1882

FIGURE 2.5. Photography was introduced in auroral science at the beginning of the twentieth century. C. Störmer (sitting) used it extensively in determining the height of the aurora.
Source: Courtesy of University Tromsø

was further refined by Yasha Feldstein and his colleagues (1961) and Bengt Hultqvist (1961), based on International Geophysical Year (IGY) data.

Based on such studies, it had been tacitly believed for more than 100 years that the auroral zone was the actual belt along which the aurora lies. It was Sydney Chapman, president of the IGY, and Chris Elvey, director of the Geophysical Institute, University of Alaska, who thought that the actual belt of the aurora should be determined photographically, not by statistics as done by Loomis,

Fritz, and Vestine. For this purpose, they took the leadership in constructing all-sky cameras (Figures 2.6a and 2.6b).

Auroral researchers in several countries responded to Chapman and Elvey by designing and constructing their own all-sky cameras. During the IGY, such cameras were operated at more than 100 locations and took photographs of the sky at one-minute intervals, regardless of sky conditions. The films were then sent to the World Data Center in Moscow and the Geophysical Institute, University of Alaska.

When I became a graduate student of the Geophysical Institute in December 1958, I had an opportunity to observe the aurora with my colleagues, including Gene Wescott and Charles Deehr. I observed that the aurora tends to appear in

FIGURE 2.6a. All-sky camera and a photometer.
Source: Courtesy of Geophysical Institute, University of Alaska

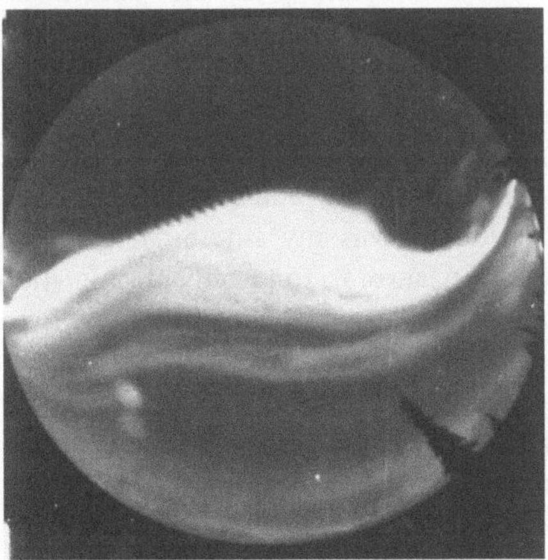

FIGURE 2.6b. An example of a photograph taken by an all-sky camera.
Source: Courtesy of Geophysical Institute, University of Alaska

the northern sky in the evening, advances toward the zenith (or even the southern sky) of Fairbanks (gm lat. 64.6°), and recedes toward the northern sky in the morning. This north-south shift of auroral arcs was a well-known fact by then (V.R. Fuller and E.H. Bramhall, 1937; Jim Heppner, 1954). I recall that I asked Elvey why this shift occurs, if auroral arcs were supposed to lie along the auroral zone. His response was that it was perhaps because auroral arcs tend to form at the centerline of the auroral zone (gm lat. 67°) and then the auroral arcs, after their formation, move equatorwards.

My question was simply that if the concept of the auroral zone was correct, we should be able to see auroral arcs near the zenith of the sky above Fairbanks at 6 p.m. when the sky becomes dark enough (actually, in Fairbanks, the sky is dark enough to observe the aurora even before 5 p.m. around the winter solstice). Instead, auroral arcs almost always appear near the northern horizon first and advance equatorward. My question to Elvey was the naïve one of a graduate student.

After this conversation with Elvey, I examined newly arrived IGY all-sky films taken at Fort Yukon, Alaska (gm lat. 66.6°), which is located at about the center line of the auroral zone. To my great surprise, auroral arcs behaved in a similar way at Fort Yukon as in Fairbanks. That is, auroral arcs appeared first near the northern horizon. Therefore, I also examined all-sky films from Barrow, Alaska (gm lat. 68.5°), well north of the centerline of the auroral zone. It was even more surprising to me that auroral arcs behaved in a similar way at Barrow. The only difference is that the local time of the first appearance in the northern sky and of the arc arrival at the zenith are earliest at Barrow, than at Ft. Yukon, than at Fairbanks. Figure 2.7a shows simultaneous all-sky camera photographs from Sachs Harbor (gm lat. 76.0°), Inuvik (gm lat. 71.0°), Fort Yukon, and College. The photographs show the equatorward shift of the aurora in the evening; see also Figure 2.8a.

It was quite obvious to me at that time that auroral arcs do not lie along the auroral zone. I realized, also, that Loomis, Fritz, and others did not and could not take into account the local time dependence of the auroral distribution (namely, only how many nights per year) in their statistical study, meaning that the instantaneous belt of auroral arcs can be quite different from the auroral zone.

All-sky films from many IGY arctic stations started to arrive at the Geophysical Institute in 1959 and 1960. It was my finding that the actual distribution of auroral arcs agrees with the auroral zone only during the midnight hours and deviates greatly from the auroral zone at the other local times. However, I could not determine the auroral distribution on the day-side of the Earth because of the lack of data at that time.

Yasha Feldstein (1963) determined the complete distribution of the aurora at all local times, using the films from Heiss Island and other sites that can observe midday auroras (Figure 2.7b). His distribution showed that the belt of the auroral zone is located at about 78° during midday hours, instead of 67° (Figure 2.3). Further, the center of the belt is shifted by about 3° from the geomagnetic pole toward the midnight sector. This belt is called the auroral *oval*.

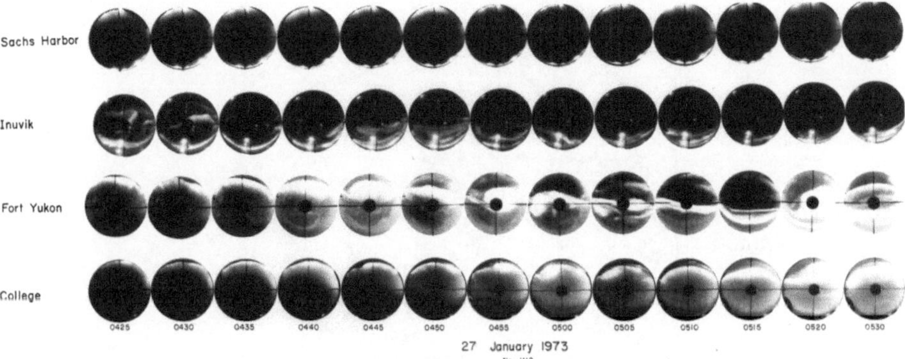

FIGURE 2.7a. All-sky photographs from the Alaska meridian chain (Figures 2.8a and 2.8b), showing the equatorward shift of the aurora. A substorm began toward the end of this series.
Source: Akasofu, S.-I., *Space Sci. Rev.,* **16**, 617, 1974

FIGURE 2.7b. From the left, S.N. Vernov, O.V. Khorosheva, and Yasha Feldstein at the University of Moscow campus at the occasion of my first visit to Russia.
Source: Akasofu, S.-I, 1968

Since the results obtained by Feldstein were basically the same as mine for the dark hours, I supported his results immediately. On the other hand, Knud Lassen, in Copenhagen, proposed once that there were two belts of aurora instead of the auroral oval.

That time was a sort of golden age of auroral spectroscopy. All-sky cameras were not considered even to be a scientific instrument for auroral spectroscopists, compared with their then-sophisticated spectroscopic instruments. In fact, some of my senior colleagues advised me that the aurora should be the same in Alaska, Siberia, Canada, and Norway, that physics of the aurora should be the same everywhere, that the distribution of the aurora is thus not a major issue, and

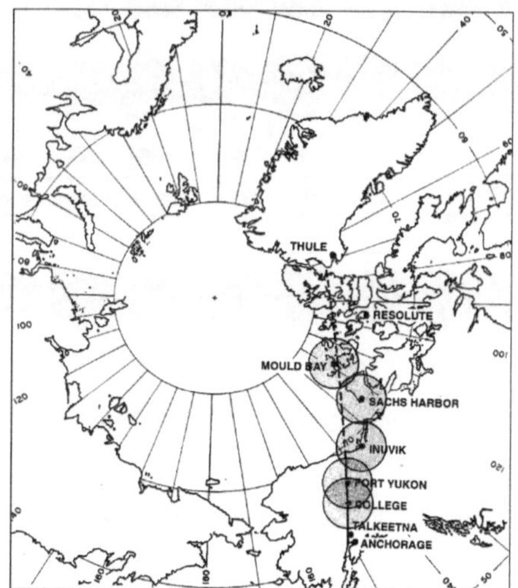

FIGURE 2.8a. Alaska meridian all-sky cameras. A circle indicates the field of view of each camera.
Source: Akasofu, S.-I, 1968

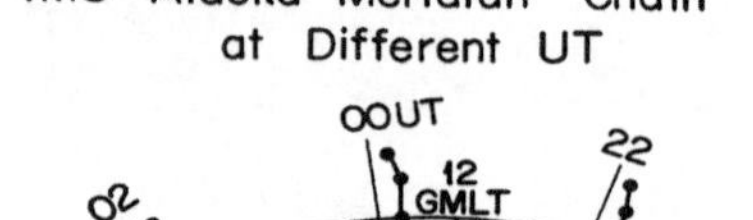

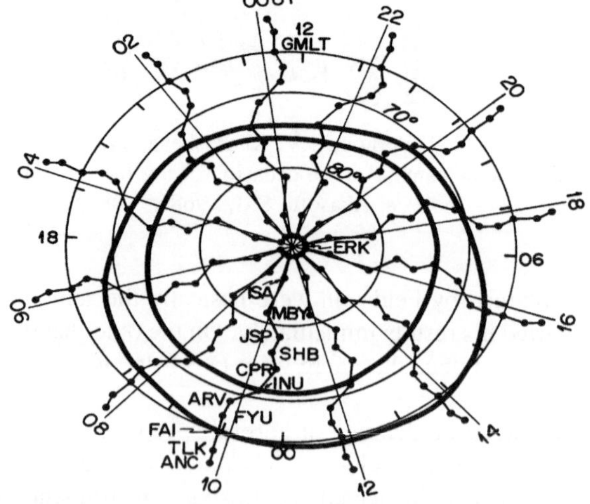

FIGURE 2.8b. Position of the Alaska meridian cameras at different universal times (UT) on geomagnetic latitude-magnetic local time (MLT) coordinates.
Source: Snyder, A.L., Ph.D. Thesis, University of Alaska, 1972

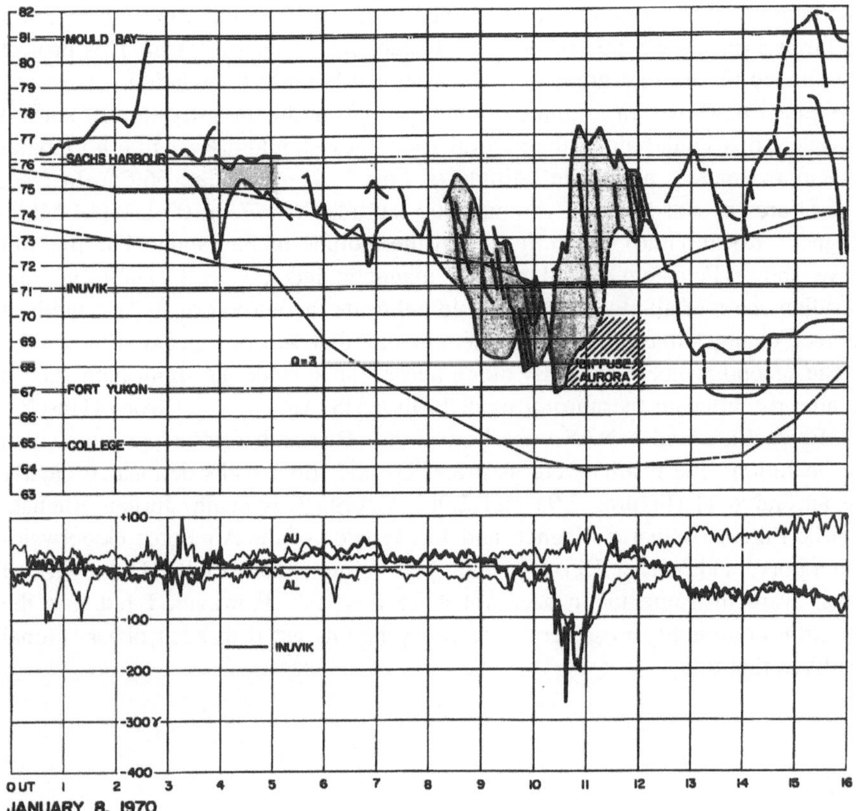

FIGURE 2.8c. Location of the aurora observed by the Alaska meridian all-sky cameras. *Source*: Snyder, A.L. and S.-I. Akasofu, *J. Geophys. Res.*, **77**, 3419, 1972

thus was a waste of time to work on it. I objected to this argument. Auroral arcs appear in a very specific belt, the auroral oval, and not along the auroral zone, and not all over the polar region. This fact tells us something about their cause and origin. Therefore, it is important to determine their actual distribution accurately.

In such an atmosphere, Feldstein's results got little attention from the scientific community. Worse, since the auroral zone had been believed to be the belt of auroral arcs for more than 100 years, it was difficult for both of us to convince our colleagues of the validity and significance of the auroral oval.

In order to convince the scientific community that Feldstein's and my views about the auroral oval were valid, I planned several projects. The first was to establish the Alaska meridian chain of all-sky cameras (Figures 2.8a). Taking advantage of the Earth's rotation, a meridian chain of all-sky cameras can scan the entire polar sky (like an azimuth-scanning radar at an airport) once a day, and delineate the auroral oval that is fixed with respect to the Sun (Figure 2.8b). As far as I am aware, this is the largest scanning device on Earth. This project was funded by my first grant from the National Science Foundation.

Figure 2.8c shows an example of the results from this investigation. If auroral arcs were distributed along the auroral zone, they should appear in a horizontal belt approximately along the latitude of Fort Yukon (gm lat. 66.6°). Instead, auroral arcs appear at about geomagnetic latitude 76°–77° at 00 UT (14 MLT, Magnetic Local Time) and shift toward the latitude of Fairbanks. The line dot curve shows Feldstein's oval for the magnetic index Q = 3. Therefore, the meridian chain of all-sky cameras could delineate the auroral oval. The width of the oval changes intermittently, a phenomenon that will be discussed later.

The second project was to fly along auroral arcs, since the flight path should be able to delineate the auroral oval. Both a US Air Force jet from Hanscom Air Force Base and a NASA jet from Ames Research Center participated in the operation. The results were as predicted: the flight paths delineated clearly the auroral oval (Figures 2.9a and 2.9b). George Gassmann, Jurgen Buchau, Charlie Pike, Rosemarie Wagner, and Jim Whalen of the Air Force Geophysics Laboratory, and Walter Heikkila and Dave Winningham of the University of Texas were instrumental in accomplishing this task. However, I felt that the scientific community in general was not very interested in such observational results at that time.

FIGURE 2.9a. A U.S. Air Force aircraft that participated in auroral research. *Source*: Courtesy of G. Gassmann

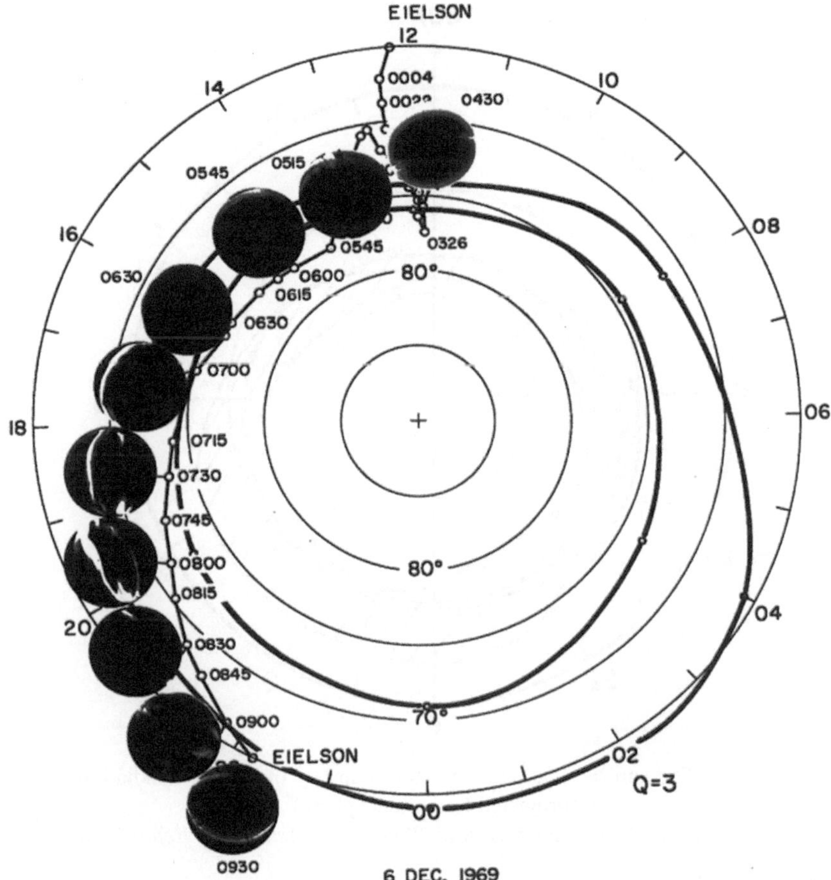

FIGURE 2.9b. All-sky photographs taken from a U.S. Air Force jet that flew along the evening half of the auroral oval.
Source: Buchau, J., J.A. Whalen, and S.-I. Akasofu, *J. Geophys. Res.*, **75**, 7147, 1970

2.3. The Auroral Oval as the Natural Coordinate System

One lesson I learned in elucidating the auroral oval is that one specific finding alone will not get much attention from the scientific community. When one finds an interesting phenomenon, it is necessary to relate it to other significant phenomena and demonstrate that a new finding is worth paying attention to. Thus, the third attempt was to find other geophysical phenomena that have a geographic distribution similar to the auroral oval. Fortunately, I had an opportunity to work with the space physics group of the University of Iowa. I found one day that Lou Frank, James Van Allen, and John Craven were plotting the outer boundary of the outer radiation belt onto the Earth's surface.

I was greatly surprised that the boundary they delineated coincided fairly well with the auroral oval (the solid curve in Figure 2.9c). This result suggested to

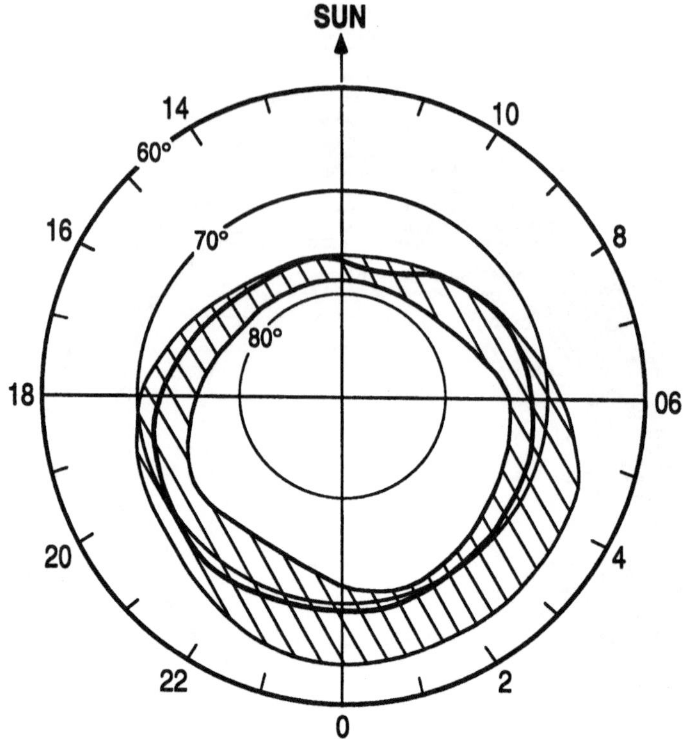

FIGURE 2.9c. The auroral oval and the outer boundary of the outer radiation belt (projected to the ionosphere).
Source: Akasofu, S.-I., *Department of Physics and Astronomy Report*, 1967

me that auroral electrons penetrate into the polar upper atmosphere by moving along the outer boundary of the outer radiation belt. I remember that I reported the results immediately to Van Allen. It was a time when the initial hope of associating auroral phenomena with the radiation belts had faded, so initially convincing my colleagues of this finding's significance was difficult. After the discovery of the plasma sheet, this result had long been forgotten, and it was only during the last few years that some researchers have been coming back to the boundary of the outer radiation belt in their search for the origin of auroral arcs.

Thus, it was fortunate that Al Zmuda and his colleagues (1966) found on the basis of TRIAD satellite data that field-aligned currents flow in or out from a belt that is basically identical to the auroral oval. He told me that he plotted the location of the field-aligned currents on my figure (Figure 2.9c) after finding it in one of the University of Iowa reports. This fact suggested that auroral electrons carry field-aligned currents. It so happened that the tape recorder aboard the satellite failed, so Zmuda asked me to help install his satellite receiving station at the top of the Geophysical Institute building. It was installed when the temperature was 50 degrees below zero Fahrenheit. Using his satellite data and

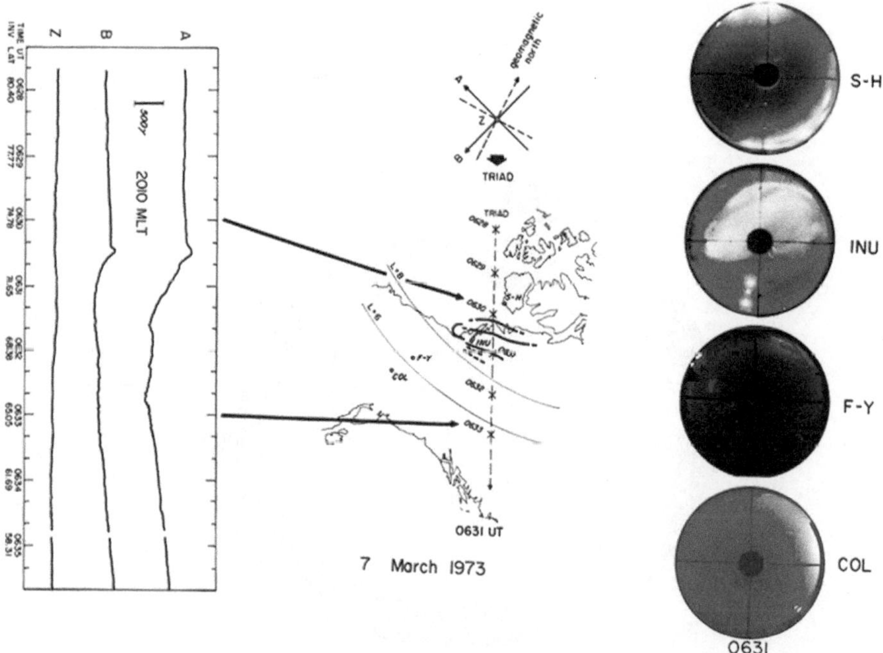

FIGURE 2.10. Simultaneous observations of the aurora (all-sky photograph) and the field-aligned currents (TRIAD satellite).
Source: Kamide, Y. and S.-I. Akasofu, *J. Geophys. Res.*, **81**, 3999, 1976

the simultaneous all-sky data, we found that auroral arcs appear where there is upward field-aligned current (Figure 2.10).

Takeshi Iijima and Tom Potemra (1976) completed Zmuda's work by showing the distribution of field-aligned currents at the ionospheric level (Figure 2.11). Further, solar protons of energies on the order of 1.5 MeV were found to penetrate uniformly over the polar region bounded by the aurora oval. Energetic solar electrons were also found in the area bounded by the auroral oval (Figure 1.16). These results indicated that the auroral oval delineates approximately the boundary of the polar cap. The field lines that originate at the polar cap are connected with the interplanetary magnetic field lines, so that they are "open" field lines. In this way, the significance of the auroral oval was firmly established. I learned thus that it is very important to find as many relevant results as possible in proving the importance of a newly observed result.

The validity and significance of the auroral oval began to be recognized toward the beginning of the 1970s. However, we had to wait for full recognition of the auroral oval until 1971, when a scanning instrument devised by Cliff Anger, and installed on the ISIS 2 satellite, imaged the entire oval (Figures 2.12a and 2.12b). Tony Lui came to Alaska as a postdoctoral fellow, starting joint projects on

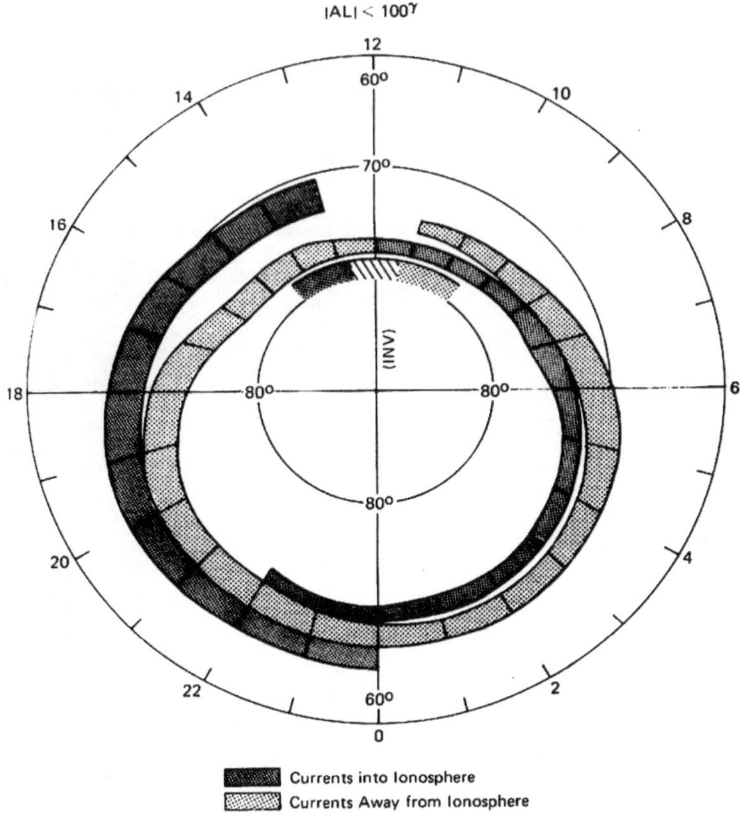

FIGURE 2.11. Distribution of the field-aligned currents.
Source: Iijima, T. and T.A. Potemra, *J. Geophys. Res.* **83**, 599, 1978

ISIS-2 data with the University of Calgary group. Their work extended Anger's observation. After this, the concept of the auroral oval was accepted as if there had been no controversy about it in the past. In any modern monograph on the aurora, one can find a simple statement that auroral arcs lie along the auroral oval. Thus, it is interesting to recognize that such a simple fact – perhaps one sentence in modern textbooks – had a long history; it took about a decade of struggle for it to be accepted by the scientific community.

Another important finding about the auroral oval is that it provides the natural coordinate in organizing a great variety of polar upper atmosphere phenomena. Before the aurora oval was established, all aurora-related phenomena were organized in terms of the geomagnetic coordinate system. However, the same or similar phenomena occur along the auroral oval, not along geomagnetic latitude circle of say 67°. The same or similar phenomena occur along the line equatorial distance from the oval, not along the same latitude circle.

FIGURE 2.12a. Cliff Anger's scanning instrument aboard the ISIS-2 Satellite.
Source: Courtesy of C. Anger

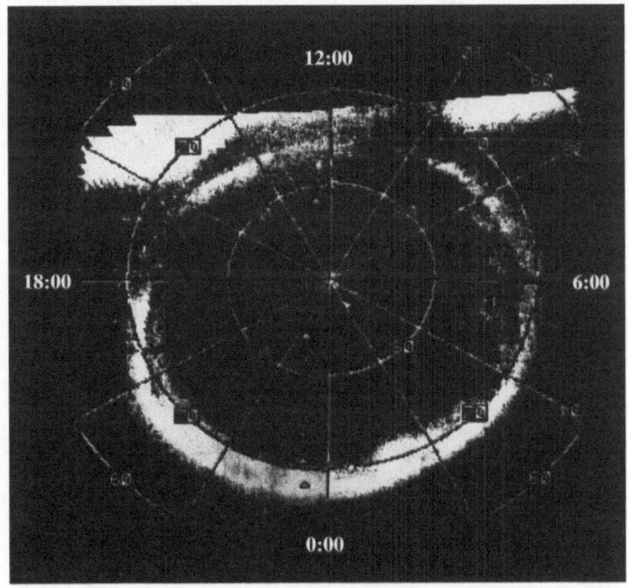

FIGURE 2.12b. First image of the auroral oval depicted by Cliff Anger's instrument.
Source: Courtesy of C. Anger

For me, it started out with a naïve question about daily auroral behavior that was well known at the time. Recalling those days, I appreciate the foresight and courage of both Chapman and Elvey for taking the leadership of the all-sky camera project in spite of the fact that auroral spectroscopists and auroral physicists in general paid little attention to it. I should note that, as far as I know, Neil Davis, Carl Gartlein, Alexander Lebedinsky, and W. Stoffregen were the first who used an all-sky camera for aurora studies.

2.4. Auroral Substorms: Fixed Pattern to Substorm Pattern

It had long been believed, on the basis of the pioneering study of the aurora by V.R. Fuller and E.H. Bramhall (1937) and also by Jim Heppner (1954), that in the evening sky, auroral arcs *always* had a quiet and homogeneous form, that auroral arcs were *always* very active in midnight hours and that arcs were broken up into patchy forms in the morning sky (Figure 2.13). In this view, the auroral activity pattern is fixed with respect to the Sun (and thus to magnetic local time). Observers at points on the earth, rotate with it, under such a fixed pattern of activity once a day. That is, a single observer at a point observes a quiet form, an active form, and a patchy form in the evening, midnight, and morning skies, respectively, during the course of a night. This is certainly true statistically.

At the beginning of the IGY (1957–1958), however, little was known about how auroras behave *simultaneously* in Siberia (in evening hours) and Canada (in morning hours), when auroras became suddenly active over the Alaskan

AURORAL RESEARCH
at the
UNIVERSITY *of* ALASKA
1930 - 1934

VERYL R. FULLER *and* ERVIN H. BRAMHALL

VOLUME III
MISCELLANEOUS PUBLICATIONS OF THE UNIVERSITY OF ALASKA
1937

FIGURE 2.13. Cover page of the report by V.R. Fuller and E.H. Bramhall. *Source*: Courtesy of University of Alaska

sky (in midnight hours). Up until then, there had not been any simultaneous observations of auroras over a long local time span up to then. At the Rasmuson Library of the University of Alaska Fairbanks, I found a letter from C. Störmer to Fuller and Bramhall, urging them to conduct, jointly, simultaneous observations of the aurora in Norway and Alaska. It must have been Störmer's dream to make such a joint observation.

As I began to examine IGY films, I found that the view commonly held on the auroral activity pattern was incorrect. Indeed, all-sky films even at a single station on the same night showed that auroral arcs can transform themselves from quiet to active and back to a quiet form two or three times (Figure 2.14). This fact suggested to me either that the fixed pattern concept was not correct or that the Earth rotated two or three times in a single night! As a graduate student, I was obviously puzzled, but was overwhelmed at that time by the firm believers in the fixed pattern. Thus, I decided to examine *simultaneous* all-sky photographs from Siberia, Alaska, and Canada, when Alaska was in the midnight sector. It was my finding that when an auroral arc is quiet in Alaska (in the midnight sky), it is also quiet over Siberia (in the evening sky), and Canada (in the morning sky), in

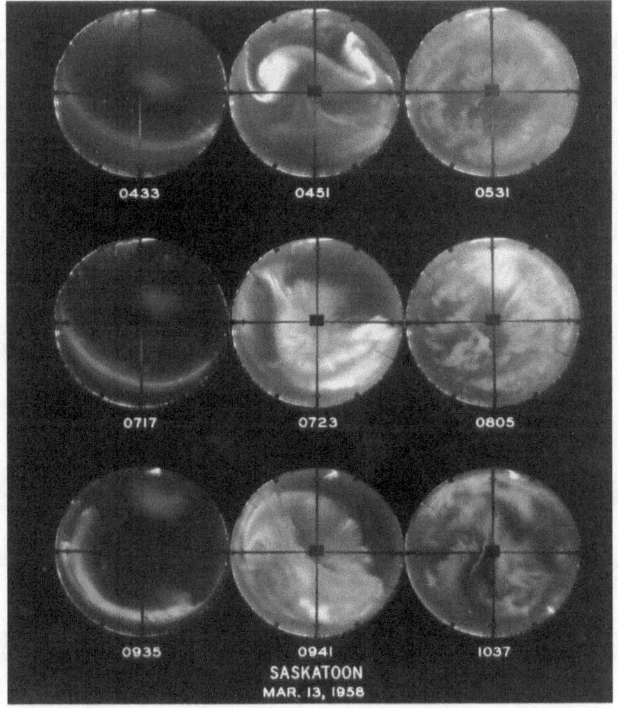

FIGURE 2.14. All-sky photographs taken during the night of March 13, 1958, at Saskatoon, Saskatchewan, Canada. Note that the aurora underwent three cycles of its activity during a single night.
Source: Akasofu, S.-I.

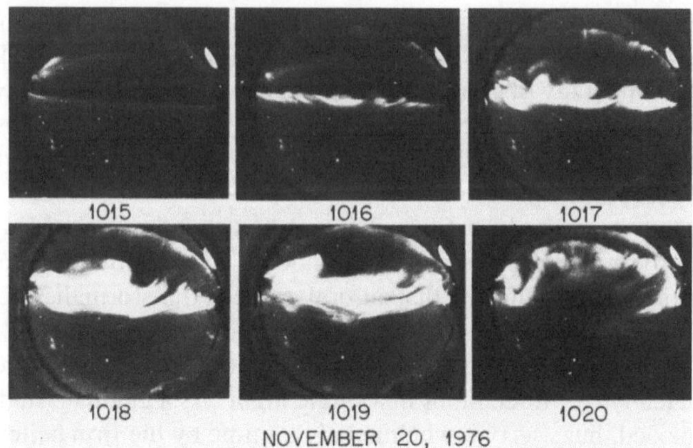

FIGURE 2.15. All-sky photographs showing a sudden brightening of an auroral arc at substorm onset.
Source: Akasofu, S.-I.

addition to the fact that the aurora over Alaska can be quiet even in the midnight hours. When an auroral arc suddenly brightens and moves rapidly poleward over the Alaska sky (Figures 2.15 and 2.16a), this activity generates a large wavy or folding structure (the westward traveling surge), which propagates along the arc toward Siberia (toward the evening sky, Figure 2.16b). This surge like activity was recorded first at the Siberia station closest to Alaska several minutes after its formation over Alaska and, subsequently, at other earlier evening stations in Siberia. This activity could propagate all the way to the dayside of the oval with a speed of a few kilometers per second. At the same time, auroras over

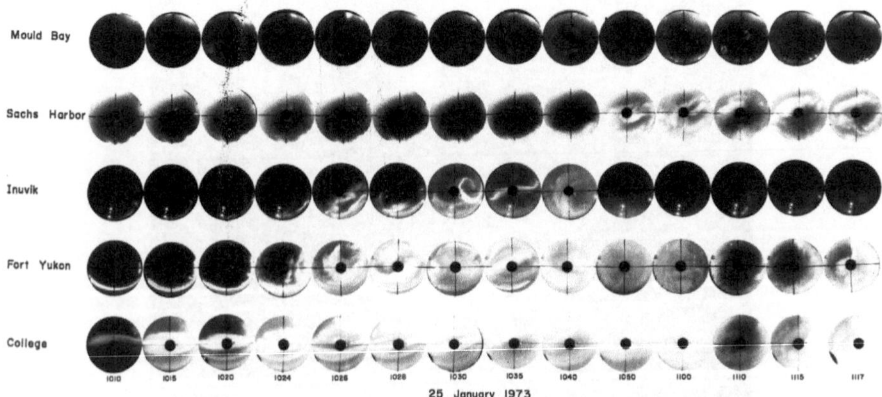

FIGURE 2.16a. Intense poleward expansion of the auroral activity depicted by the Alaska meridian all-sky camera from the zenith of College (Fairbanks) to the northern sky of Sachs Harbor (see Figure 2.8a).
Source: Akasofu, S.-I., *Space Sci. Rev.,* **16**, 617, 1974

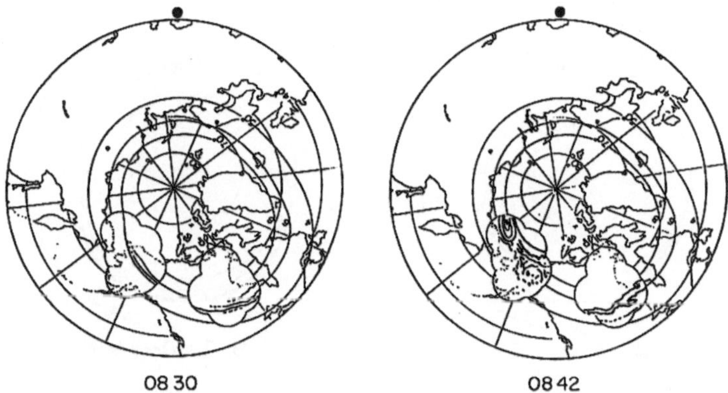

FIGURE 2.16b. Simultaneous auroral activity observed in Alaska, Siberia, and Canada. *Source*: Akasofu, S.-I., *J. Geophys. Res.*, 68, 1667, 1963

Canada became active, often forming an inverted Ω shaped form (called the omega band). To the south of the omega band, auroral arcs became folded in a very complicated way. Folded portions of an arc appear as shafts of light, or patchy forms, scattering all over the sky.

More important, when auroras over Alaska in the midnight sector became quiet again, in about 2–3 hours after an active period, auroras over Siberia and Canada also became quiet. Further, such activity often repeated two or three times during an active night. Chapman coined the term auroral substorm for this transient phenomenon. My paper on this subject initially had a title of *Auroral Activation*. Chapman refused to review my paper unless I changed the title to *Auroral Substorm* (Akasofu 1964). There was little mention of such auroral features in the current and most authoritative book, published by Joe Chamberlain (1961). Therefore, I sent a paper to the *Journal of Geophysical Research* reporting on our findings. The paper was rejected on the basis that there was nothing worth reporting, so I decided to analyze simultaneous all-sky films from a large number of stations. As I did, I became more convinced of the validity of my findings. A new paper was then sent to the late Sir David Bates, the editor of *Planetary and Space Science*, who accepted it without review (Figure 2.17). I assumed this because I received his acceptance letter only about 10 days after sending the paper to him.

However, I found it very difficult at first to convince my colleagues of my auroral substorm findings at first (although this paper was later recognized as one of the most cited papers by the *Science Citation Index* in 1979). This was particularly the case for those who were experienced in observing the aurora. This was because a single observer, standing at a point on the Earth, is carried by the Earth's rotation with a speed of 15° (in longitude) per hour, so that the observer gets the impression statistically that the fixed pattern was correct. Elvey was a firm believer in the fixed pattern concept. Many auroral scientists who have actually little experience in observing the aurora simply followed the

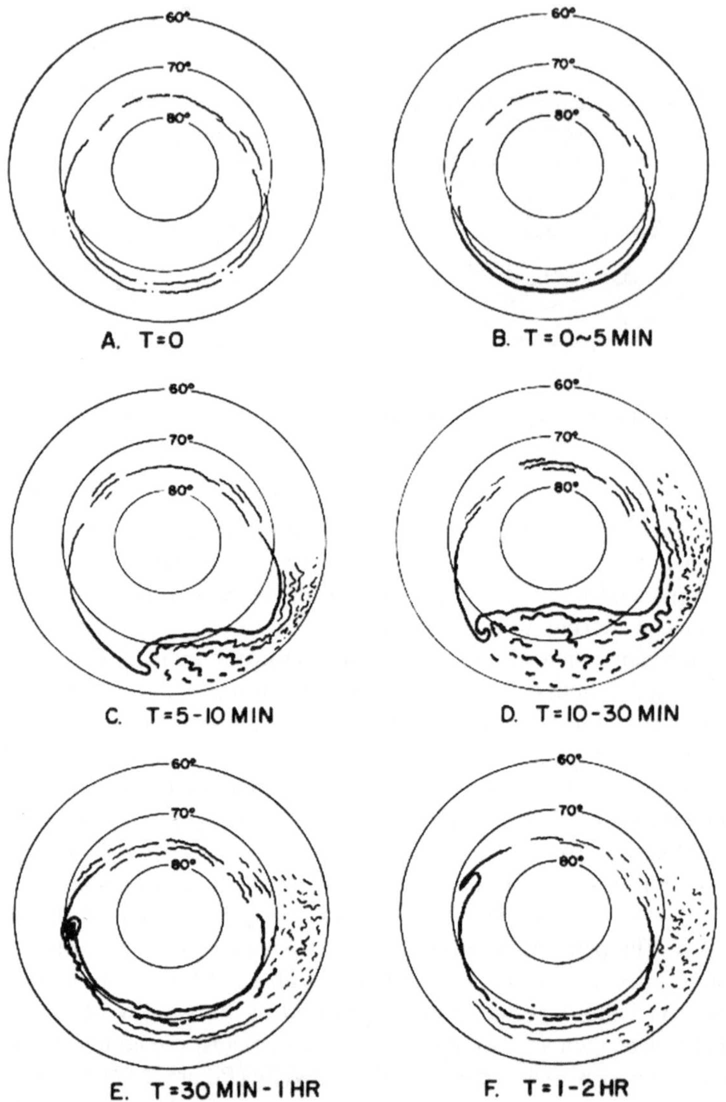

FIGURE 2.17. Schematic view of the auroral substorm as seen from above the geomagnetic pole.
Source: Akasofu, S.-I., *Planet Space Sci.*,**12**, 273, 1964

experienced ones. Thus, it was hard to convince anyone about the validity of the concept of the auroral substorm. The only exception at that time was Feldstein, who strongly supported my findings.

I had to devise a scheme to prove the validity of the concept of the auroral substorm. The best way would have been to observe the aurora from a fixed point (with respect to the Sun) well above the North Pole for many hours, as the

Dynamic Explorer satellite did in the 1980s. In the middle of the 1960s, this was nothing but a dream. One method I conceived was to fly on a jet plane westward, under the aurora, along the latitude circle of 65° or so. Because the speed of a jet plane is about the same as the rotation speed of the Earth at such a high latitude, it can cancel the effects of the Earth's rotation. Thus, a jet plane can stay in the midnight sector for about 6 hours by flying at midnight from the East Coast of the U.S. to Alaska. Both NASA and Air Force jet planes contributed to the so-called *constant local time (midnight) flights* many times for this study (Figures 2.18 and 2.19).

On my way back from one of my trips to Hanscom Air Force Base in Massachusetts, I learned that Elvey, who had since retired in Tucson, Arizona, was critically ill. I decided to visit him. Resting at his hospital bed, Elvey was waiting for my results. We sat together at his bedside to scan the all-sky film obtained on one of the constant local time (midnight) flights, which clearly registered intermittent auroral activities in the midnight sector. We shook hands firmly. He said, "Syun, you did a good job." I believe that I had finally convinced him of the validity of the concept of the auroral substorm. As I shook his hand, I noticed that his arms were just skin and bones. He died about 10 days later.

However, in spite of such an effort, the confirmation of the concept of the auroral substorm had to wait for images from satellites. The first images were obtained from the ISIS-2 satellite in 1971; they showed the clear pattern of the westward traveling surge (Figure 2.20), which is very similar to the auroral substorm pattern I constructed (Figures 2.16b and 2.17). I recall I was naturally excited in examining those images. This observation began to convince some researchers, both believers and nonbelievers alike.

The second set of convincing images came from the Defense Meteorological Satellite Program (DMSP) satellites (Figure 2.21) and showed different stages of the development of auroral substorms. Both polar-orbiting satellites scanned the polar region every 100 minutes or so, and were unable to obtain a sequence

FIGURE 2.18. The NASA jet plane Galileo, which participated in auroral research. *Source*: Ames Research Center, NASA (see also Akasofu, S.-I., *Planet. Space Sci.*, **16**, 365, 1968)

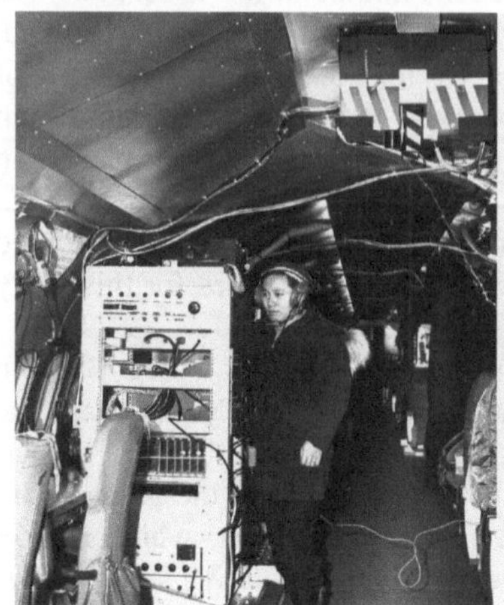

FIGURE 2.19. Left: Operating instruments aboard the NASA jet Galileo: Geophysical Institute, University of Alaska. Right: Explaining instruments on the NASA Galileo to Sydney Chapman and Jerry Romick.
Source: Courtesy of Geophysical Institute, University of Alaska

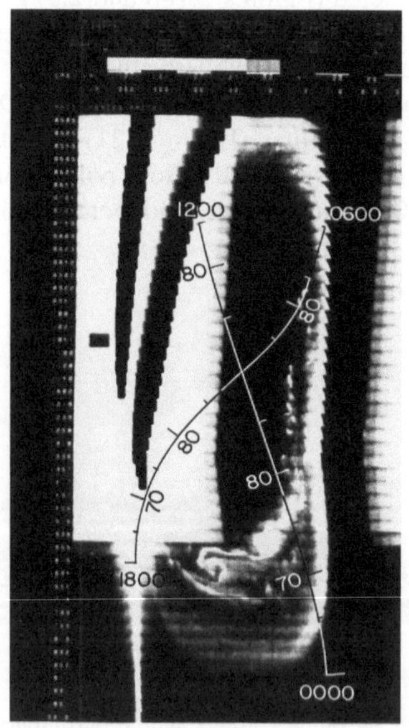

FIGURE 2.20. An auroral image depicting an auroral substorm.
Source: Anger, C.D., A.T.Y. Lui, and S.-I. Akasofu, *J. Geophys. Res.*, **78**, 3020, 1973

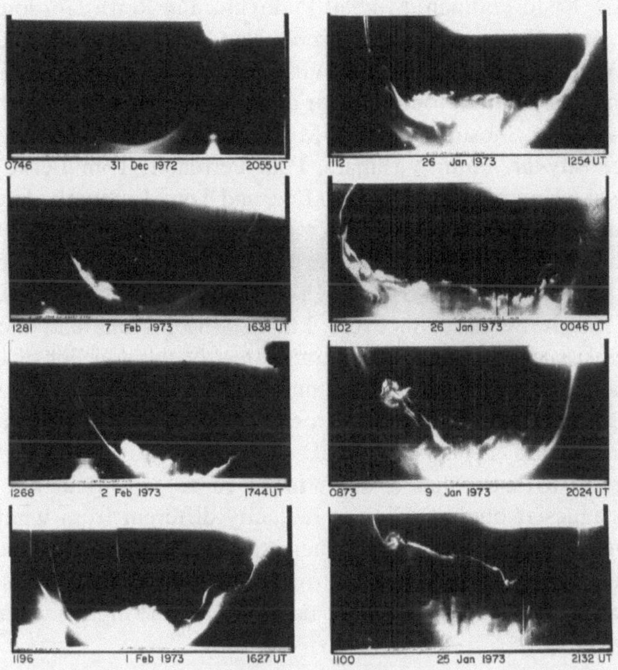

FIGURE 2.21. Snapshots of auroral substorm images by the DMSP satellite. *Source*: Akasofu, S.-I., *Planet. Space Sci.*, **24**, 1349, 1975

of images for a single substorm (as illustrated in Figure 2.22a and 2.22b). Nevertheless, many of my colleagues were surprised by the similarity of such images to each of the sequences of my substorm pattern (Figures 2.17 and 2.21). Ching Meng, Lee Snyder, Don Kimball, Jurgen Buchau, Jim Whalen, and many others joined me in a study of DMSP images.

When the aurora was detected by the DMSP satellite for the first time, the images were classified as "Top Secret" data. I was asked to visit Hanscom Air Force Base to identify "strange lights" in the images. Not being a U.S. citizen then, they allowed me to see only pencil sketches of auroral images at that time. With Lee Snyder, we worked hard to declassify DMSP data.

During the next decade, I was fortunate in that many people realized they could understand and interpret their observational results better in terms of the concept of the auroral substorm rather than of the fixed pattern. Among the early participants were Roger Arnoldy, Dan Baker, Peter Banks, Wolfgang Baumjohann, J. Birn, Asgeier Brekke, Jim Burch, Ferd Coroniti, Stan Cowley, Don Fairfield, Carl-G. Fälthammer, Yasha Feldstein, Lou Frank, Yu Galperin, Ray Greenwald, Don Gurnett, Gerhard Haerendel, Walter Heikkila, H. Herman, Ed Hones, Charlie Kennel, Richard Lundin, W.B. Lyatsky, Larry Lyons, Carl McIlwain, Bob McPherron, V.M. Mishin, Atsusi Nishida, J. Opgenoorth, George

Parks, R. Pellat, Risto Pellinen, Mikhail Pudovkin, Pat Reiff, Gordon Rostoker, Chris Russell, V.A. Sergeev, George Siscoe, Dan Swift, O.A. Troshichev, N.A. Tsyganenko, Vytenis Vasyliunas, Jack Winckler, and Dave Winningham. Later, waves of the new generation joined in our effort, particularly during the International Conference on Substorms (ICS). My former students Ching Meng, Koji Kawasaki, Lee Snyder, Fumi Yasuhara, Paul Perreault, Tom Berkey, and my associates Yosuke Kamide, Joe Kan, Lou Lee, and Tony Lui worked very closely with me on substorm research.

Finally, long-awaited images from the Dynamic Explorer satellite began to arrive (John Craven and Lou Frank, 1983). I visited my colleagues at the University of Iowa to witness this event. I thanked Lou Frank and congratulated him on this great success. It was the ultimate test of the concept of the auroral substorm because the auroral substorm must be the same seen from below and above, see Figures 2.22a and 2.22b. Auroral morphology was further advanced by the Canadian group (Elphinstone et al., 1996).

It is important to learn that it takes much more time than one thinks to convince colleagues if one's finding is radically different from what has been believed for years. Figure 2.23 shows schematically the auroral features at about the maximum epoch of a typical substorm. The visible feature consists of three parts, as shown on the left-hand side, the dayside part, the nightside part, and the

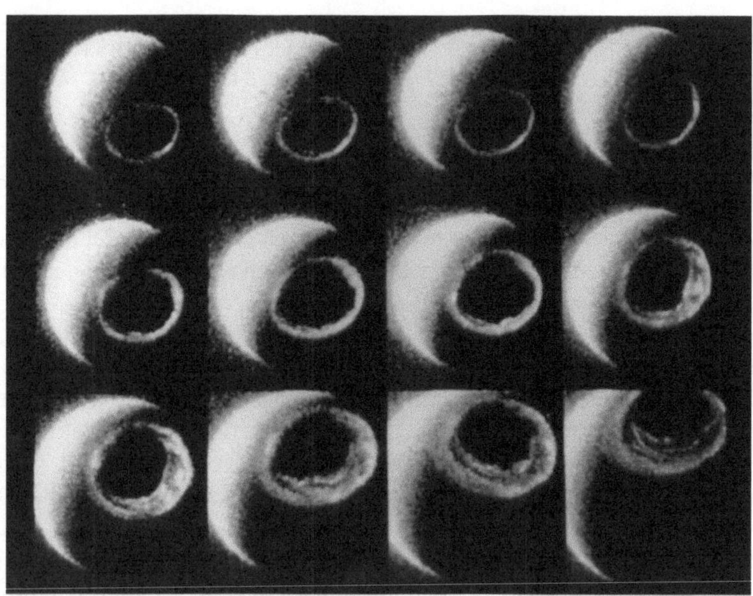

FIGURE 2.22a. Auroral images taken from the DE satellite that depict the development of an auroral substorm.
Source: Craven, J.D., Y. Kamide, L.A. Frank, S.-I. Akasofu, and M. Sugiura, Magnetospheric Currents, ed. by T.A. Potemra, *Geophysical Monograph*, 28, AGU, Washington D.C., 1983

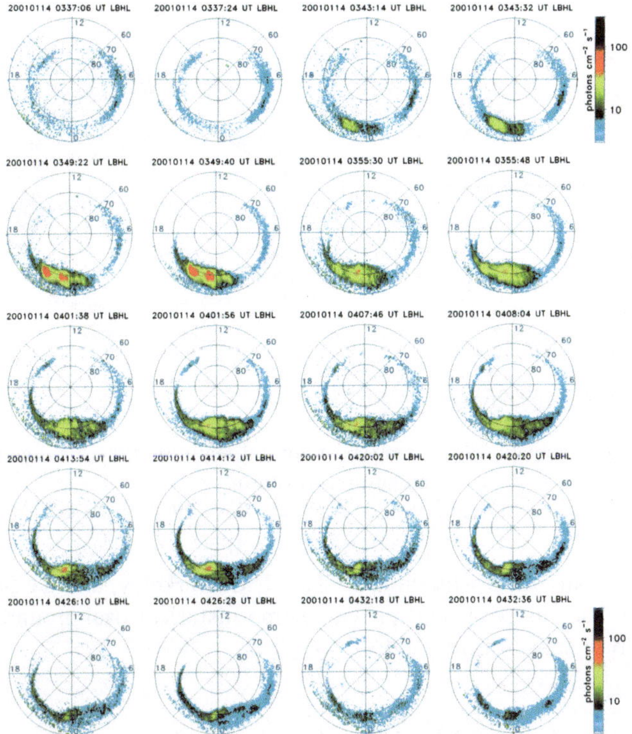

FIGURE 2.22b. A typical substorm depicted by the POLAR satellite.
Source: Courtesy of G. Parks

diffuse aurora, which is located equatorward of the first two arc structures. Note the diffuse aurora evolves into many arcs that develop further complex folds.

As the study of substorms had progressed by the work of a large number of researchers in the 1980s, we thought that an organized effort was needed to advance it further. Joe Kan was instrumental in establishing the ICS. The first conference was held in 1992 under the leadership of Bengt Hultqvist at the Swedish Institute of Space Physics, in Kiruna, Sweden. The second conference was held in 1994 at the University of Alaska Fairbanks, commemorating the publication of my 1964 paper on auroral substorms. The conference was blessed by active auroral displays over Fairbanks. The ICS brought many younger researchers who have considerably advanced the study of magnetospheric substorms. I also wish to express great appreciation for the close interaction with the following groups: the Swedish group in Stockholm, headed by Carl-Gunne Fälthammer; the Norwegian group in Oslo, headed by Alv Egeland; the Danish group, headed by Knud Lassen; the Canadian group headed by Cliff Anger; the Russian groups at Apatity, Moscow, Irkutsk, and Petersburg; and many U.S. groups, including Aerospace Corporation, Boston University, Johns Hopkins University, University of New Hampshire, Rice University, Southwest Research

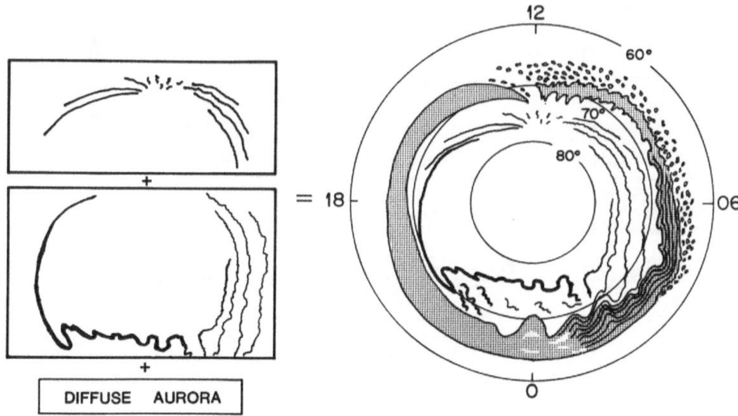

FIGURE 2.23. A schematic representation of auroral features at the maximum epoch of an auroral substorm.
Source: Akasofu, S.-I., *Space Sci. Rev.*, **19**, 169, 1976

Institute, University of Washington, UCLA, U.C. Berkeley, and UCSD. It is my sincere hope that this book will provide some historical background and some new direction in substorm research.

2.5. Publication of My First Monograph

In about 1966, I decided to write a book, summarizing a great variety of auroral and geomagnetic phenomena could be synthesized in terms of the substorm concept.

The main reason for writing the book was that I found that the concept of auroral substorms could provide at least the time frame of reference in organizing a great variety of aurora-related phenomena. Polar magnetic disturbances (Chapter 3), ionospheric disturbances, X-ray bursts, VLF emissions, geomagnetic micropulsations, and other phenomena occur in harmony with auroral substorms. Further, the pattern of auroral substorms was useful in analyzing and understanding the above phenomena. Obviously, based on such analysis, all these phenomena are found to be closely related. In this way, the concept of auroral substorms provided the basic frame of reference in synthesizing all polar atmospheric phenomena.

The book was published by D. Reidel Publishing Company in 1968 under the title *Polar and Magnetospheric Substorms*. It was dedicated as follows:

TO SYDNEY CHAPMAN, who unbeknown to most scientists, has encouraged and inspired the world's magnetic and auroral observatories to maintain the essential records upon which our understanding of geomagnetism and the aurora rests.

Chapman indeed had visited many magnetic observatories in the world and encouraged them to continue the recording. He told me that in the 1950s some prominent scientists were of the opinion that magnetic observatories were no longer necessary, because they thought that the geomagnetic daily variations and storms were already well understood. Some of them later regretted their premature judgment after noting the great development of geomagnetism, as it led to the discovery of the magnetosphere.

2.6. Auroral Storms

In 1962, when I began to study the great geomagnetic storm of February 11, 1958 (one of the most intense storms in the twentieth century, with the maximum Dst decrease being as large as 450 nT at 10:00 UT; Figure 2.24a), I was greatly surprised to find that the aurora can be *very quiet,* even when the main phase of a great storm is reaching its maximum epoch; it is natural to assume that the intensity of auroral displays would also reach highest at the maximum epoch. In Figure 2.24b, the aurora at 10:20 UT, on February 11, 1958, was quiet, in spite of the fact that the accompanying geomagnetic storm was at about the maximum epoch; the auroral oval (both the northern and southern boundaries) expanded greatly towards the equator. I thank Carl Gartlein who installed a number of all-sky cameras near the US–Canada border that accurately located the expanded oval on that day.

However, such a quiet condition did not last too long, and a great auroral substorm activity began soon afterward, resulting in one of the most spectacular poleward expansions of the auroral oval. There were similar displays a few hours before and after the event that began at 10:20 UT. That is, similar auroral activity repeated several times during the great storm. The onset of the poleward expansion at 10:20 UT was observed at geomagnetic latitude as low as $54°, L = 2.7$ (Pullman, Washington).

This event also became the basis for Chapman and me to consider that an auroral storm consists of a number of distinct impulsive phenomena, namely the auroral substorms. That is to say, an auroral storm consists of a number of auroral substorms; an auroral substorm is the element of the auroral storm.

In my 1964 paper on auroral substorms I reported how individual auroral substorms develop (Figure 2.17). In high latitudes, a geomagnetic storm also consists of a number of impulsive changes (Figure II in the Prologue), coinciding with the auroral substorm, which is called the polar magnetic substorm. This idea developed into the concept of a magnetospheric storm that consists of a number of magnetospheric substorms. Thus, it was concluded that it is necessary to study magnetospheric substorms in order to understand a magnetospheric storm. It was in this way that the magnetospheric substorm became one of the main topics of research in magnetospheric physics.

Chapman used to tell me how lucky I was as a student of the aurora, since the IGY provided my generation with a great wealth of auroral data. In our pre-computer days, the data analysis by hand was laborious, but there was enough time to consider what the data were trying to tell us. Compared with present data

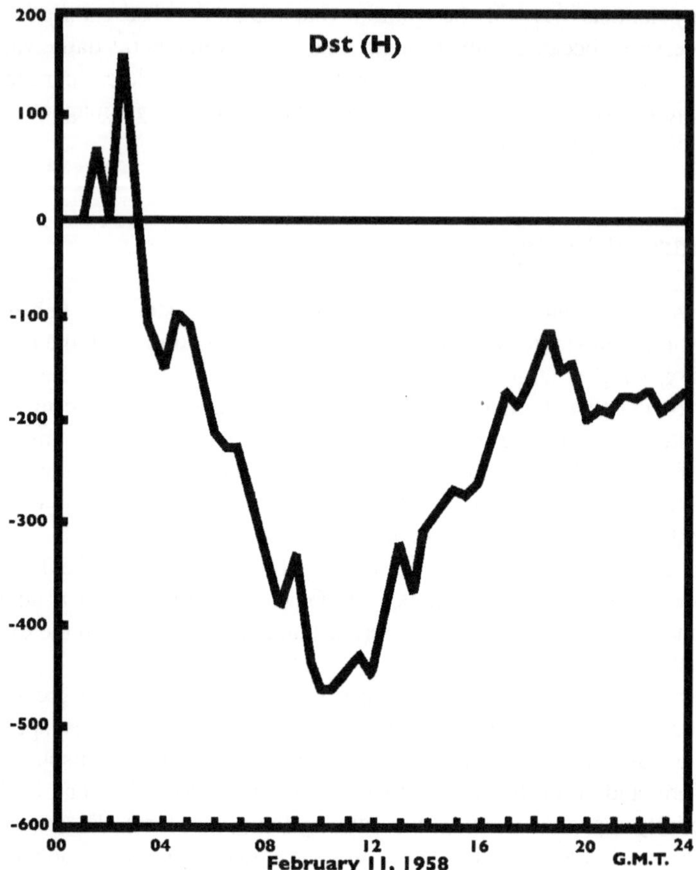

FIGURE 2.24a. The Dst index during the geomagnetic storm of February 11, 1958. *Source*: Akasofu, S.-I. and S. Chapman, *Planet. Space Sci.*, **24**, 785, 1962

gathering, our days were almost like the Stone Age. Further, most of the data one needs, including real-time auroral images from the POLAR satellite, can now be obtained instantly by clicking on a computer; in the 1960s, 1970s, and even in the 1980s, it took several years to gather necessary data. My only hope is that the new generation of researchers would not have indigestion because of the present wealth of data.

2.7. Auroral Rays

The curtain-like structure of the aurora frequently develops vertical striations called the auroral rays. Tom Hallinan, of the Geophysical Institute, developed a high-sensitivity TV camera and with it captured these striations clearly for

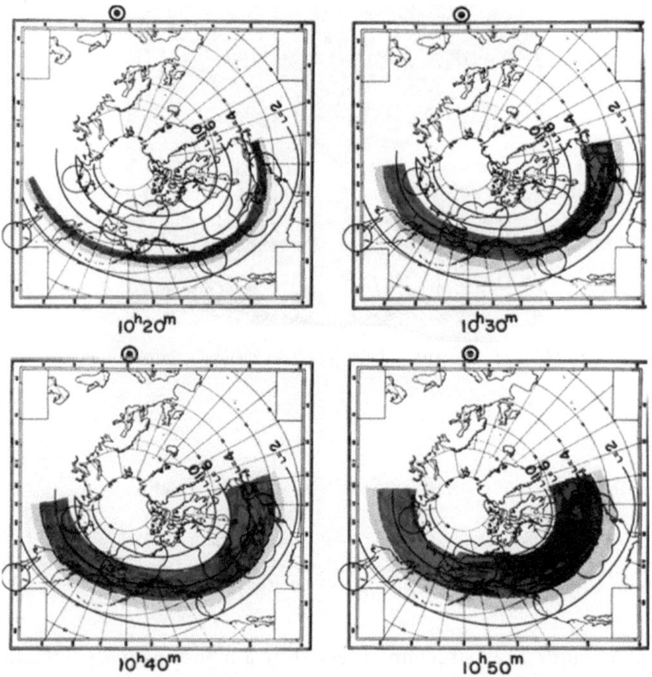

FIGURE 2.24b. The distribution of the aurora at about the maximum epoch of the main phase of the geomagnetic storm of February 11, 1958.
Source: Akasofu, S.-I. and S. Chapman, *Planet. Space Sci.*, **24**, 785, 1962

the first time (1983). He found that the rays are a sort of fine pleating of the curtain-like structure of the aurora.

At that time, I had a graduate student, John Wagner, who wanted to study the ray structure by a computer simulation method. The problem I faced was that the Geophysical Institute did not have a high-speed computer for such a project and I knew nothing about a plasma simulation method. This was one of the most difficult problems I faced as a professor, but I was determined to solve it. Thus, I contacted Walter Orr Roberts, the founder of the National Center for Atmospheric Research and my other mentor, to seek advice on how I might obtain a high-speed computer; he arranged to fund $500,000 from the Fleishmann Foundation for this purpose. It was in this way that the Geophysical Institute obtained the first modern high-speed computer. Then, I sent Wagner to work with UCLA's John Dawson for a year to learn about a plasma simulation method. Wagner satisfied my expectations and simulated the ray structure that was successfully imaged by Hallinan. Wagner showed that the ray structure develops as a result of the counter-streaming plasma flow across an auroral curtain (Figure 2.25).

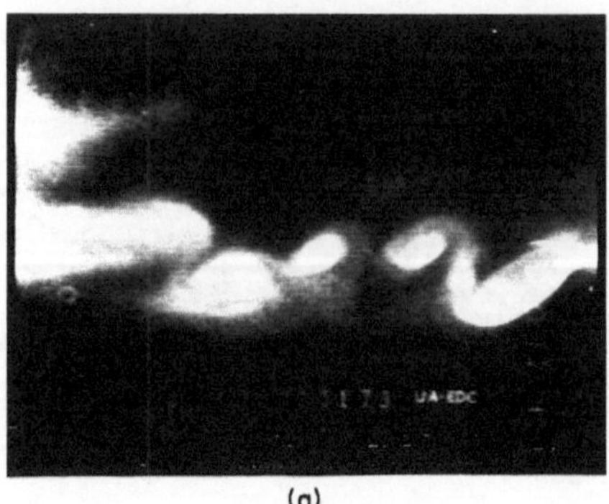

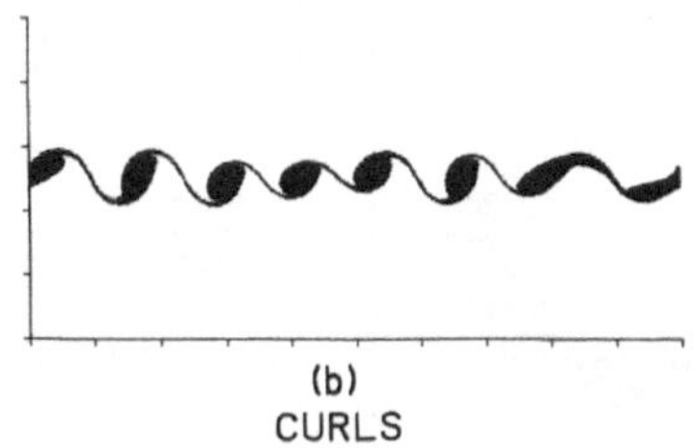

FIGURE 2.25. An auroral ray structure observed by a high-sensitivity TV camera and a simulated one.
Source: Wagner, J.S., R.D. Sydora, T. Tajima, T. Hallinan, and S.-I. Akasofu, *J. Geophys. Res.*, **88**, 8013, 1983

2.8. Thickness of an Auroral Curtain

As far as I am aware, I was the first to determine accurately the thickness of auroral curtains. The corona-type aurora is observed when an auroral curtain is located near the magnetic zenith. In some of the aurora photographs I took, the bottom edge of the curtain was clearly captured.

Using a star constellation map, I could measure the thickness to be about 500 m. Chapman encouraged me to publish the result, and I wrote a short paper that appeared in the *Journal of Atmospheric and Terrestrial Physics*. It should be noted that the reason for the thin curtain-like form of the aurora is still one of the long-standing unsolved problems. The question remains why the field-aligned currents occur in the form of thin sheets (Chapter 4).

John Wagner also simulated the auroral potential structure, which may be responsible for accelerating auroral electrons and for generating the radiation. However, there is still no agreed-upon mechanism for the acceleration process of auroral electrons. There have been a number of efforts to learn about the individual curtain-like structure of the aurora and precipitating electrons (Figure 2.26). The acceleration of charged particles in the magnetosphere and the solar atmosphere, perhaps even in galaxies, is one of the most fundamental issues in cosmic electrodynamics.

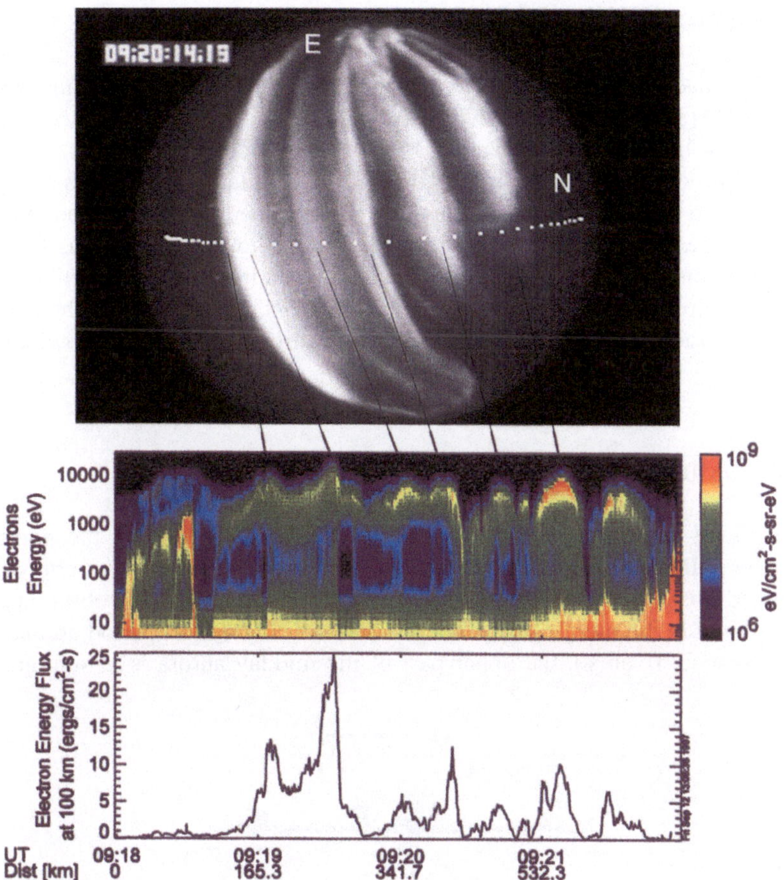

FIGURE 2.26. The Fast Auroral Snapshot Small Explorer (FAST) satellite passing through multiple auroral arcs at 0920:14 UT, February 6, 1997. The 110-km conjugate to the satellite is shown in the all-sky image at 10s intervals as FAST passed across from left (south) to right (north). The center panel is a "normal" format of the electron energy spectrum (integrated overall pitch angles) and shows a number of inverted-V structures. The bottom panel is the precipitated energy flux on a linear scale. The individual auroral arcs are clearly displayed here. The auroras, in particular the two arcs to the right (north), did change over the 4 minutes of the pass, so a detailed comparison between all-sky image and the particle data could not be attempted.
Source: Kimball, C. Chaston, J. McFadden, G. Delory, M. Temerin, and C.W. Carlson, *Geophys. Res. Lett.*, 25, 2073, 1998

I salute Duncan Bryant's devotion to this subject. He has pursued the processes associated with the acceleration of auroral electrons for more than 20 years, in spite of the fact that his idea was not widely accepted. Bryant's efforts are well described in his recent book *Electron Acceleration in the Aurora and Beyond* (1999).

2.9. Auroral Kilometric Radiation

In 1973, I had an opportunity to visit the University of Iowa with the newly acquired DMSP images in hand. I also was able to talk to Don Gurnett who told me that his radio detector aboard the IMP-6 satellite had observed a new type of radio emission and was wondering if it was related to auroral activity. It was fortunate that the period his data covered coincided with the period that my DMSP images covered. It became immediately obvious to both of us that the radio emission occurred when DMSP images showed intense auroral activity. We published a joint paper on this subject (Figure 2.27). It is likely that the auroral kilometric radiation is emitted from the auroral electrons' acceleration region.

2.10. Auroral Observation at the South Pole

It is difficult to observe the midday part of the auroral oval in the northern hemisphere. Because the geomagnetic pole is located in the northwestern part of Greenland (instead of at the geographic pole), the midday aurora is observable at Svalbard and in the middle of the Arctic Ocean during a short period around the winter solstice. Even so, the upper part of the midday aurora is in sunlight. In

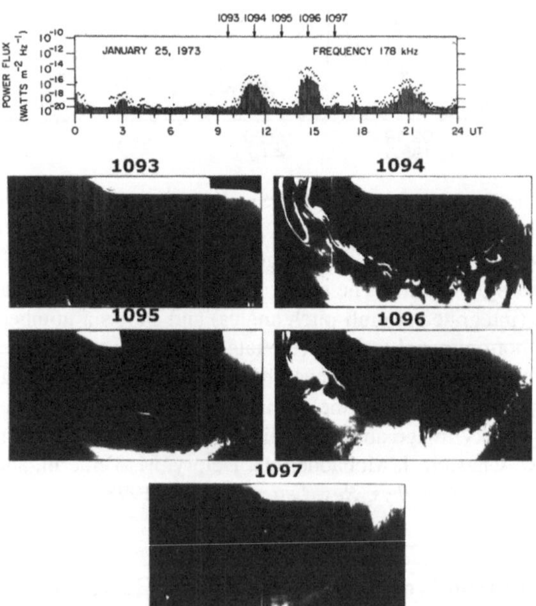

FIGURE 2.27. Auroral kilometric radiation observed by the IMP-6 satellite (D. Gurnett) and DMSP images.
Source: Voots, G.R., D.A. Gurnett, and S.-I. Akasofu, *J. Geophys. Res.*, **82**, 2259, 1977

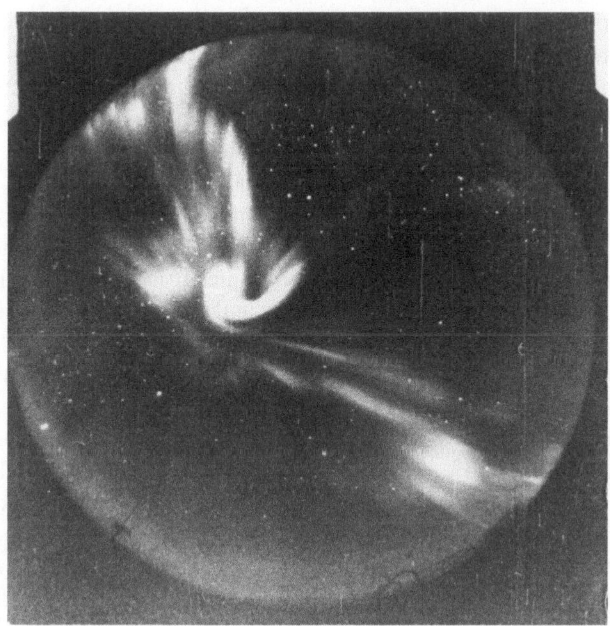

FIGURE 2.28. Midday aurora observed at the South Pole station.
Source: Akasofu, S.-I., Upper Atmosphere Research in Antarctica, Antarctic Research Series, vol. 29, ed. by L.J. Lanzerotti and C.G. Park, *AGU*, Washington, D.C., 1978

1970, we had a NASA airborne expedition over the Greenland Sea and confirmed the presence of the midday aurora in spite of the twilight. I clearly remember Bob Eather's excited voice on the intercom, describing that the spectrum was rich in the oxygen-red emission.

It so happens that the South Pole is located at gm lat. 78°, and the shadow height is highest on Earth. Therefore, one can clearly observe the entire midday part of the auroral oval. It is the most ideal location to observe the midday aurora. Merritt Helfferich was sent to install a Fairchild all-sky camera at the South Pole and high-resolution images of the midday aurora were obtained (Figure 2.28). The films provided us with many new results on the midday aurora.

2.11. Auroral Spectra as Tools for Detecting Extraterrestrial Life

One of the most prominent emissions from the aurora is the greenish-white light from oxygen atoms, while the Jovian aurora contains atomic hydrogen emissions (Clarke et al., 1989). Most of the processes leading to the production of oxygen atoms are directly or indirectly related to molecular oxygen produced near ground level. Thus, the oxygen emission, the so-called *green line* (557.7 nm) of the terrestrial aurora, arises mostly because plants release abundant free oxygen into the atmosphere by photosynthesis.

Thus, *an intense green line emission suggests that plant life exists on Earth.* It is expected that the green emission from oxygen atoms dissociated from CO_2 may also exist, but its contribution is very small. This is because even on the ground level, the amount of CO_2 is about 1/1000 that of O_2.

It was recently reported that Upsilon Andromedae, which is a solar-type star, has three planets. This discovery is significant because it shows the planetary system, like the solar system, is not quite unique. It is expected that a number of stars are accompanied by several planets, and it may not be too long before the aurora on such planets can be discovered.

One possible way to detect plant life on such planets is to examine their auroral emissions. If strong oxygen emissions can be detected among other emissions in the planetary auroral oval, the possibility of the presence of plant life is high. Further, if plant life exists, animal life, whether lower or higher, can also exist there.

The Earth-like auroral processes leading to the green light emission from the auroral oval require, in addition to plant life, both stellar wind and planetary magnetism. It is highly probable that solar-type stars have stellar wind. If such a planet does not have a strong dipole-like magnetic field, the stellar wind can cause atmospheric glow, in which oxygen emissions may be present. In any case, if *strong* oxygen emissions are detected in the planetary auroral spectra, the possibility of plant life there is high.

There is no doubt that the detection of the oxygen lines is technically a very challenging problem, particularly from ground level. However, the planets expose their full dark sides to the Earth once during their revolution around their parent stars. Further, there are a number of prominent oxygen emissions in the infrared and far ultraviolet ranges that can be detected by satellites. In any case, this is only a technical problem to be solved.

Auroral science will evolve in a variety of ways in the future. It would be a great boon for auroral science if it could contribute to the search for extraterrestrial life, one of the ultimate human endeavors.

2.12. Emperor Showa and the Aurora

On October 3, 1985, I gave a special lecture on the aurora for the Emperor of Japan in his palace in Tokyo. It seems the Emperor, a marine biologist, had an unusual interest in the aurora and prepared a large number of questions before my lecture. After my slide presentation, he asked how we could confirm ancient sighting reports of the aurora in Japan. He was not satisfied with my response that anomalous events in the sky were well documented in an ancient publication titled *Japanese Meteorological Data*, and he asked further how one could confirm such sightings as auroral events. He was visibly pleased to learn that the dates of these sightings coincided with those recorded elsewhere in the world.

Many people still believe that the aurora occurs more frequently as we approach the north magnetic pole. The Emperor was not an exception. Thus, when I told him that this is not the case, he was very puzzled. However, he was delighted when I showed him an image of the ring-shaped aurora taken from the Dynamics Explorer satellite (provided to me by Lou Frank, University of Iowa) and I explained that at the geomagnetic pole, which is located near the center of the auroral oval, the aurora is located well beyond the southern horizon most of the time. He was pleased to learn that the aurora appears along an annular ring. This is because on his flight from Anchorage to London, he observed the aurora above Alaska and expected to see more at higher latitudes. However, he could not see the aurora farther north. Obviously, he was very puzzled by this experience, but the satellite image I presented solved his puzzle. He also wanted to know about the auroral spectral composition. "I want to make sure that the auroral green line comes from atomic oxygen, not from molecular nitrogen," he told me.

A videotape of a spectacular auroral display that was recorded at our Poker Flat Research Range, Geophysical Institute, University of Alaska Fairbanks, fascinated the Emperor. It was projected on a 150-inch screen, which was kindly provided by Panasonic. The Emperor was pleased to know that a Japanese company produced the high-sensitivity video camera. He asked how often we could observe such a display in Fairbanks (how many days per week and then how many hours on active nights, and so on). He was interested in astrophysical implications and practical applications of auroral studies, and he asked me a few basic and technical questions about them. I also recall that on the occasion when I received the Japan Academy Award in 1977, he asked me if there is any relationship between the solar corona and the aurora. I will never forget my meeting with the Emperor.

2.13. Exciting New Developments

The auroral dynamo requires both the solar wind and a planetary magnetic field to be available. Thus, it was expected that both Jupiter and Saturn would have the auroral oval, while Venus and Mars would not. The Hubble Space Telescope depicted a clear image of the auroral oval on Jupiter and Saturn (Figure 2.29), while both the Venus orbiter and the Mars orbiter did not find any indication of an auroral oval similar to that of the Earth. These results confirm that the solar wind-planetary magnetic field interaction is essential in providing the power for auroral discharge, as we learned in the case for the Earth.

Furthermore, there is some indication that Saturn's magnetosphere has substorm activities. It is of great interest to extend our knowledge of Earth's substorms to those of Saturn and vice versa (Figure 2.30).

FIGURE 2.29. The aurora on Earth, Jupiter, and Saturn, NASA Hubble Space Telescope Project.
Source: Bhardwaj, A. and G.R. Gladstone, *Rev. Geophys*. **38**, 295, 2000, Hubble Space Telescope Project

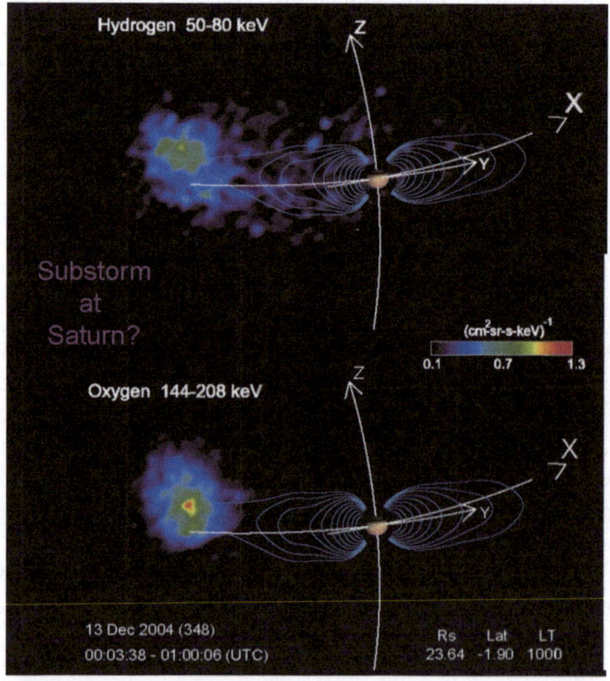

FIGURE 2.30. Energetic neutral atom (ENA) imaging of substorm-like activities in the magnetosphere of Saturn.
Source: Courtesy of D. Mitchell, Earth-Sun System Exploration: Energy Transfer, January 16–20, 2006, Kona, Hawaii

The solar wind causes both the aurora and the comet's tail.
Source: Geophysical Institute, University of Alaska Fairbanks

With Cheng Meng, my first Ph.D. graduate (1968).
Source: Geophysical Institute, University of Alaska Fairbanks, 1968

With Charlie Kennel and Vytenis Vasyliunas, at the Kiruna Geophysical University.
Source: Kiruna Geophysical Observatory

From left to right, Mrs. Olson, Mrs. Kathy McPherron, Hiroshi Fukunishi, Bob McPherron, Atsushi Nishida, Gordon Rostoker, and Syun Akasofu at the Grenoble, France, IUGG General Assembly meeting (1975).
Source: Akasofu, S.-I.

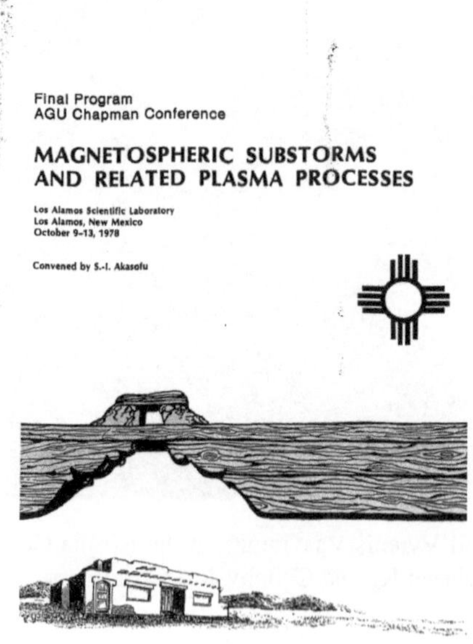

The cover of the program of the *AGU Chapman Conference on Magnetosphere substorms and Related Plasma Processes,* held at Los Alamos Scientific Laboratory, October 9–13, 1978.
Source: Akasofu, S.-I.

On the occasion of the Los Alamos Conference (1978), several attendees went to Pueblo to dance, from left Syun Akasofu, Carl McIlwain, unidentified, Jack Winckler, and Cheng Meng (1978).
Source: Akasofu, S.-I.

Lecturing at the Institute of Space Physics as an invited guest by the People's Republic of China.
Source: Akasofu, S.-I.

Japanese Academy Award (front and back of award).
Source: Akasofu, S.-I.

3
Realizing the Dream of Our Pioneers: Polar Magnetic Substorms and the Associated Current System

In modern terms, the Earth and its magnetic field are permanently enclosed in a comet-shaped cavity, called the magnetosphere, which is carved in the solar wind, a high temperature plasma flow from the Sun. Geomagnetic disturbances can be defined as the magnetic manifestation of an increased level of the solar wind-magnetosphere interaction, resulting in an increased electric power output and currents from the solar wind-magnetosphere dynamo process. Magnetic fields generated by the resulting increased electric currents are defined as geomagnetic disturbance fields $\Delta \boldsymbol{B}$. The disturbance magnetic fields are recorded as variations superimposed on the main field $\boldsymbol{B}_0$, namely $\boldsymbol{B} = \boldsymbol{B}_0 + \Delta \boldsymbol{B}$. More precisely, $\Delta \boldsymbol{B}$ includes the solar quiet-day daily variation that is caused by solar tidal effects on the ionosphere and the Chapman–Ferraro current on the magnetopause (Section 1.2).

One of the important purposes of the discipline of geomagnetism is to examine the configuration of the electric current systems that cause geomagnetic disturbances $\Delta \boldsymbol{B}$, and to elucidate their driving processes in terms of the solar wind-magnetosphere interaction. This long-term effort began with the pioneering effort of scientists such as K. Birkeland, C. Störmer, Sydney Chapman, and Hannes Alfvén at the beginning of the twentieth century. Before satellite observations gave views of the three-dimensional space around the Earth, the early studies were limited to inference of the current systems $\boldsymbol{J}$ (r, θ, λ, t) in space around the Earth based on records of magnetic variations $\Delta \boldsymbol{B}$ (r = a, θ, λ, t) on the Earth's surface (r = a), where a, θ, and λ denote the Earth's radius, latitude, and longitude, respectively. One of the early debates in this discipline was whether one could determine uniquely $\boldsymbol{J}$ (r, θ, λ, t) on the basis of observed variations $\Delta \boldsymbol{B}$ (r = a, θ, λ, t) at a number of places on the Earth's surface. As physicists, Birkeland and Alfvén attempted to determine a unique three-dimensional current configuration. Meanwhile Chapman, as a mathematical physicist, limited his study to a mathematical equivalent current on a spherical shell concentric to the Earth, avoiding the uniqueness issue.

3.1. The Three-Dimensional Current System

3.1.1. The Uniqueness Problem

The non-uniqueness of the solution to the problem of determining the distribution of $\boldsymbol{J}$ was recognized by both Birkeland and Chapman independently. Chapman (1935) noted:

> It is, of course, in principle, impossible to infer uniquely, purely from observations of a magnetic field (of external origin) at the earth's surface, the location of the current system, which is the source of the field.

It is interesting to note that by facing this non-uniqueness problem, Birkeland and Chapman took contrasting approaches. Being a physicist, Birkeland (1908) attempted to construct a three-dimensional current system by stating:

> If we assume, as from a physical point of view we might legitimately do, that the current is of a cosmic nature, and consists of negatively and positively charged corpuscles, the trajectories of the separate corpuscles must, as already stated, more or less approximately, follow the magnetic lines of force, moving in spirals about them.

On the other hand, as a mathematical physicist, Chapman limited himself to a two-dimensional equivalent current system on a spherical shell. Chapman (1935) wrote:

> The current distribution over a spherical sheet can easily be represented by a diagram using any projection of the sphere upon a plane. This is one method of representing the potential of the field graphically.

Figure 3.1 shows schematically the equatorial view and the polar view of the current systems considered by Birkeland and Chapman.

3.1.2. Chapman's Equivalent Current System

As a mathematical physicist, Chapman considered that the magnitude of the storm field $|\Delta \boldsymbol{B}| = \mathrm{D}$ consists of two components, the component independent of longitude or local time (Dst) and the component dependent on longitude or local time (DS). Both depend on the storm time t,

$$\mathrm{D}(\theta, \lambda, t) = \mathrm{Dst}(\theta, t) + \mathrm{DS}(\theta, \lambda, t).$$

The equivalent current for the Dst component is a zonal current on a spherical surface, since it does not depend on longitude. It can be considered as equivalent to the ring current. The ring current index Dst originates from this definition. The equivalent current for the DS component has a more complicated distribution (Figure 3.2). The polar view of the DS component shows two concentrated

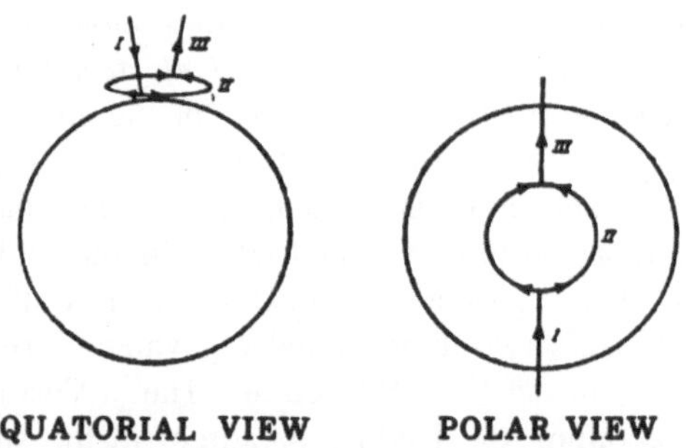

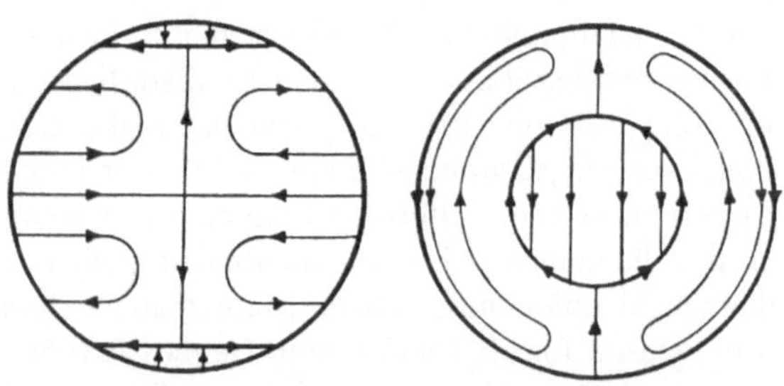

FIGURE 3.1. Birkeland's 3-D current system and Chapman's 2-D equivalent current system, both equatorial and polar views.
Source: Chapman, S., *Terr. Magn.*, **40**, 349, 1935

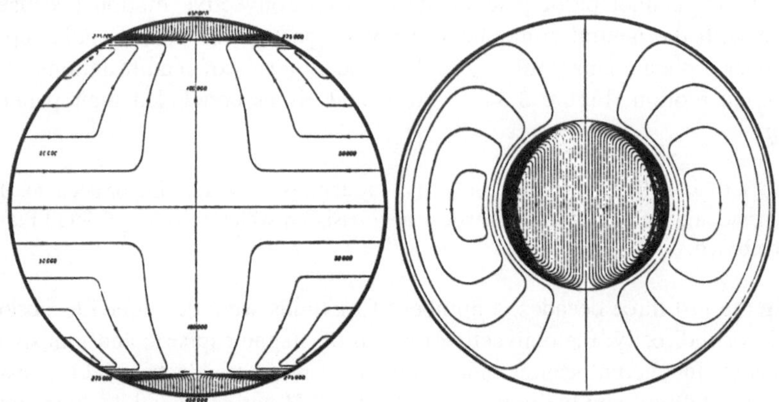

FIGURE 3.2. Details of Chapman's 2-D equivalent current system for the DS component, the equatorial (left) and polar (right) views for the DS component.
Source: Chapman, S., *Terr. Magn.*, **40**, 349, 1935

currents of equal intensity, but of opposite directions, along the classical auroral zone. Chapman coined the terms the *eastward electrojet* for the current in the evening sector, and the *westward electrojet* for the current in the morning sector. Each electrojet has a return current in the polar cap and the lower latitude. Note that this analysis of D is basically Fourier analysis for a given latitude θ; Dst is the first constant term and DS is in sinusoidal terms, namely $DS(\lambda) = a_1 \sin\lambda + a_2 \sin 2\lambda \ldots$ for a given θ. Unfortunately, after the ionosphere was discovered, Chapman's two-dimensional equivalent current was accepted as the real ionospheric current (H.C.Silsbee and Harry Vestine, 1942; Naoshi Fukushima, 1953), becoming the leading paradigm in the study of magnetospheric current systems. Thus, field-aligned currents were not considered by most researchers for a few decades.

Chapman told me that he thought that there were an infinite number of possible current systems for a given distribution of magnetic disturbance fields observed on the ground, choosing just one arbitrarily did not make sense. Instead, he thought that he could remain accurate so long as he dealt with the equivalent (two dimensional) current system. Although Chapman had many deep insights into physical processes, he tended to become an applied mathematician when he encountered mathematical uniqueness issues. Mathematical rigor was his life, and it was part of the reason for his friction with Hannes Alfvén, who tended to be intuitive in interpreting physical phenomena. Later, Chapman told me that he went a little too far in avoiding the non-uniqueness.

Chapman's DS component has a return current from each electrojet in the polar cap, constituting two current cells in the polar ionosphere, one located in the morning sector and the other in the evening sector. Ian Axford and Colin Hines (1961) thought that Chapman's two current cells are the Hall current and thus that they are a manifestation of large-scale convective ($\boldsymbol{E} \times \boldsymbol{B}$) motions of plasma in the magnetosphere. In the E region of the ionosphere, the Hall current arises from the flow of electrons along the streamlines of the convection flow; positive ions cannot participate in the ($\boldsymbol{E} \times \boldsymbol{B}$) convective motion because of friction with the neutral component. They suggested that various polar upper-atmospheric phenomena could be understood in terms of manifestations of the convective motion (Figure 3.3a). Axford and Hines concluded their paper by stating:

> We are led to believe that the convective system is of major importance to these phenomena, and we expect it to provide a new basis on which theories of detail may, in time, be based.

During the last three decades, a number of methods were developed to observe, directly or indirectly, the convective motion of magnetospheric and ionospheric plasmas by incoherent scatter radars, chemical releases, electric field measurements by satellites, and balloons. As Axford and Hines predicted, the convection of magnetospheric plasma has become one of the most important processes and paradigms in our present understanding of magnetospheric processes as a whole.

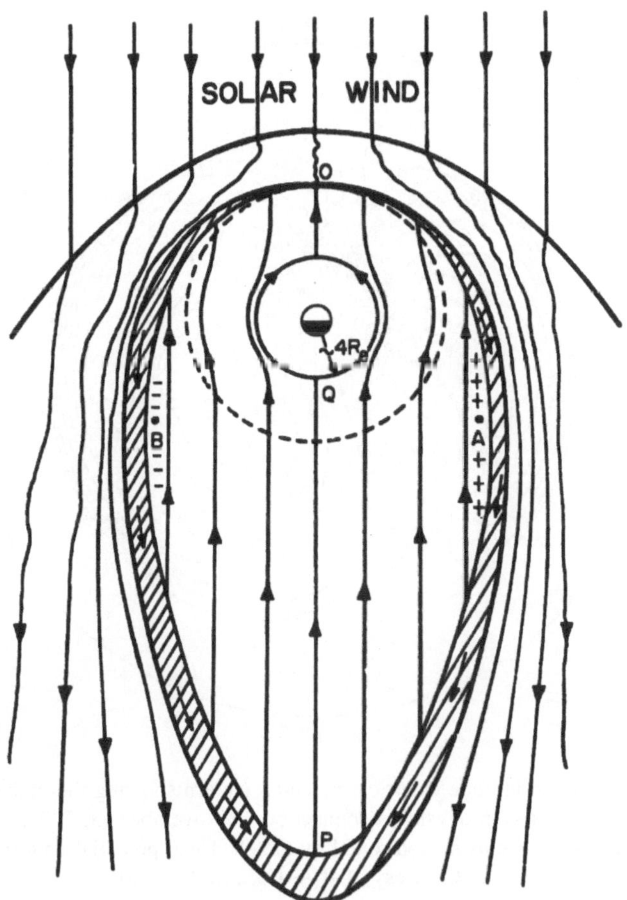

FIGURE 3.3a. The $(\boldsymbol{E} \times \boldsymbol{B})$ convective motion of magnetospheric plasma on the equatorial plane inferred by Ian Axford and Colin Hines who projected Chapman's two-dimensional flow pattern onto the equatorial plane of the magnetosphere. They suggested that a viscous-like interaction between the solar wind plasma and the magnetospheric plasma drives the convective motion. It is more likely that the electric field generated by the solar wind-magnetosphere dynamo process is the direct source. *Source*: Axford, W.I. and C.O. Hines, *Can., J. Phy.*, **39**, 1433, 1961

Thus, Chapman's equivalent current system contributed to magnetospheric physics in this important and interesting way, akin to how we learned earlier that Störmer's study of motions of electrons from the Sun is not applicable to magnetosphere formation but has helped us understand motions of trapped particles in the radiation belt and the ring current belt.

It is now possible to observe the convection on a real-time basis, by a network of HF radars, the SuperDARN network. Figure 3.3b shows an example of

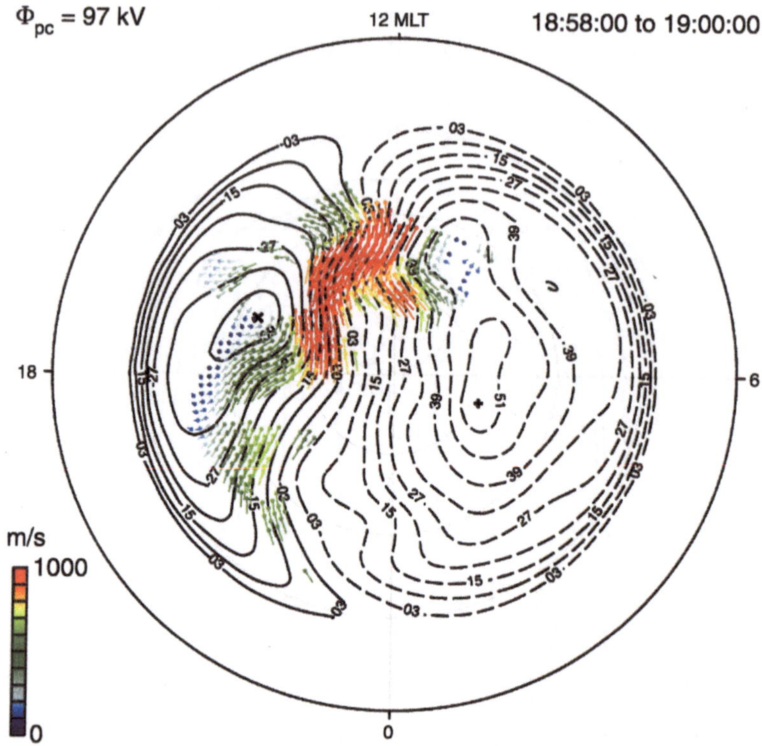

FIGURE 3.3b. The high-latitude electrical potential pattern showing the global response to a significant increase in dayside reconnection on November 13, 1996. Color bars show speed of plasma controlled by the electrical potential pattern.
Source: Courtesy of R.A. Greenwald, 2001

the convection pattern about 10 minutes after the dayside reconnection (Ray Greenwald, 1999). It remains to be seen whether or not the corresponding large-scale, systematic convection exists on the equatorial plane.

3.1.3. Birkeland–Alfvén Model

Most researchers took Chapman's equivalent current system as the true current system for many decades. Meanwhile, Hannes Alfvén (1950) demonstrated with a wire model and a search coil that the local time-dependent part of the geomagnetic storm field (DS) defined by Chapman can be reproduced by a combination of field-aligned currents, the auroral electrojet and the connecting equatorial currents. However, this work did not receive the attention it deserved. It was unfortunate that Alfvén could not attract the attention of the scientific community to the merit of his three-dimensional model against the two-dimensional equivalent current model.

Koji Kawasaki, Ching Meng, and I decided to examine whether the observed distribution of magnetic disturbance vectors can be reproduced by a model three-dimensional current system, which was developed by Alfvén and modeled by C.B. Kirkpatrick (1952). Kirkpatrick's three-dimensional model is not too much different from the currently accepted one. It was a great surprise to us that Kirkpatrick's model reproduced the observations very well (compare Figure 3.4 and 3.5). When I showed my results to Alfvén during one of my visits to Stockholm, he was almost irritated. He said that I was too slow to recognize the validity of his three-dimensional current system. I could well understand his impatience.

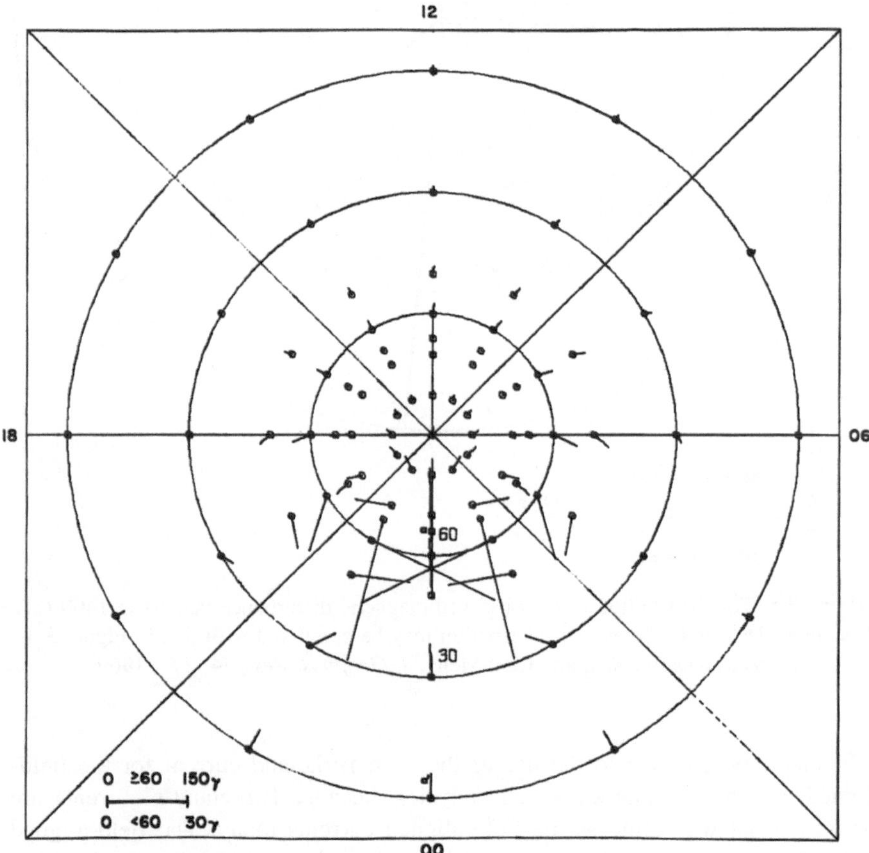

FIGURE 3.4. Distribution of the magnetic disturbance vectors for the Kirkpatrick-Alfvén model.
Source: Kawasaki, K. and S.-I. Akasofu, *Planet. Space Sci.*, **19**, 543

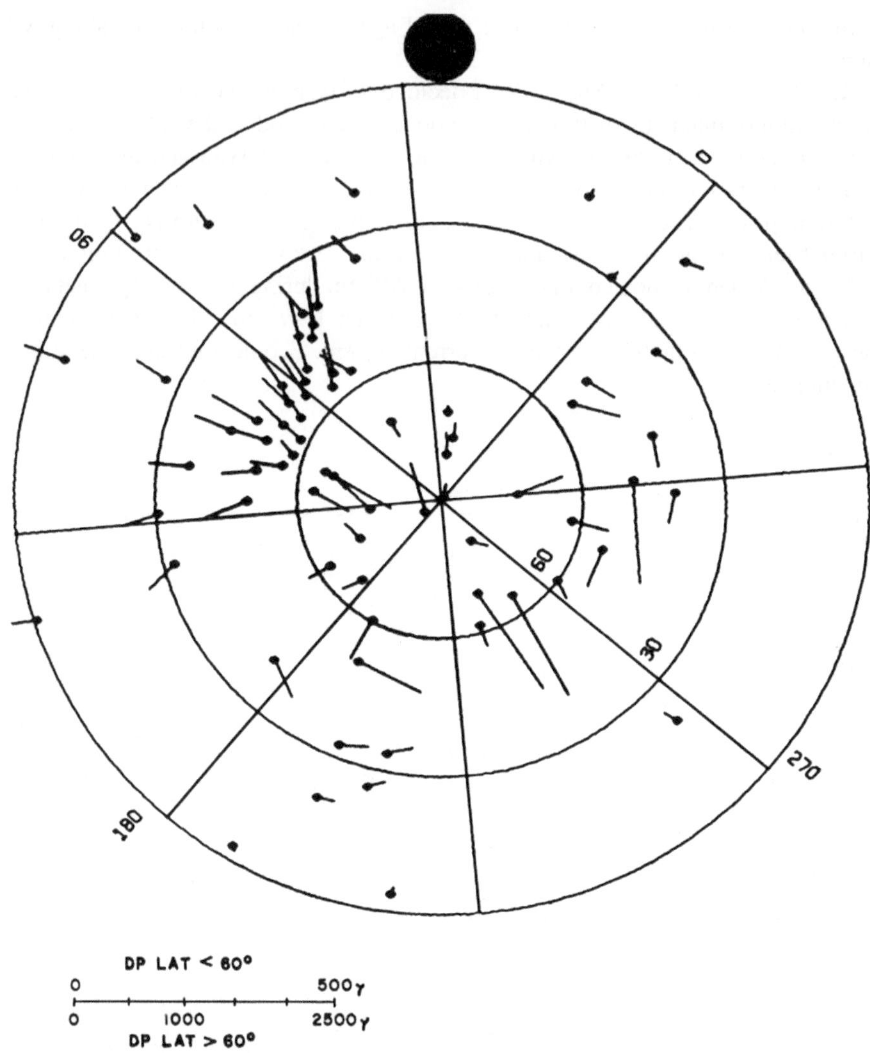

FIGURE 3.5. The distribution of the observed magnetic disturbance vectors at 1400 UT, December 16, 1964. The vector distribution may be compared with that in Figure 3.4. *Source*: Akasofu, S.-I. and C.-I. Meng, *J. Geophys. Res.*, **74**, 293, 1969

Incidentally, I object to the use of the term Birkeland current for the field-aligned currents in magnetospheric physics, because Birkeland's currents are far from what we define as the field-aligned currents today. The field-aligned currents flow between the magnetosphere and the ionosphere as a result of the magnetosphere–ionosphere coupling, and they are not *extraterrestrial* currents from the Sun, as Birkeland thought. I also object to statements that imply that Chapman was wrong in rejecting Alfvén's paper on magnetic storms. Note that neither Birkeland nor Alfvén could conceive of the magnetosphere in the way we envision it today. In their first paper on the formation of what we now call the magnetosphere, Chapman and Ferraro (1931) obtained an equation similar to the Debye length and concluded that the solar gas flow must be treated as what we now call *plasma*. Birkeland, Störmer, and Alfvén treated the solar gas as if it were composed of solitary particles. Alfvén's magnetosphere is quite different from what we know today.

Since this point is so basic in the Chapman–Ferraro theory, it was difficult for Chapman to entertain Alfvén's theory, in which protons and electrons in the stream are semi-independent. This point became one of the most serious controversies of the 1960s. During many international conferences, Alfvén (1954) confronted Chapman and emphasized the importance of the electric field in the solar plasma by stating, "The solar plasma flow must satisfy $\boldsymbol{E} + \boldsymbol{V} \times \boldsymbol{B} = 0$." Chapman thought that Alfvén's theory was not appropriate for dealing with the solar plasma as Alfvén ignored the strong electrostatic coupling between protons and electrons in the solar plasma. Chapman was correct in emphasizing that the solar gas be treated as plasma, but his theory could not account for the solar wind-magnetosphere coupling because his plasma was diamagnetic. Alfvén was correct in introducing the concept of the interplanetary magnetic field $\boldsymbol{B}$ and electric field $\boldsymbol{E} = -\boldsymbol{V} \times \boldsymbol{B}$, but failed to treat the solar gas as plasma. Both Chapman and Alfvén were emphasizing the importance of two different electric fields in the solar wind. This should be the way by which science develops constructively, even though controversies tend to flare up from time to time, although it was unfortunate that Chapman and Alfvén could not communicate better.

I am not trying to criticize the monumental pioneering work of the Scandinavian school of Birkeland, Störmer, and Alfvén. What I emphasize here is that we must be cautious in carelessly commenting on the works by our great pioneers. We must give credit where it belongs. We also have to be careful in assigning the nomenclature with the name of the appropriate scientist.

I used to tell my students and colleagues that when two groups have very different views on the same phenomenon, it is very likely that they are observing two different aspects on the same phenomenon. I used to use a pencil to explain such a situation. One end of a pencil is sharp and hard, while the other end, the eraser, is round and soft. Each of the two groups observes only one end, and thus a controversy erupts. Eventually, a third group finds that the two groups are looking at the different ends of the same object, a pencil. An irony may be that the third group gets most of the credit, at least in some cases.

One lesson here is that when a serious controversy erupts, there may be a way to integrate two controversial views and find an epoch-making advance. That is to say, in such a situation, the problem may not be that one is right and the other is wrong, but how to integrate two seemingly contradicting views (see Epilogue).

3.1.4. *Nikolsky's Spiral*

This is one of the early stories of the study of the current distribution in the polar region. The concept of the auroral zone (a circular belt in the geomagnetic coordinate system, centered around the geomagnetic pole) had greatly influenced the study of geomagnetic disturbances. The SD current system obtained by Chapman (1935) was an example of this influence. He suggested that the auroral currents consist of a pair of concentrated currents along the auroral *zone* (not the auroral *oval*, which was not known then). There were the westward auroral electrojet in the morning sector and the eastward electrojet in the afternoon sector, and their return currents in the polar cap and in lower latitudes (Figure 3.2). It was thought that a magnetic observatory rotates under such a fixed current system, registering the daily magnetic variations. Under the eastward electrojet (in the afternoon sector), there occurs positive (poleward) magnetic disturbances in the horizontal (H) component, while the westward electrojet produces negative (equatorward) magnetic disturbances in the morning sector. The SD current system had become the standard model and thus a major paradigm for a few decades.

However, Nikolsky (1947) found an interesting phenomenon: geomagnetic disturbances recorded at high-latitude stations have three activity peaks during a day. He denoted three peaks: A (afternoon), N (night) and M (morning) (Figure 3.6a). Further, he found that the peak tends to occur earlier in time at

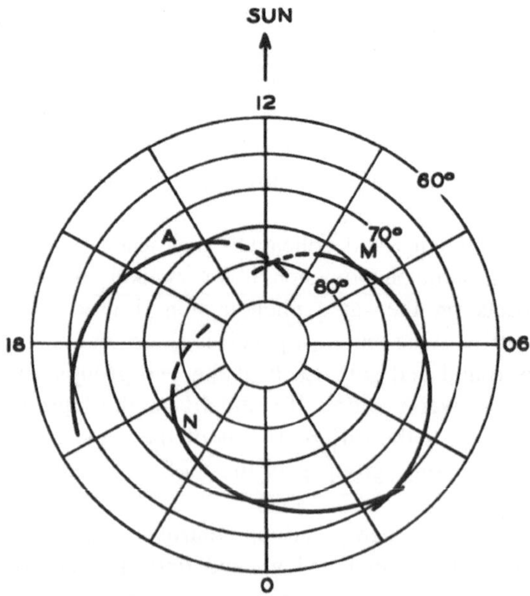

FIGURE 3.6a. Nikolsky's A, N, and M spirals. Both N and M constitute the auroral oval. *Source*: Akasofu, S.-I. and C.-I. Meng, *J. Atm. Terr. Phys.*, **29**, 965, 1015, 1967

higher latitudes for the A and N peaks, later for the M peaks. Thus, in a polar plot, the peak occurrence times for A, N, and M delineate three spiral curves. I was fascinated by Nikolsky's results, but had no idea as to how to interpret them. This is because it is not possible to understand his results in terms of the SD current system. However, one day I recognized that the combination of the N and M spirals delineates the auroral oval; the A peak spiral indicates the eastward electrojet. The results suggested to me that the westward electrojet does not stop in the midnight sector (as indicated by the SD current), but continues to flow westward, with the westward-traveling surge along the auroral oval in the evening sector where auroral arcs actually lie. Thus, the westward electrojet is located at latitudes higher than 65°–70° in the evening sector, not along the auroral zone (Figure 3.6b). When I reported this result in Moscow in 1968, Nikolsky was very happy. He wrapped me in a typical Russian bear hug, saying that I was his son's age.

I thought that my interpretation on Nikolsky's results was reasonable, because the auroral ionization takes place along the auroral oval, not along the auroral

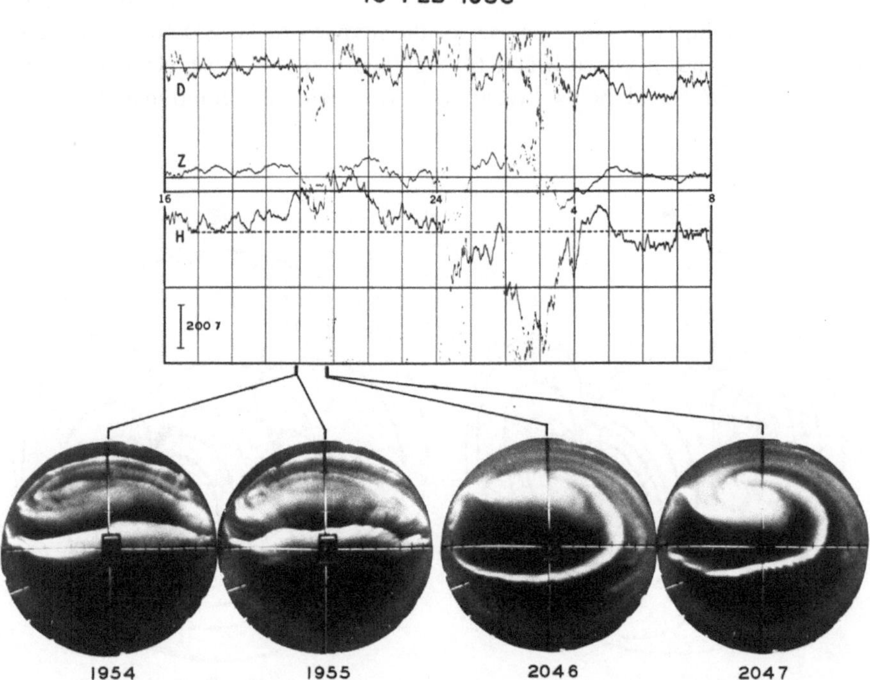

FIGURE 3.6b. College magnetic record and all-sky photographs from Fort Yukon. When positive changes in the H component were observed at College, an intense westward-traveling surge was passing over Fort Yukon; the surge is the leading edge of the westward electrojet.
Source: Akasofu, S.-I. and C.-I. Meng, *J. Atm. Terr. Phys.*, **29**, 965, 1015, 1967

zone. Therefore, there is no reason why the westward electrojet must stop in the midnight sector as the SD current system indicates. Fortunately, the Alaska meridian chain of all-sky camera stations (Figures 2.8a and 2.8b) were also equipped with magnetometers. An examination of both all-sky photographs and magnetic records indicated clearly that the westward electrojet extends into the evening sector along the auroral oval with the westward-traveling surge. During the passage of westward-traveling surges to the north (Figure 3.6b), an auroral zone station (Point A in Figure 3.7) registered positive changes in the H component, while at a station of gm lat. 70°–75° (Point B in Figure 3.7), negative changes with greater magnitudes were observed. Therefore, it became obvious that the westward electrojet flows along the oval, not the auroral zone.

Thus, Meng and I found that Chapman's SD current in the polar region must be revised. As the right side of Figure 3.7 shows, the westward electrojet during substorms forms a single vortex, not a double vortex as the SD current system suggests. As described in Section 1.10, the double vortex pattern is prominent during the growth phase. A strong single vortex pattern appears at substorm onset. It was hard for Chapman to realize that his SD current system was not correct. However, after examining my analysis, he was convinced about its validity, becoming a strong supporter of the revision of his SD current system. On the other hand, it was difficult to convince many of my colleagues of the results in the 1960s and 1970s. I recall that there were even emotional objections

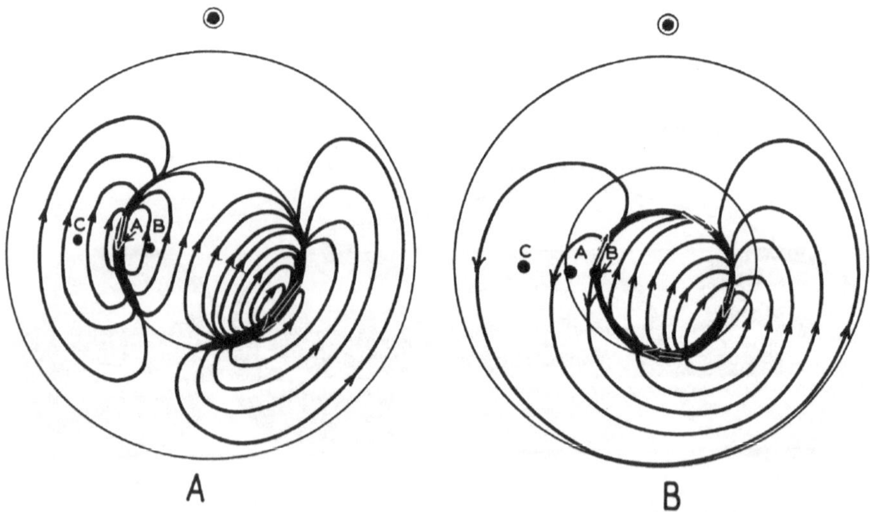

FIGURE 3.7. Left: Chapman's SD current system. Right: Revised equivalent current system.
Source: Akasofu, S.-I. and C.-I. Meng, *J. Atm. Terr. Phys.*, **30**, 227, 1968

FIGURE 3.8. With Yosuke Kamide and Yasha Feldstein (Prague).
Source: Akasofu, S.-I

to the revision. Again, Yasha Feldstein was one of the strong supporters of my work (Figure 3.8).

3.2. Alaska Meridian Chain of Magnetic Observatories

The early study of the current systems in the polar region suffered from insufficient magnetic records, since there were very few possible locations for the observatories in high latitudes. As a first step to improve the situation, I equipped the entire Alaska meridian chain of all-sky camera stations with magnetometers (Figures 2.8a and 2.8b). Like the meridian chain of all-sky cameras, the meridian chain of magnetometers scans the polar magnetic variations once a day. This was the first attempt in the history of geomagnetism to obtain geomagnetic data systematically as a function of latitude. The College Observatory, operated by Jack Townshend of the USGS, was the key observatory in this operation.

When the Alaska meridian chain of magnetometers became operational, I was very surprised that we could obtain a fairly systematic magnetic vector distribution over the entire polar region by averaging the data for only a few months (Figure 3.9a). Yosuke Kamide from the University of Tokyo joined my group at that time as a post-doctoral fellow. I suggested to him that there may be a way to obtain the true current system, not the equivalent current system, using such a systematic data set by knowing that the currents can flow only in the ionosphere and along the geomagnetic field lines. After moving to the National Center for Atmospheric Research, in Boulder, Colorado, he developed a computer algorithm with Art Richmond and Sadami Matsushita for this purpose. This code is called the KRM code and has been most useful in studying the high-latitude current configuration.

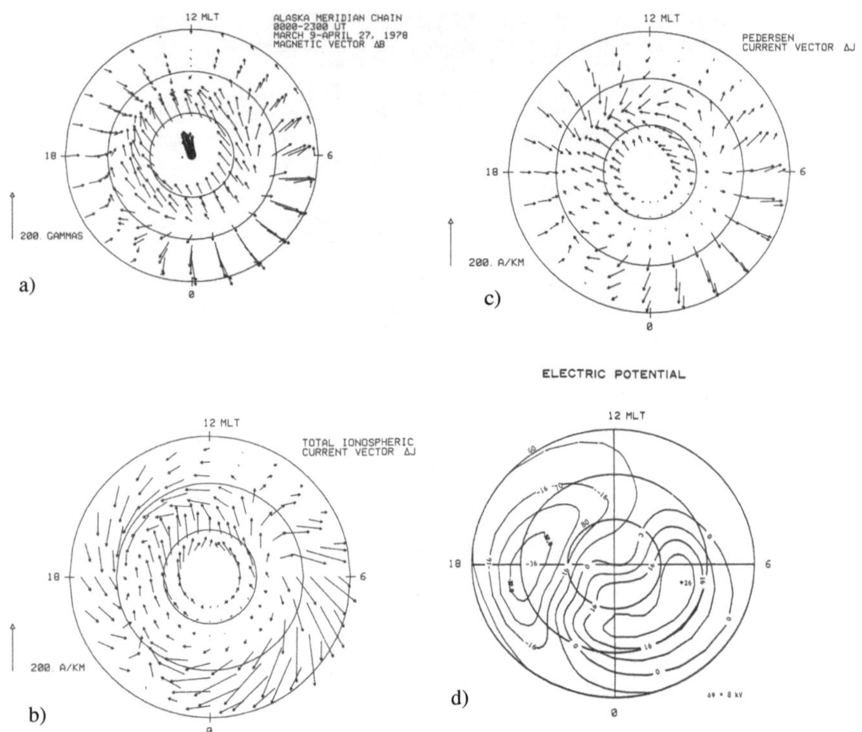

FIGURE 3.9. (a) Hourly average horizontal magnetic disturbance vectors obtained by the Alaska meridian chain between March 9 and April 27, 1978; (b) The total current vectors are obtained from Figure 3.9a by applying the KRM algorithm; (c) The Pedersen component of Figure 3.9b; (d) The potential distribution obtained from Figure 3.9a. *Source*: Akasofu, S.-I. and Y. Kamide, *J. Geophys. Res.*, **103**, 14,939, 1998

The KRM code was successfully applied to the data obtained from the Alaska meridian chain, although it was necessary to assume the distribution of the conductivity for both the Pedersen and Hall currents. Byung-Ho Ahn, one of my graduate students, developed an excellent conductivity model for this purpose. Thus, we could obtain the daily average current pattern over the entire polar region. The pattern of both the westward electrojet and the eastward electrojet was clearly elucidated (Figure 3.9b). Figure 3.9c shows the distribution of the Pedersen component. Figure 3.9d shows the distribution of the electric potential. All the results are self-consistent.

3.3. The IMS Meridian Chains of Observatories

When I was operating the Alaska meridian chain of observatories in the 1970s, Gordon Rostoker was also establishing a meridian chain of magnetometers in Canada. We agreed to operate the Inuvik, Northwest Territories, station jointly. We flew to Inuvik in small planes to install a magnetometer there.

Our operation of the meridian chains inspired our colleagues in Europe and Russia to establish four other chains during the International Magnetosphere Study (IMS). Thus, six meridian chains of magnetometers, consisting of 71 magnetometer stations, became operational during the IMS (Figure 3.10). The KRM computer code had further been developed to deal with instantaneous current patterns based on the simultaneous magnetic records from all six IMS meridian chains of magnetometers. Thus, both the data from the six meridian chains and the powerful KRM code enabled Kamide and his colleagues (1982) to study the development of the three-dimensional substorm current system with a time resolution of 5 minutes. A great wealth of knowledge on the ionospheric currents, electric fields, potential field, field-aligned currents, and the Joule heating rate was obtained over the entire polar region. This was a great joint effort by many of our colleagues, which made the dream of our pioneers a reality after more than half a century. They were Yosuke Kamide, Byung-Ho Ahn, Wolfgang Baumjohann, Eigil Friss-Christensen, Herb Kroehl,

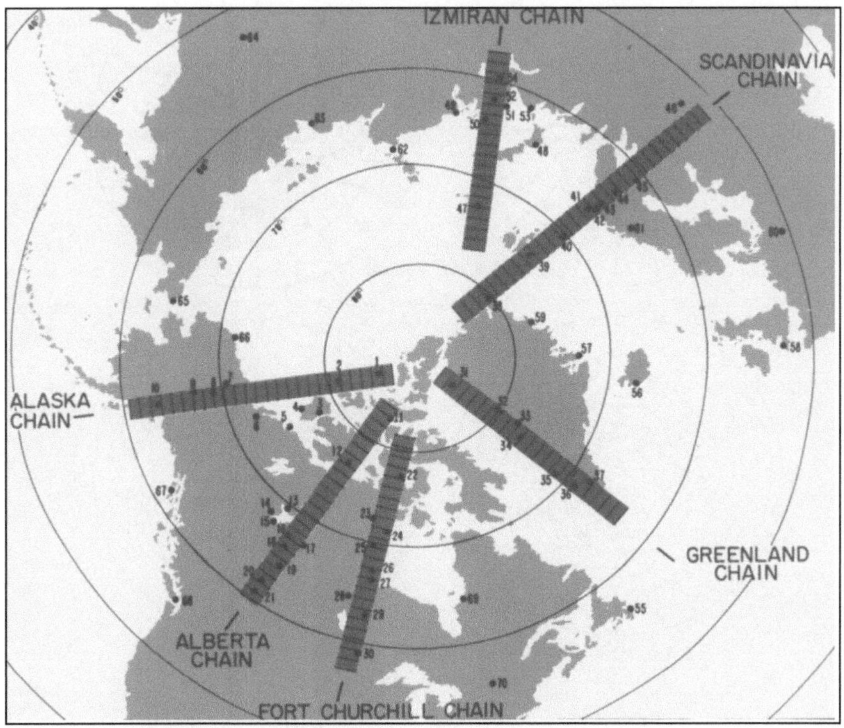

FIGURE 3.10. Six IMS meridian chains of magnetometers.
Source: Ahn, B.-H., Y. Kamide, and S.-I. Akasofu, *J. Geophys. Res.*, **89**, 1613, 1984

H. Maurer, Art Richmond, Gordon Rostoker, R.W. Spiro, John Walker, and Alex Zaitzev.

Figure 3.11 shows the current distribution at 0900 and 1000 UT, on March 18, 1978. At 0900, the condition was fairly quiet, although a very weak current pattern can be seen. The electric potential pattern at the same time shows the two-cell pattern (the directly driven component in Section 1.10). At 1000 UT, a strong westward current developed in the night sector. The potential pattern shows a large single cell pattern.

Figure 3.12 shows the Joule dissipation rate at 0700, 0800, 0900 and 1000 UT on the same day. During a quiet period (0900 UT), a weak Joule heat dissipation was seen in the afternoon sector. During a disturbed period (1000 UT), a large Joule dissipation occurred in the night sector, particularly in the late evening sector; maximum dissipation rate was close to 50 m Watt/m^2.

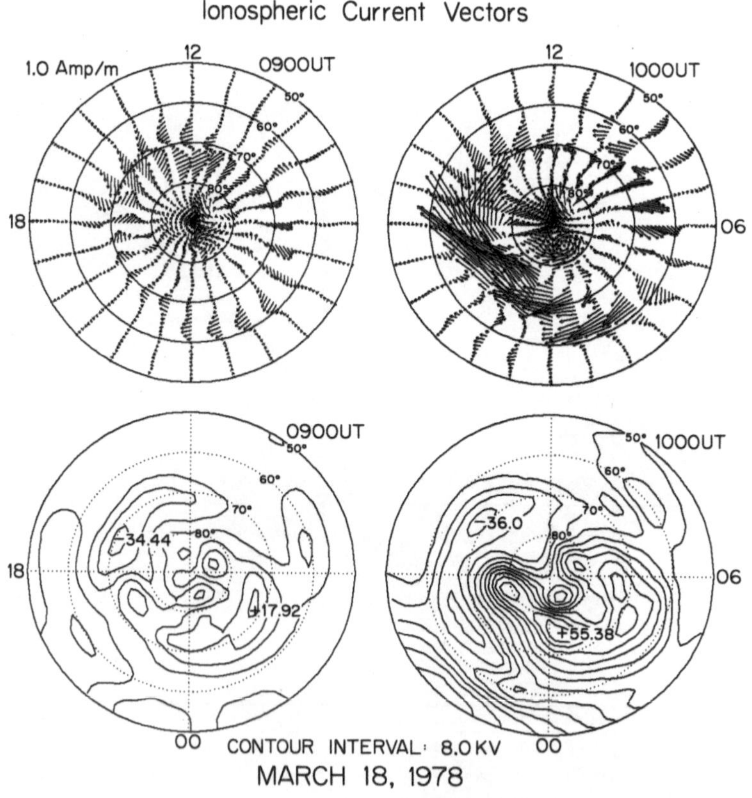

FIGURE 3.11. Example of the results obtained by the KRM method on magnetic data that was obtained from the IMS meridian chains.
Source: Akasofu, S.-I., W. Sun and B.-H. Ahn, 2000

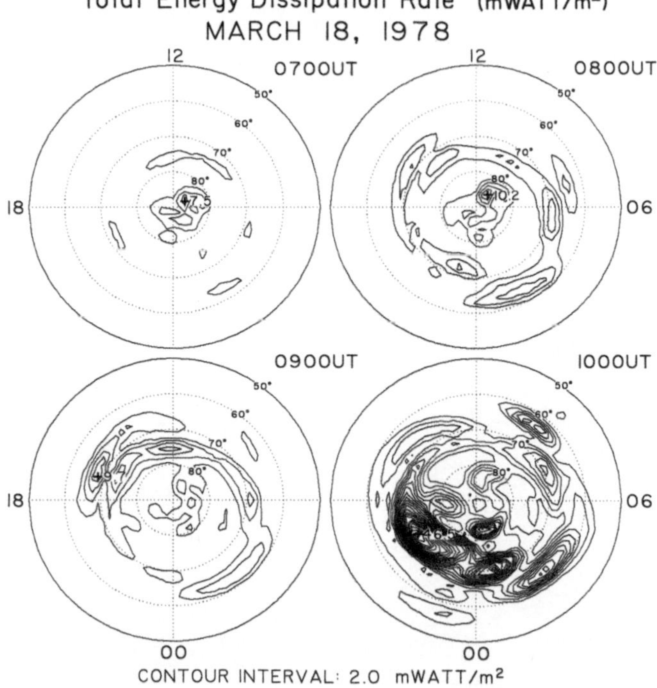

FIGURE 3.12. The Joule heat production rate at 0700, 0800, 0900, and 1000 UT on March 19, 1978.
Source: Ahn, B.-H., S.-I. Akasofu, and Y. Kamide, *J. Geophys. Res.*, **88**, 6275, 1983

It is in this way that the ground-based observation $\Delta \boldsymbol{B}$ (r = a, θ, λ, t) has finally enabled us to determine $\boldsymbol{J}$ (r at the ionospheric level, θ, λ, t), realizing the dream of our pioneers, although our results were still only the first approximation.

4
Synthesis of Magnetospheric Substorm Phenomena

The concept of substorms was a product of the earlier efforts of syntheses on many polar upper atmospheric phenomena, such as auroral activities, magnetic and ionospheric disturbances with a great variety of observation methods based on different instruments.

In this chapter, an attempt is made to synthesize some of the basic aspects of magnetospheric substorms. This is not intended to be a review of substorm literature. This is an example of synthesis, regardless of correctness of the conclusion I attempt to arrive at. My interest is to combine four basic aspects of substorm onset, namely the sudden brightening of an auroral arc, development of the westward electrojet, the so-called "dipolarization," and the poleward expansion in an attempt to reach a simple explanation.

4.1. Boström's Current Loops

4.1.1. Proof of Boström's Current Loops

In spite of much progress in understanding the electric current system during substorms, we encountered a serious problem in confirming the validity of our results described in Chapter 3. An independent method may be found by analyzing satellite observations. However, most satellite-based observations of electric current $\boldsymbol{J}$ must be based on measurements of $\Delta\boldsymbol{B}$ (r, θ, λ) at single points and at particular times. Thus, except for a rather simple geometrical configuration of the field-aligned currents just above the ionosphere, it is not possible to determine $\nabla \times \Delta\boldsymbol{B} = \boldsymbol{J}$ at a particular point and at a particular instant, because it takes at least one year of data to determine the average distribution of currents on the entire polar region or the equatorial plane by a satellite (Iijima et al., 1990). For this very reason, it is not possible to obtain the distribution of electric currents in the magnetosphere at a particular time, $\boldsymbol{J}$ (r, θ, λ, t) by a satellite or two. Thus, our problem was that neither ground-based nor satellite-based observations alone would allow us to confirm each other's results.

After much thought, an interesting idea emerged based on one theoretical insight by Rolf Boström. As early as 1964, Boström had suggested that ionospheric currents and equatorial currents are connected by two types of field-aligned loop currents, the meridional loops and the azimuthal loops (Figure 4.1). Thus, one way to test both the ground-based and satellite-based results, as well as Boström's loops, is to project the *average* ionospheric currents onto the equatorial plane and compare the results with the satellite results obtained by Takeshi Iijima and his colleagues (1990). The left-hand side of Figure 4.2 shows the projected *average* Pedersen component of the ionospheric current (Figure 3.9c) onto the equatorial plane (the insert shows the geometry of a meridian loop). Note that the ionospheric Pedersen current during substorms is basically the north–south current ($\boldsymbol{J}_\mathrm{p}$ in Figure 4.1). This projected distribution may be compared with the distribution of the satellite-based average radial currents on the equatorial plane on the right-hand side of Figure 4.2 (Iijima et al., 1990). I was greatly surprised by this unexpected agreement between the two.

In spite of the fact that both results are obtained by entirely different methods and at different times, the agreement shown in Figure 4.2 is quite satisfactory, testifying that both the KRM method and satellite observation can obtain the *average* ionospheric and equatorial currents reasonably well. Further, the results indicate the general validity of Boström's meridional loop currents. It took several more years to identify the azimuthal loops. This was because satellites measure both the ring current and the cross-tail current, which do not close in the

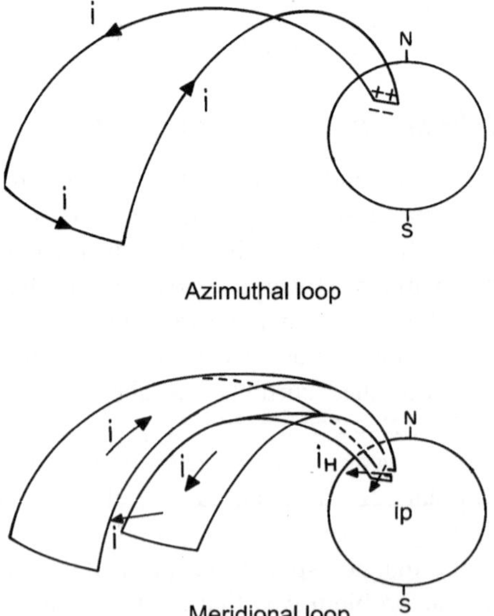

FIGURE 4.1. Two types of the magnetospheric current system proposed by Boström. *Source*: Boström, R., *J. Geophys. Res.*, **69**, 4983, 1964

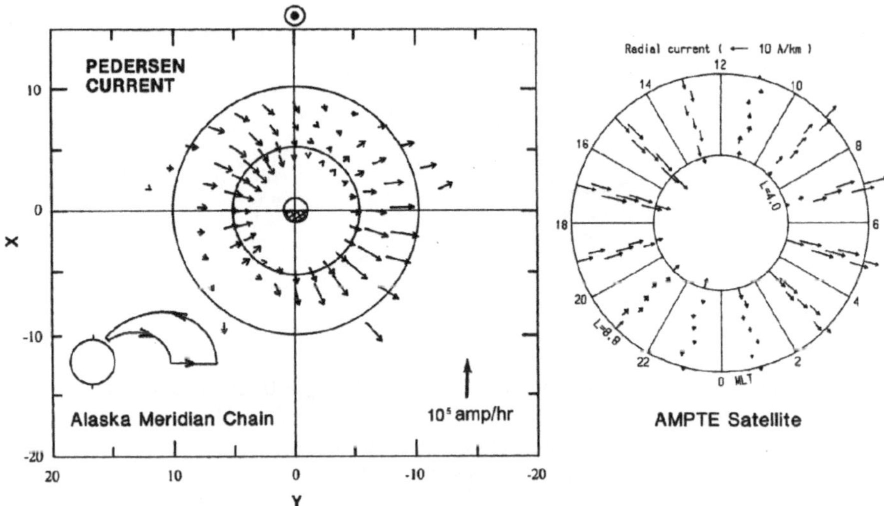

FIGURE 4.2. Left: The Pedersen current vectors are projected on the equatorial plane as indicated by the insert (Akasofu, 1992). Right: The distribution of the radial currents on the equatorial plane was obtained by Iijima et al. (1990).
Source: Akasofu, S.-I., *Geophys. J. Int.*, **109**, 191, 1992

ionosphere, so that we had to rely on modeling efforts to extract the azimuthal loop currents from the satellite observations (Sun et al., 1996, 2000).

One of my colleagues had never trusted our ground-based study of inferring the current system. It was his strong opinion that satellite measurements ($\nabla \times \Delta \boldsymbol{B} = \boldsymbol{J}$) are the only way (although satellites do not carry a current meter!). Thus, I asked him to compare the two diagrams in Figure 4.2. After this comparison, I have not heard him remark on our ground-based study of the magnetospheric currents.

4.1.2. The Pedersen Current

It has long been known that the Pedersen current $\boldsymbol{J}_p$ in the polar ionosphere flows southward in the early morning sector, see Figure 3.9c. By definition, it flows along the direction of electric field $\boldsymbol{E}$, so that it indicates the direction of the driving electric field. In fact, radar observations show that the southward directed Pedersen current is associated with the southward directed electric field (see Figure 4.6; Brekke et al., 1974). It is thus important to note that the electric fields $\boldsymbol{E}$ and $\boldsymbol{J}_p$ are related by $\boldsymbol{E} \cdot \boldsymbol{J}_p > 0$ in the ionosphere. Therefore, the Pedersen current must be driven externally from outside the ionosphere. Since the Pedersen current flows mostly in the north-south direction, it is expected that it is connected to radial currents $\boldsymbol{J}_r$ on the equatorial plane. It was most fortunate that Iijima et al. (1990) obtained the distribution of the radial current

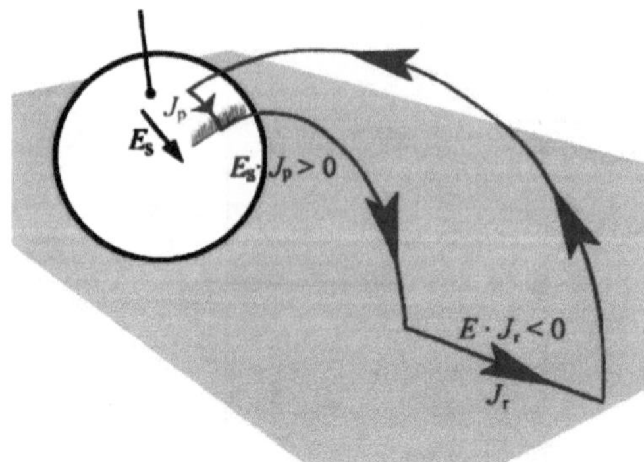

FIGURE 4.3. The Pedersen current circuit. Because $\boldsymbol{E} \bullet \boldsymbol{J}_p > 0$, the dynamo must be found in the equatorial plane $\boldsymbol{E} \bullet \boldsymbol{J}_r < 0$. The auroral arc is formed at the feet of the upward field-aligned current.
Source: Akasofu, S.-I.

in the equatorial plane using the AMPTE satellite data set that is shown on the right hand side of Figure 4.2.

The fact that the distributions on the left and right sides in Figure 4.2 are strikingly similar suggests that the Pedersen current is connected to the radial current $\boldsymbol{J}_r$ where the electromotive forces $\boldsymbol{E}$ and $\boldsymbol{J}_r$ must be related by $\boldsymbol{E} \bullet \boldsymbol{J}_r < 0$ in order to drive the meridional current system, including the Pedersen current and the upward and downward field-aligned currents (Figure 4.3). The fact that the Pedersen current is connected to the field-aligned currents, as shown in the insert of the right-hand side of Figure 4.2, is demonstrated by the distribution of the field-aligned currents in Figure 4.4a. The field-aligned current associated with meridional currents is a *sheet* current. The upward sheet current, which is carried by downward flowing electrons, must be associated with an auroral arc.

Thus, the ionospheric and equatorial observations are combined to indicate that the current element in Figure 4.3 constitutes a sheet current circuit in Figure 4.4b, and that there must be a process within a distance of 10 Earth radii to drive $\boldsymbol{J}_r$ (Figure 4.3). Further, the upward field-aligned current in a sheet form, carried by the downward streaming electrons, causes the curtain form of an auroral arc.

4.1.3. The Westward Electrojet is the Hall Current

It has long been known that the westward electrojet is the Hall current, which is driven by a southward oriented electric field (Akasofu, 1960).

In 1971, at the occasion of the International Symposium on Solar-Terrestrial Physics, held in St. Petersburg, Russia, May 12–19, I presented an idea that both the poleward expansion of the bulge and the poleward shift of the westward electrojet were associated with the disruption/diversion of the cross-tail current and its tailward expansion. My Figure 12 is reproduced here as Figure 4.5. Its popularized version was later promoted by a UCLA group (McPherron

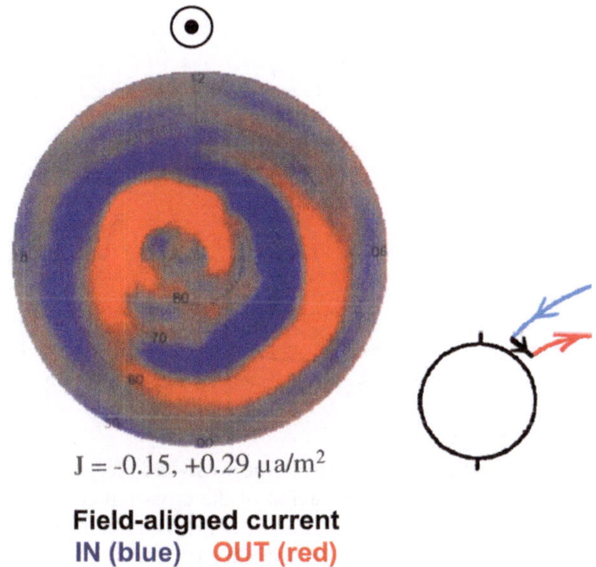

FIGURE 4.4a. The distribution of field-aligned currents determined by the Iridium satellite.
Source: Waters, C.L., B.J. Anderson, and K. Liou, *Geophys. Res. Lett.*, **28**, 2165, 2001

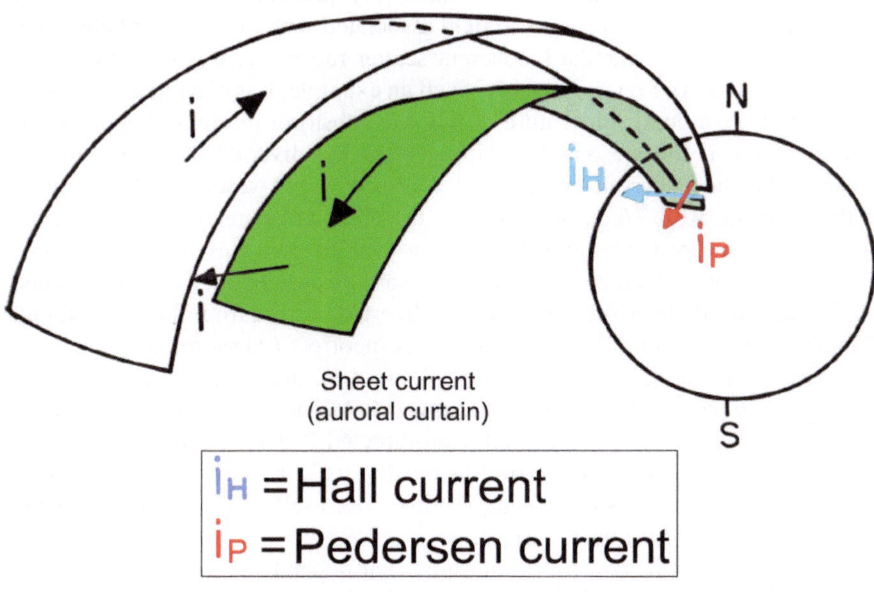

FIGURE 4.4b. The upward sheet current must be related to an auroral arc.
Source: Akasofu, S.-I.

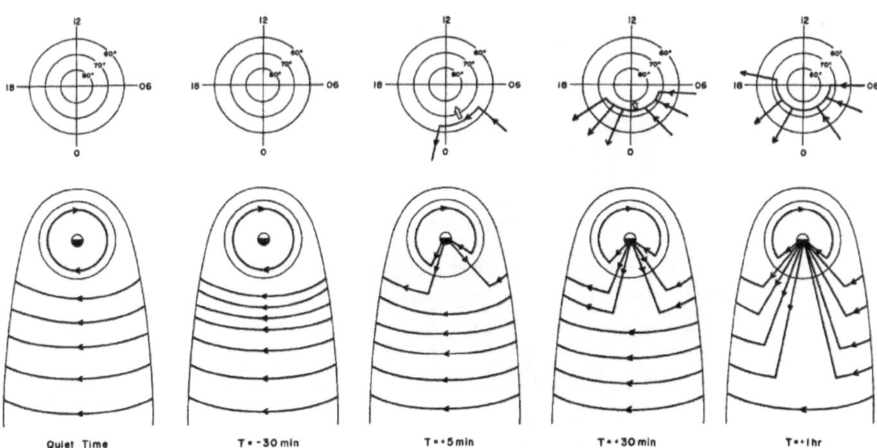

FIGURE 4.5. An attempt to explain the cause of the westward electrojet and its poleward advance in terms of the diversion of the cross-tail current.
Source: Akasofu, S.-I., *Solar Terr. Phys.*, Part III, E.R. Dyer and J.R. Roederer, ed., D. Reidel Publishing Co., 1972

et al., 1973) in terms of "wedge current." Since then, it had long been believed that the westward electrojet was the diverted current from the magnetotail.

However, in 2001, about 30 years after my publication on the westward electrojet, I re-examined the electric field associated with the westward electrojet on the basis of the Chatanika incoherent scatter radar data that was studied by Brekke et al. (1974). Figure 4.6 shows such an example. It is clear that southward electric field is a dominant feature during the substorm in the midnight sector. I realized that if the westward electrojet were the diverted cross-tail current, it had to be the Pedersen current and was driven by westward electric field of the order of 50 mV/km, arising from the cross-tail potential. However, the electric field is directed southward (driving the Pedersen current), so that there is no observational evidence of such a westward electric field. Thus, I concluded that the westward electrojet *couldn't* be a diverted current from the magnetotail. Therefore, the concept of "wedge current" is incorrect (Akasofu, 2002).

Boström's azimuthal current system (Figure 4.1), hereafter referred to as the *triangular current*, is driven by the southward electric field, which drives both the westward electrojet and the southward-directed Pedersen current.

The westward electrojet must find its closing path. However, since the polar ionosphere is not conductive enough, it must close itself on the equatorial plane, forming Boström's azimuthal current.

The intensity of this equatorial return current depends on the intensity of the westward electrojet. It is expected that it can be weaker or more intense than the cross-tail current (Figure 4.7).

The longitudinal extent of the southward electric field can be examined by obtaining its electric potential contours. An example is given in Figure 4.8a.

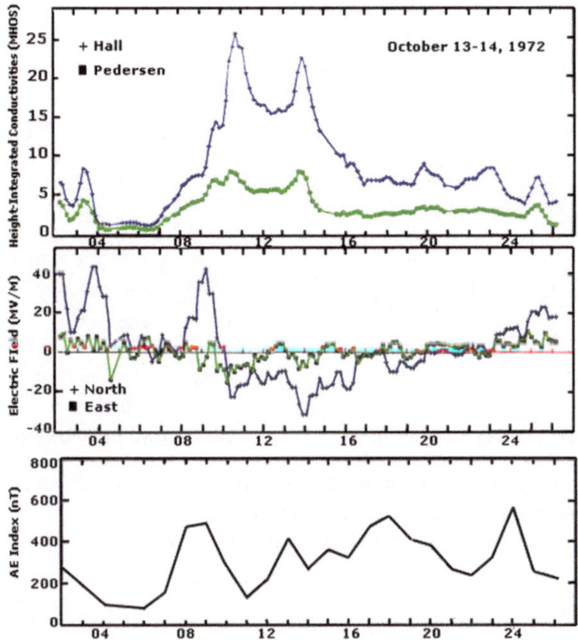

FIGURE 4.6. Top: The Hall and Pedersen conductivities during a substorm on October 13–14, 1972. Middle: North-South and East-West electric fields. Bottom: AE index. *Source*: Brekke, A., J.R. Doupnik, and P.M. Banks, *J. Geophys. Res.*, **79**, 3773, 1974

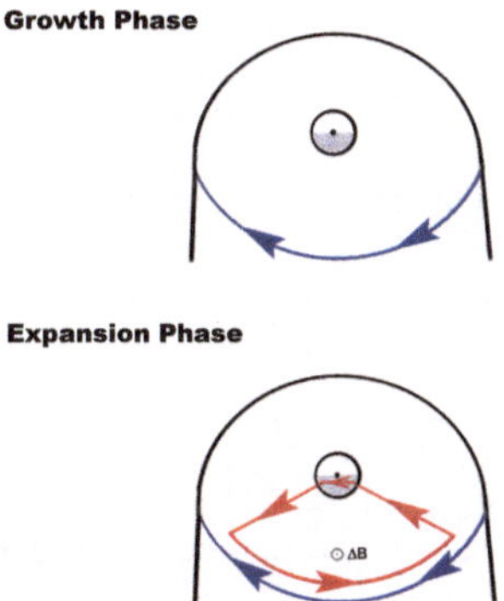

FIGURE 4.7. In the past, it has been believed that the enhanced cross-tail current during the growth phase is disrupted and diverted to cause the westward electrojet. However, the westward electrojet is the Hall current, which closes on the equatorial plane (red line in the lower figure). This return current can cause the "over-dipolarization," when it is more intense than the cross-tail current.
Source: Akasofu, S.-I., *J. Geophys. Res.*, **108**, 8006, 2003

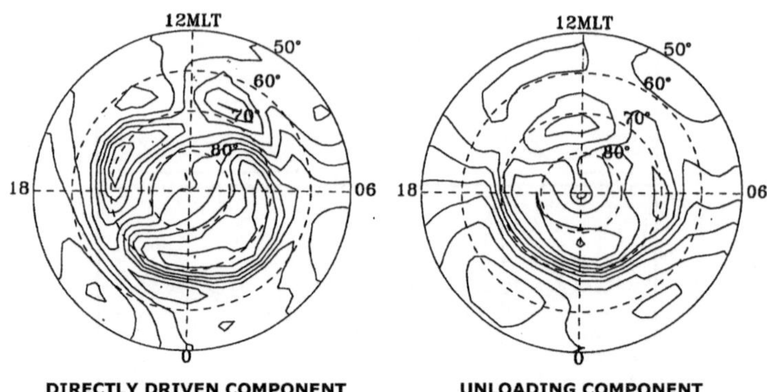

FIGURE 4.8a. Directly driven and unloading components of the equivalent current. The current lines are similar to the flow lines of ionospheric plasmas.
Source: Sun, W., W.-Y., Xu, and S.-I. Akasofu, *J. Geophys. Res.*, **103**, 11695, 1998

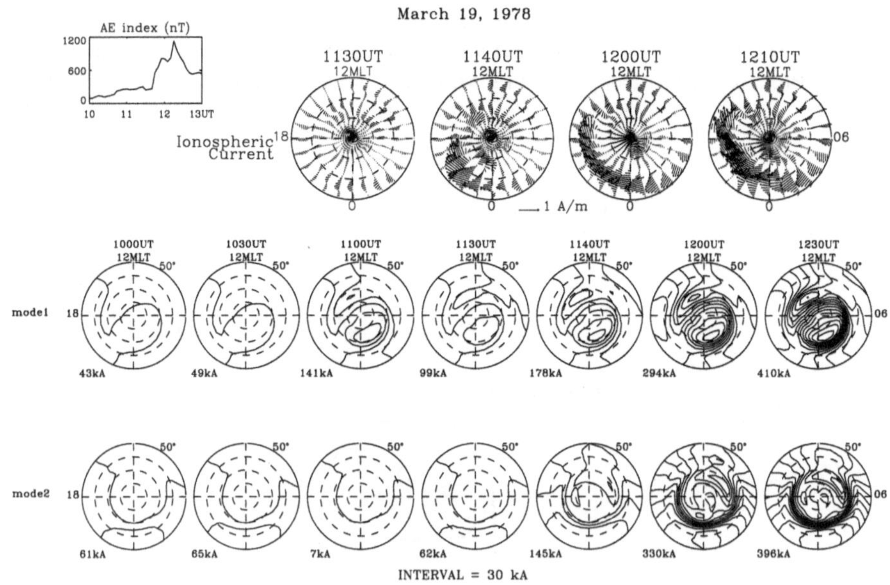

FIGURE 4.8b. Top: The ionospheric current distribution between 11:30 and 1210 UT on March 19, 1978. Middle: The potential contours associated with the two-cell pattern (mode 1). Bottom: The potential contours associated with the single-cell pattern (mode 2).
Source: Sun, W., W.-Y. Xu, and S.-I. Akasofu, Geospace Mass and Energy Flow, *Geophysical Monograph*, 104, *AGU*, 1998

The potential pattern can be divided into two modes. The first mode (mode 1) is the so-called "two-cell" pattern (often referred to as the DP2 component) and its intensity fairly well follows the dynamo power ε (Chapter 1). It is for this reason that it is called the directly driven component; it is the dominant feature during the growth phase. The second mode (mode 2) is the so-called "single-cell" pattern (often referred to as the DP1 component) and it grows suddenly at substorm onset (Figures 4.8b and 4.8c). Thus, it is called the unloading component (Figures 1.13a and 1.13b). The potential pattern shows that the equipotential lines are confined in a narrow belt along the 65° latitude line. The associated electric field must be directed southward.

Our success of projecting the ionosphere current on the equatorial plane had encouraged us to extend our method to infer the distribution of substorm electric currents on the equatorial plane with a time resolution of five minutes by projecting the ionospheric currents, using Boström's two elementary loops. There is no way to obtain such a pattern, even with hundreds of satellites.

Figure 4.9 shows the distribution of ionospheric and magnetospheric currents from 11:30 to 12:10 UT on March 19, 1978. It is a substorm event. The top row shows the distribution of ionospheric currents; magnetic records from 71 high-latitude observatories were used for this particular analysis. The current pattern at 11:30 UT shows the distribution just prior to the onset, at 11:35 UT, of

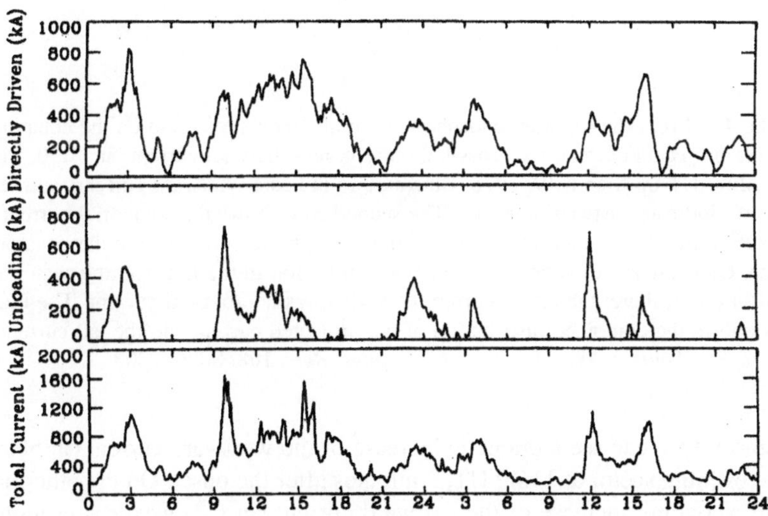

FIGURE 4.8c. Top: Time variations of the directly driven (mode 1 = the two-cell pattern) current. Middle: Time variations of the unloading (mode 2 = the single-cell pattern) current. Bottom: Time variations of the total (mode 1+ mode 2) current. Note, in particular, an impulsive growth of the mode 2.
Source: Sun, W., W.-Y. Xu, and S.-I. Akasofu, *J. Geophys. Res.*, **103**, 11695, 1998

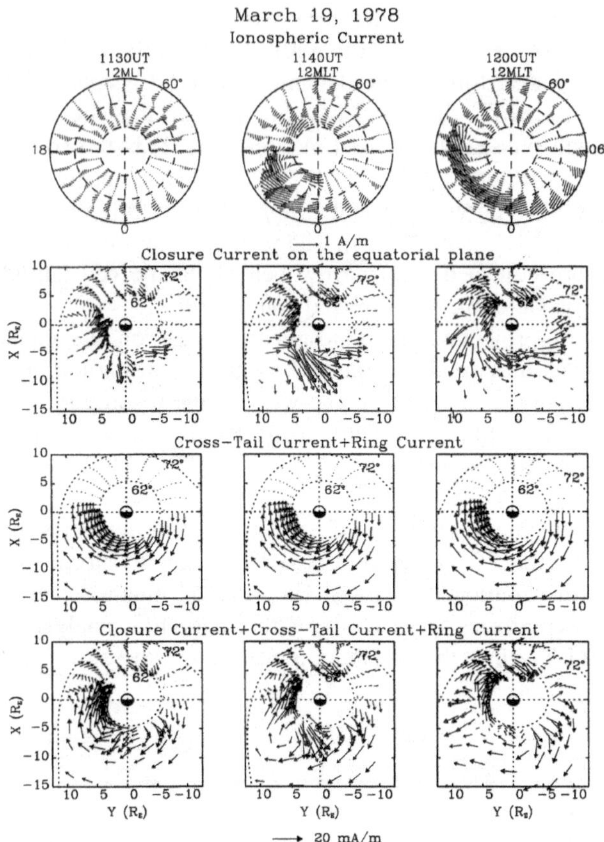

FIGURE 4.9. From the top, the ionospheric currents, their projection on the equatorial plane, the projected current, the cross-tail current, and the total current, at 11:30, 11:40, and 12:00 UT on March 19, 1978. Blue arrows indicate westward currents and red arrows indicate eastward currents. The second row shows the equatorial current distribution that is obtained by projecting the ionospheric currents onto the equatorial plane. One can see a major change of the distribution in the late evening sector at substorm onset; there, the return current overwhelms the cross-tail current. The fourth row shows the combined distribution of the projected current and the tail current. *Source*: Akasofu, S.-I., *J. Geophys. Res.*, **108**, 8006, 2003

a substorm. One can see a dramatic increase of the westward current (in blue) in the late evening sector at 11:40 UT, 5 minutes after the onset. On the other hand, the corresponding increase of the eastward current (in red) was less prominent. The substorm reached its maximum epoch at 12:10 UT.

The ionospheric current vectors are projected onto the equatorial plane (in the second row), as explained in Figure 4.2. The projected current vectors in the night sectors have a large eastward component. The assumed cross-tail current is shown in the third row. In the fourth row, the return current vector (the second row) and the cross-tail current (the third row) are combined. The cross-tail current was reduced at many points. In the late evening sector, the current direction was even reversed.

4.2. The So-called "Dipolarization"

It was unfortunate that there was no satellite to measure $\Delta \boldsymbol{B}$ on the equatorial plane for the event of Figure 4.9. On the other hand, since satellite measurements cannot provide $\boldsymbol{J}$, but instead $\Delta \boldsymbol{B}$, we can examine the validity of our results by comparing the well-established observational result of $\Delta \boldsymbol{B}$ and the computed $\Delta \boldsymbol{B}$ based on our results. Figure 4.10a shows the results. One can see that the stretched field lines contract and become dipolar after substorm onset at 11:30 UT. Thus, the inferred currents on the equatorial plane can reproduce the well-known change of $\Delta \boldsymbol{B}$. This phenomenon is often referred to as *dipolarization* and is a common feature at substorm onset, so that our results are consistent with the

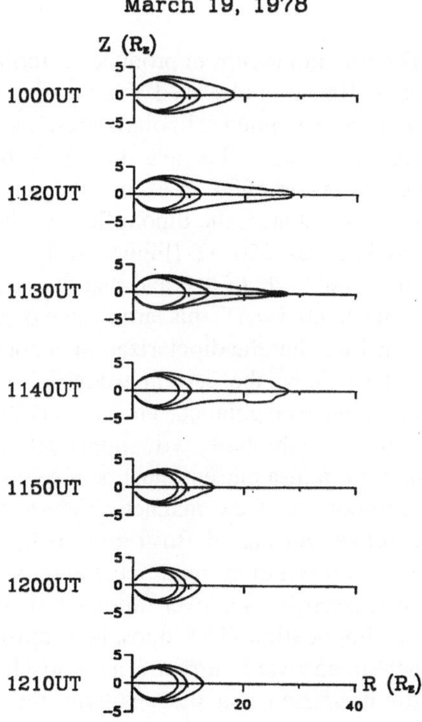

FIGURE 4.10a. Computed configuration of the magnetic field lines at the time of the substorm in Figure 4.9.
Source: Akasofu, S.-I. and W. Sun, *EOS*, 81-361, 2000

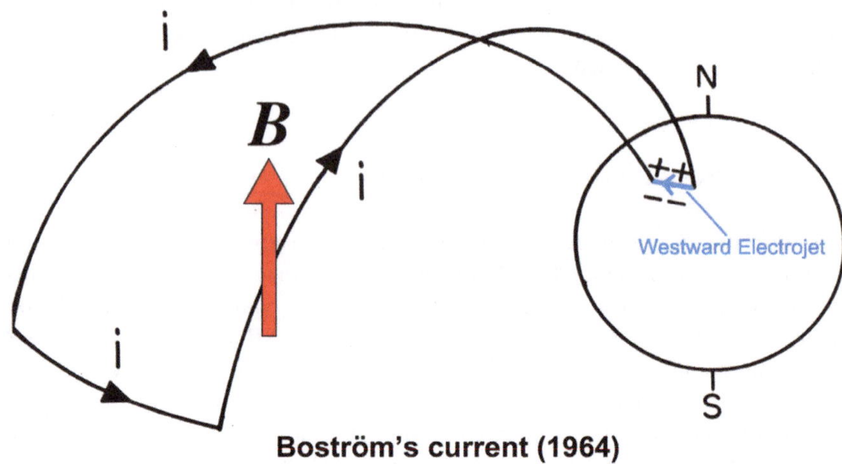

FIGURE 4.10b. The triangular current produces a northward-directed magnetic field inside it.
Source: Akasofu, S.-I.

satellite observations. The triangular current produces a northward $\Delta \boldsymbol{B}$ inside the triangular current, Figure 4.10b. It is observed as a positive bay on the ground.

It should be noted that the so-called "dipolarization" is not necessarily the recovery of the stretched field lines to become the dipole field lines. The field magnitude of the dipolarization can often far exceed the magnitude of the dipole field. At the geosynchronous distance, the dipole field is about 100 nT, but the observed field can be as large as 150 nT (Figure 4.11). Considering that the electron pressure is comparable with $B^2/8\pi$ and that the tail current can reduce the dipole field from 110 nT to 80–90 nT, this large value of 150 nT is surprising. Thus, this observation confirms that the dipolarization is not caused by diverting the westward tail current (namely, the current wedge concept). If the current wedge concept were correct, the triangular current stretched field, by an enhanced cross-tail current during the growth phase, will simply return back to the dipole field. The eastward equatorial return current must be responsible for the positive change of the vertical component of the magnetic field at the geosynchronous distance. The eastward return current of Boström's azimuthal current can be weaker or stronger than the cross-tail current, depending on the intensity of the westward electrojet. If it is stronger, an "over-dipolarization" will result.

This fact has a greater implication. The ionosphere must be responsible for the driving force for such an eastward current. The ionosphere must be playing an active role in the dipolarization and perhaps even the over-dipolarization. Thus, the ionosphere actively participates in substorm processes, rather than passively responding to magnetospheric processes. As mentioned in Section 1.10, regardless how important the magnetotail is in terms of plasma physics, the magnetotail is simply the "tail" of the ionosphere (dog).

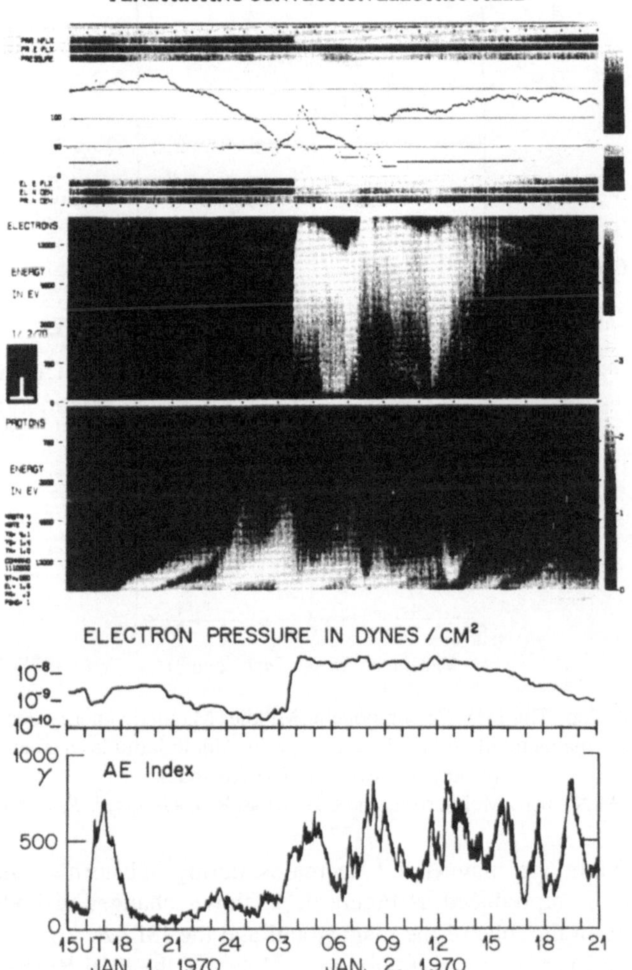

FIGURE 4.11. The "over-dipolarization" and injection of electrons and protons, observed at the geosynchronous satellite ATS-5 on January 2, 1970. From the top, the H component magnetic variations, the electron and proton spectrograms, plasma pressure and the AE index.

Source: DeForrest, S.E. and C.E. McIlwain, *J. Geophys. Res.*, **76**, 3587, 1971

4.3. Changes of Magnetic Energy in the Magnetotail

Proponents of the magnetic reconnection hypothesis have shown that the magnitude of the magnetic field B in the magnetotail decreases sharply at substorm onset as proof of magnetic reconnection.

The source of the firm belief that magnetic energy in the magnetotail decreases is the oft-quoted paper by Caan et al. (1975), which demonstrates that the

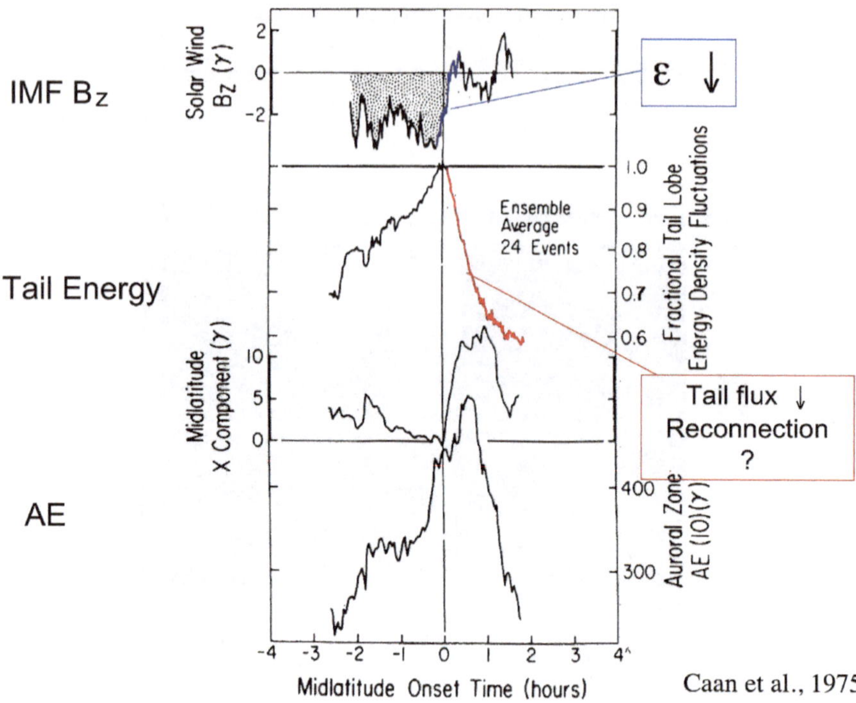

FIGURE 4.12. Top: The IMF *Bz* component. Middle: Magnetic energy density in the lobe of the magnetotail and positive change in middle latitude magnetograms (modified). Bottom: The AE index.
Source: Caan, M.N., R.L. McPherron, and C.T. Russell, *J. Geophys. Res.*, **80**, 191, 1975

magnetic field in the magnetotail decreases during substorms. One of their figures, which is reproduced as Figure 4.12, shows changes of both magnetic energy density in the lobe of the magnetotail and the IMF *Bz*.

However, this is not necessarily the case. Note that the IMF *Bz* sharply turned northward when the magnetic energy density in the lobe began to decrease, indicating that the power of the dynamo ε decreased, indicating that the magnetotail energy also decreased. Figure 4.13a shows that the ε function, magnetic energy density ($B_T^2/8\pi$), and AE index had similar time variations during two successive substorms. Wolfgang Baumjohann (1996) showed also that there is no distinct change in $B_T^2/8\pi$ for isolated substorms.

Actually, a better parameter to examine changes of magnetic energy in the magnetotail is the size of the open region, which should be approximately proportional to magnetic flux in the magnetotail, instead of the field magnitude *B* at a single point in the magnetotail.

In Figure 4.13b, the open region is defined as the highest latitude region that is free from auroral electrons except for the *polar rain* that consists of the high-energy tail of electrons in the solar wind. (The term "polar rain" was coined by Walter Heikkila of the University of Texas, where snow does not fall!)

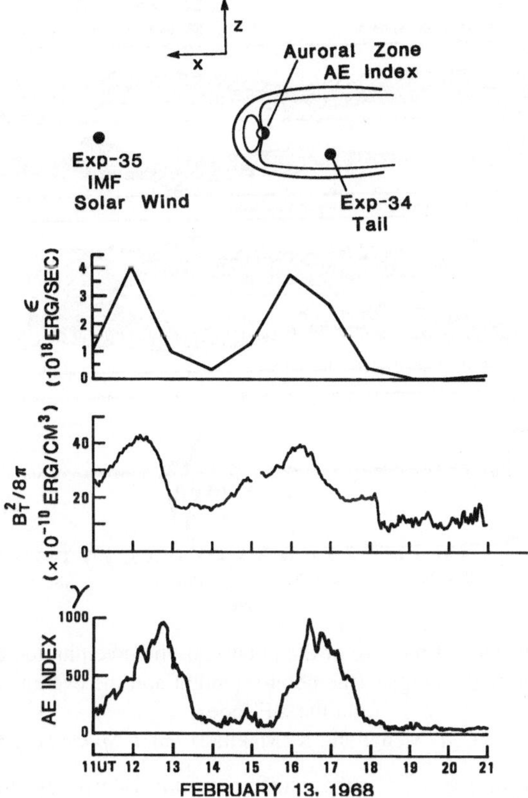

FIGURE 4.13a. Changes of ε, $B^2/8\pi$, and the AE index for two successive substorms. *Source*: Akasofu, S.-I., *Nature*, **284**, 248, 1980

Figure 4.14 is an example in which the dimension of the polar cap thus defined was monitored simultaneously by both the noon-midnight and evening-morning sector satellites for a four-day period. The auroral oval (the belt of visible auroras) is located in the cross-hatched region, while the belt of soft electrons (<500 eV) lies in the dot-shaded region. During the first 10 hours or so, until about 10:00 UT on October 30, 1984, the substorm activity was subsiding, and the noon-midnight and the evening-morning dimensions of the polar cap were decreasing. This resulted from an increase in the width of the precipitation belt of soft electrons (<500 eV) during this period, except in the night sector. These soft electrons must be a trapped population; otherwise, they will escape into interplanetary space in less than one second.

The second substorm activity began at about 20:00 UT on October 31. Both the noon-midnight and evening-morning dimensions *increased*. Substorm activity intensified at about 06:00 UT on November 1, and lasted until about the end of that day. Thus, the open region expanded during this substorm activity. First,

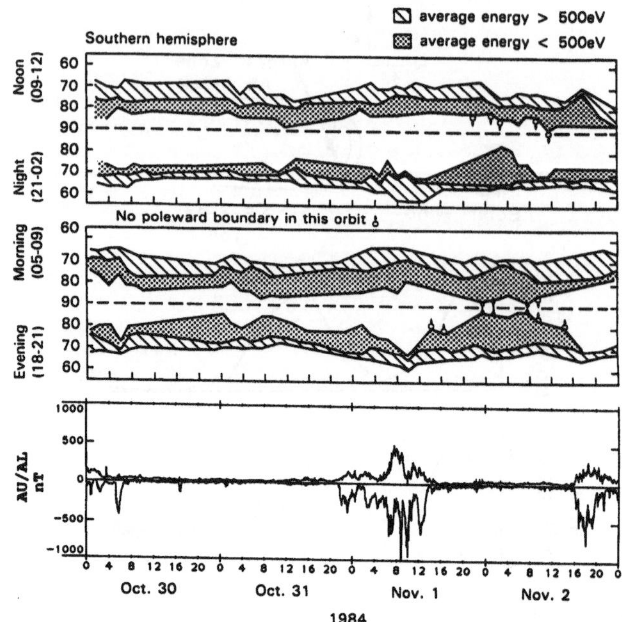

FIGURE 4.13b. Changes of the size of the polar cap, the precipitation belt of auroral electrons of low and high energies (the noon-midnight and the evening-morning orbits) and the AE index.
Source: Akasofu, S.-I., C.-I. Meng, and K. Makita, *Planet. Space Sci.,* **40**, 1513, 1992

the dayside precipitation (the cusp) shifted equatorward, as much as 10° as reported by many researchers. On the night side, there was a poleward shift of the precipitation boundary and a large expansion of the hard electron precipitation region at about 09:00 UT, indicating a large poleward expanding and bulge. After 10:00 UT on November 1, the width of the soft precipitation belt began to increase rapidly, although the hard precipitation region remained in the same latitude range; the AE index shows that the substorm activity was subsiding at that time. The open region contracted gradually until about 08:00 UT on November 2, but began to expand rapidly afterwards. This expansion was associated with the last substorm activity.

This and many other examples show that the dimension of the open region varies roughly in harmony with the AE index. Both the noon-midnight and evening-morning dimensions of the open region are substantially larger during substorm activity than those during a quiet period, by a factor of about two to three. Thus, the open magnetic flux is at least four times greater during substorm activity than during a quiet period. Actually, the expansion begins soon after the IMF Bz component becomes negative and the contraction proceeds gradually well after the IMF Bz component becomes positive.

Figure 4.14 shows also changes of the open flux, estimated on the basis of the size of area surrounded by the oval during a substorm. There is no indication of the decrease of the flux at substorm onset observed by a satellite (Frank et al., 1998).

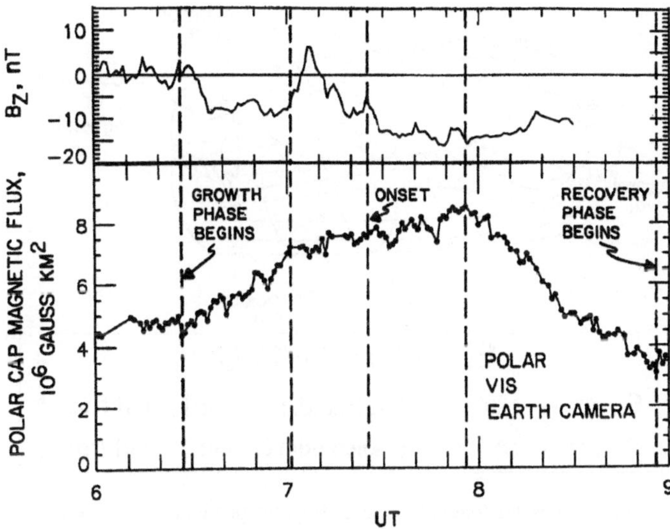

FIGURE 4.14. Upper: The IMF *Bz* component. Lower: The magnetic flux in the polar cap (modified).
Source: Frank, L.A., J.B. Sigwarth, and W.R. Patterson, *Substorms-4*, ed. by Kokubun S. and Kamide, Y., Terra Scientific Pub. Co., Tokyo, 1998

Thus, there can be more magnetic flux in the magnetotail during substorms than during a quiet period. This situation is illustrated in Figures 1.13a and 1.13c; if the input is large enough, the water level of the pitcher can be above the upper spout. If the sudden conversion of magnetic energy in the magnetotail were solely responsible for the expansive phase, the dimension of the open region should decrease suddenly at onset. This is not necessarily the case. The fact that the dimension of the open region is greater during an active period than during a quiet period shows that the magnetosphere is highly driven throughout substorm activity. We shall see in Section 7.5 that magnetic energy in an active region of the Sun tends to increase at flare onset.

This misinterpretation by Caan et al. (1975) in Figure 4.12 can be avoided by assuming that the magnetotail consists of two solenoids, together with the Earth's dipole field, its image dipole field, and the IMF (Figure 4.15). One can reproduce qualitatively much of the magnetotail features by assuming that the intensity of the solenoidal current changes in harmony with ε with a short time delay. Thus, it is quite natural for the tail field to decrease as ε decreases.

Therefore, the decrease of the lobe-field need not be a result of magnetic reconnection and the subsequent transfer back of the tail flux as generally believed. Indeed, it makes perfect sense to consider that the decrease occurred because the power ε, and subsequently the intensity of the solenoidal currents, began to decrease. Considering the fact that the lobe-field can increase or decrease during substorms, it is reasonable to consider that the lobe-field increases when the power ε is greater than the total loss rate (the energy dissipation rate) in the magnetosphere, while the lobe field decreases when the loss rate is greater than ε.

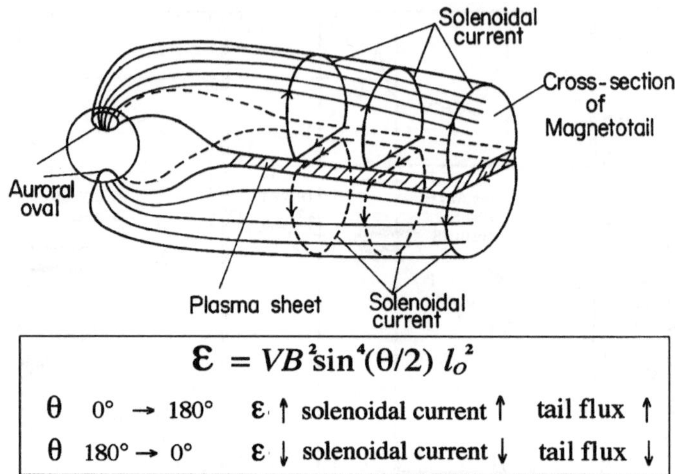

FIGURE 4.15. The solar wind-magnetosphere dynamo produces the power given by ε. It generates two solenoidal currents, one in each hemisphere of the magnetotail. The intensity of the solenoidal current varies with ε.
Source: Akasofu, S.-I.

Based on the above observation, instead of the idea of transferring magnetic flux from the dayside of the magnetosphere to the tail lobe during the growth phase and transferring back from the tail lobe to the dayside after substorm onset, it is possible to interpret the magnetic field observations in the magnetotail without contradictions in terms of increase and decrease of the solenoidal current and the cross-tail current as ε increases and decreases. A complicated MHD simulation is not needed to understand changes of the lobe field intensity.

4.4. Substorm Onset

4.4.1. An Example of Integration/Synthesis

In the earlier sections of this chapter, the essential ingredients for considering processes associated with substorms are assembled. When theoretical progress stagnates, it is best to go back to the fundamental observed facts. There are three distinct and well-established phenomena at substorm onset, as well as an enhanced convective flow after the so-called southward turning of the IMF or an enhancement of the solar wind-magnetosphere dynamo power ε:

1. A sudden brightening of an auroral arc over a distance of 1000 km at the poleward boundary of the diffuse aurora in the late evening or the midnight sector; the arc is located just the poleward side (~gm lat. 65°) of the diffuse aurora (caused by energetic electrons from the outer radiation belt), Figure 4.16.

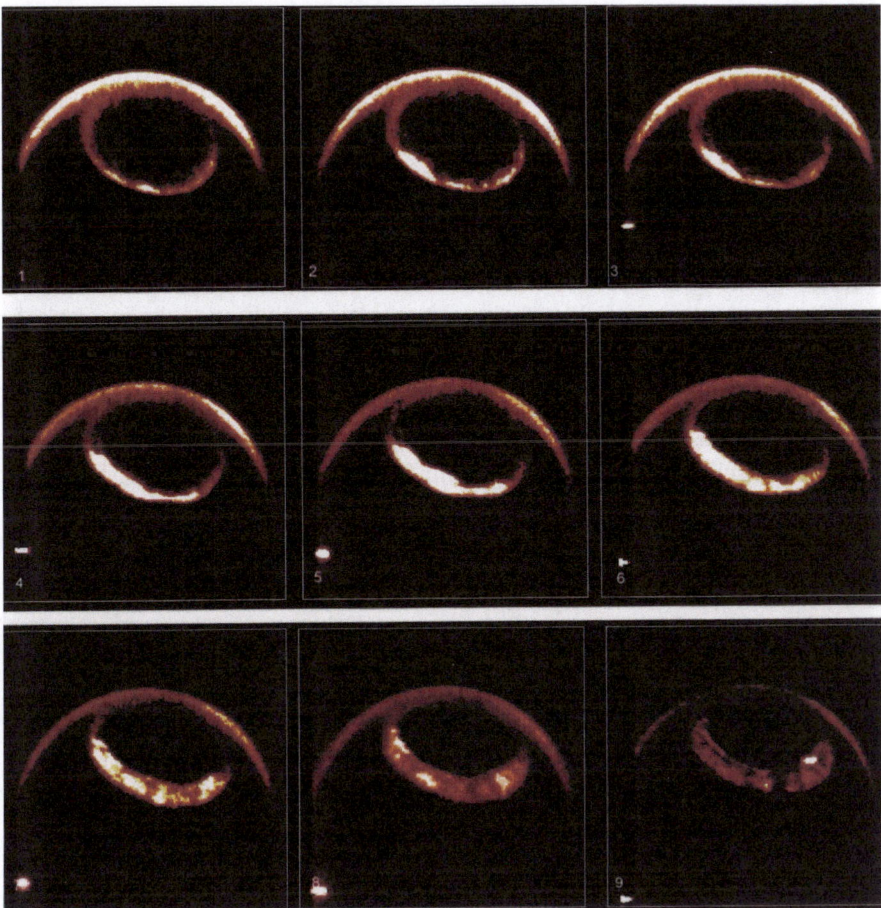

FIGURE 4.16. An example of substorm onset.
Source: Courtesy of L. Frank

2. The westward electrojet develops suddenly along the brightening arc.
3. The magnetic field structure changes suddenly from a tail-like configuration to the dipolar structure in the magnetotail. This phenomenon is often referred to as dipolarization and is known to propagate outward.

The three phenomena have the following physical meanings:

Sudden Brightening:

This phenomenon must be associated with an increase of the energetic electron flux into the existing arc, carrying the upward field-aligned current over an east–west extent of the order of 1000 km within the narrow width of an auroral arc (see Figure 4.4b).

Westward Electrojet:

The westward electrojet is essentially the Hall current carried by an eastward flow of ionospheric electrons. For this to happen, a southward electric field must be imposed along a narrow east–west strip in the ionosphere. An increased ionospheric conductivity caused by the electron precipitation will also enhance the current.

Dipolarization/Over-dipolarization:

Prior to substorm onset, the dipolar field lines at 5 Earth radii and beyond are greatly stretched by an intensification of the westward cross-tail current. The sudden dipolarization indicates that the cross-tail current is suddenly reduced. In some instances, an over-dipolarization can take place, indicating the reversal of the westward tail current.

Any successful theory of substorm onset must explain at least these three processes, which follow after the magnetosphere is driven for 30–40 minutes. The arc, which brightens first at substorm onset, is located just on the poleward side of the diffuse aurora in the late evening or the midnight sector. The diffuse aurora is caused by the precipitation of high-energy electrons from the outer trapping (Van Allen) belt. Therefore, the processes associated with substorm onset must be found at the region of transition from a dipolar field regime to the stretched field regime, not deep in the magnetotail. It is likely that this particular region is located at a distance of 5–10 Earth radii (Frank and Sigwarth, 2000, not as far as 20–30 Earth radii.

It is pointed out here that the three observations can be explained quantitatively by the single fact that Boström's current system (Figure 4.1) is activated at substorm onset.

One of the possible processes that can produce such an electric field is a sudden reduction of the cross-tail current. As the stretched field lines start to contract, electrons in the plasma sheet, being magnetized, will move with the contracting field lines, but positive ions will not follow such a motion, being left behind by the electrons because they are not magnetized. As a result, an earthward electric $\boldsymbol{E}_r$ field is produced, as Figure 4.17 shows. This possibility was suggested to me by Tony Lui. The earthward flow of electrons can be identified as $\boldsymbol{J}_r$ in Figure 4.3. The earthward electric field $\boldsymbol{E}_r$ induces an eastward plasma flow ($\boldsymbol{E} \times \boldsymbol{B}$) in the magnetosphere and the ionosphere, which generates the westward electrojet in the E region of the ionosphere.

An important question is what causes the reduction of the cross-tail current. One fact we know is that some fraction of substorms occur at the time when the IMF Bz component is suddenly reduced, causing the reduction of the cross-tail current intensity. Figure 4.18 shows the occurrence of a substorm when the IMF *Bz turned northward* (Lyons et al., 2001), indicating a decrease of ε and the cross-tail current. This case may be generalized to infer that any process that reduces the cross-tail current rather suddenly might trigger substorms after the magnetosphere is driven for 30–40 minutes. We need to find the processes responsible for the reduction of the cross-tail current in addition to the reduction of ε. This is because substorms often occur without the northward turning of the IMF. Some plasma instability associated with the cross-tail current may also

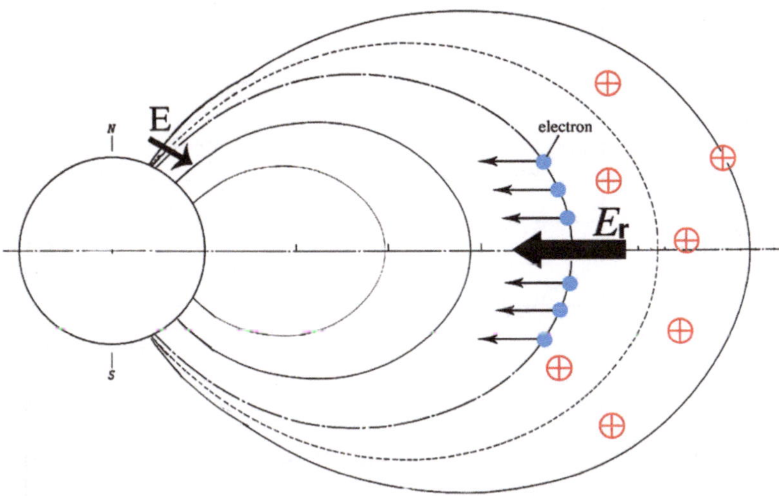

FIGURE 4.17. When the stretched field lines contract, they will bring electrons with them, but not protons. As a result, an earthward electric field develops.
Source: Akasofu, S.-I.

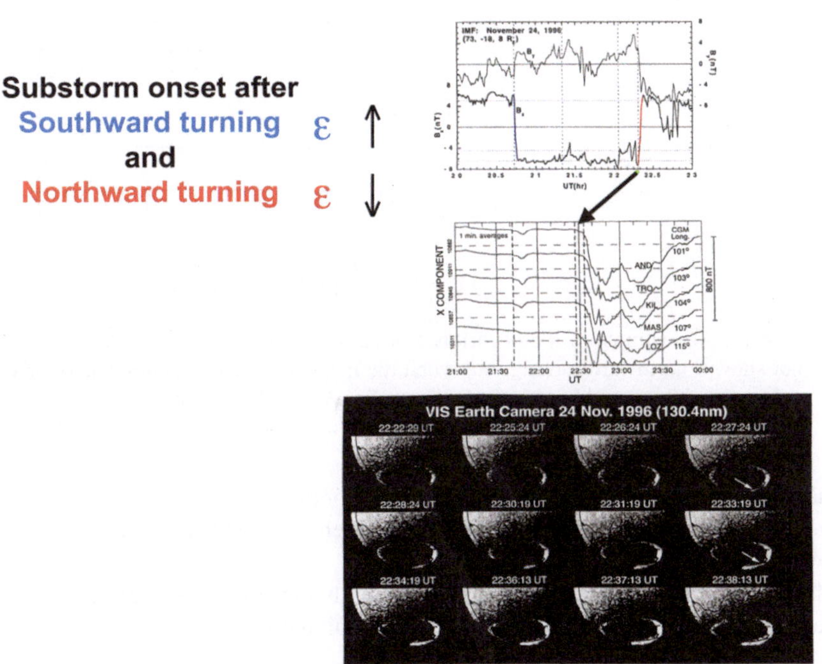

FIGURE 4.18. An example of substorms that occurred when the IMF *Bz* turned northward (modified).
Source: Lyons, L.R., R.L. McPherron, and E. Zesta, *J. Geophys. Res.*, **106**, 349, 2001

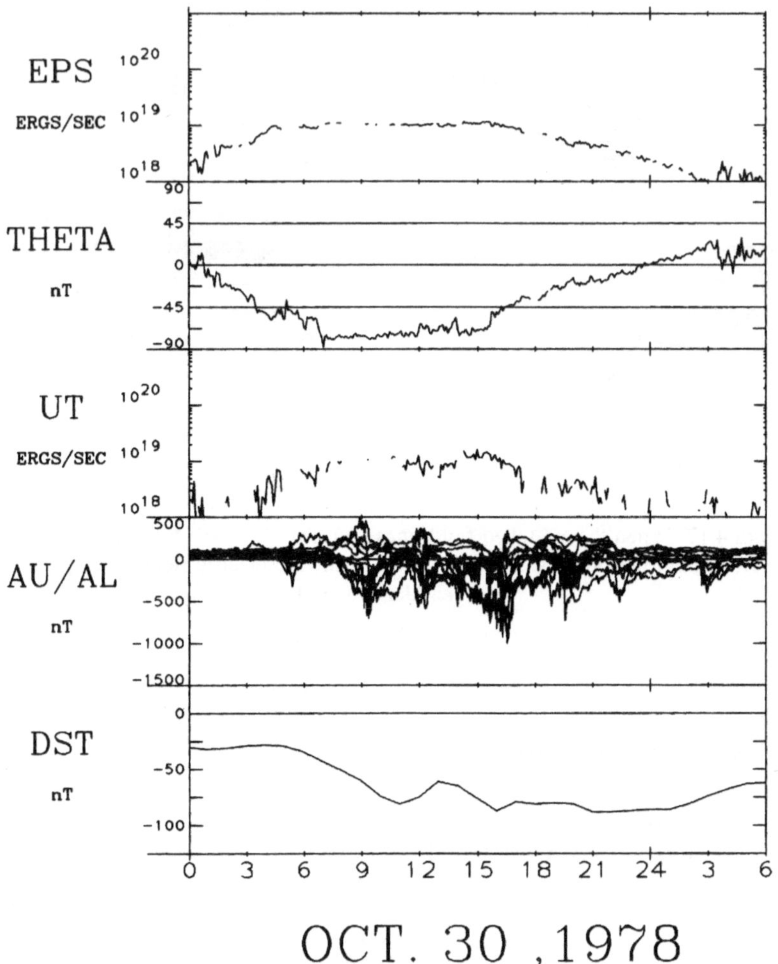

FIGURE 4.19. An example of the occurrence of substorms when the IMF *Bz* component does not show a northward turning, but when the magnetosphere is continuously driven. *Source*: Akasofu, S.-I., *Space Sci. Rev.*, **113**, 1, 2004

cause substorms when the magnetosphere is continuously driven (Figure 4.19). Figure 4.20 summarizes schematically the sequence of the events, which are also shown in a block diagram in Figure 4.21.

The three phenomena that are associated with substorm onset are likely to be directly related to each other and to occur simultaneously:

(a) After an enhancement of the cross-tail current during the growth phase, some plasma instability suddenly reduces the cross-tail current, causing the contraction of the magnetotail field, the "dipolarization."

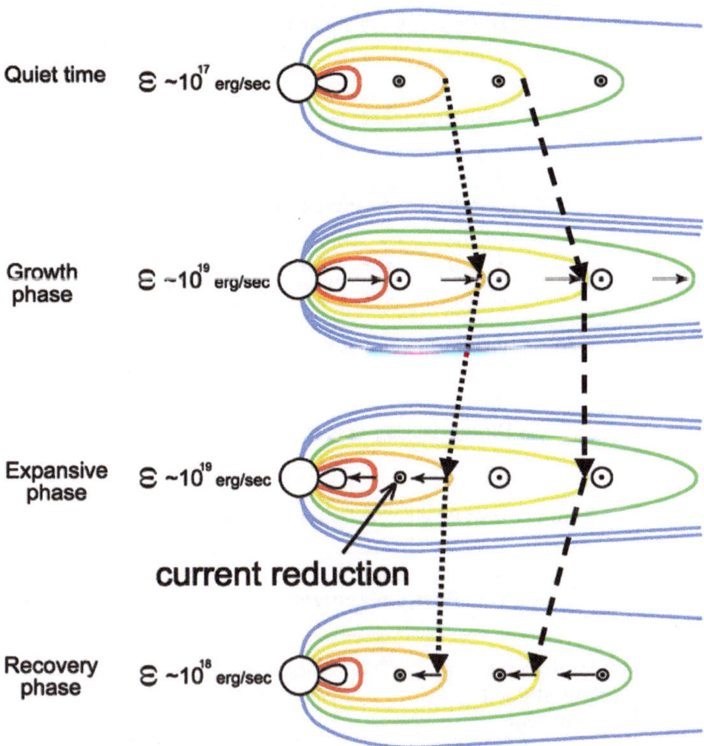

FIGURE 4.20. The summary of one possible sequence of the events from the quiet time to the recovery phase. It shows changes of the intensity of the cross-tail current and the magnetic field configuration.
Source: Akasofu, S.-I.

(b) Processes associated with the "dipolarization" result in a sudden activation of Boström's current system at a distance of 5–10 RE, causing the meridional currents. The upward portion of the sheet current will brighten the aurora.
(c) The same process will bring the equatorward electric field in the ionosphere, driving the westward-directed Hall current, the westward electrojet.
(d) The westward electrojet must close its circuit on the equatorial plane, disrupting the cross-tail current. This process further causes the "dipolarization." This is a positive feedback process that is needed for the sudden growth of substorms. In some cases, the eastward current becomes stronger than the cross-tail current, causing over-dipolarization.

4.4.2. *The Poleward Expansion*

The most prominent feature of auroral substorms is the poleward expansion of the auroral system in the midnight sector (Chapter 2). In the past, this feature has been explained by piling up of the reconnecting closed field lines. However, magnetic

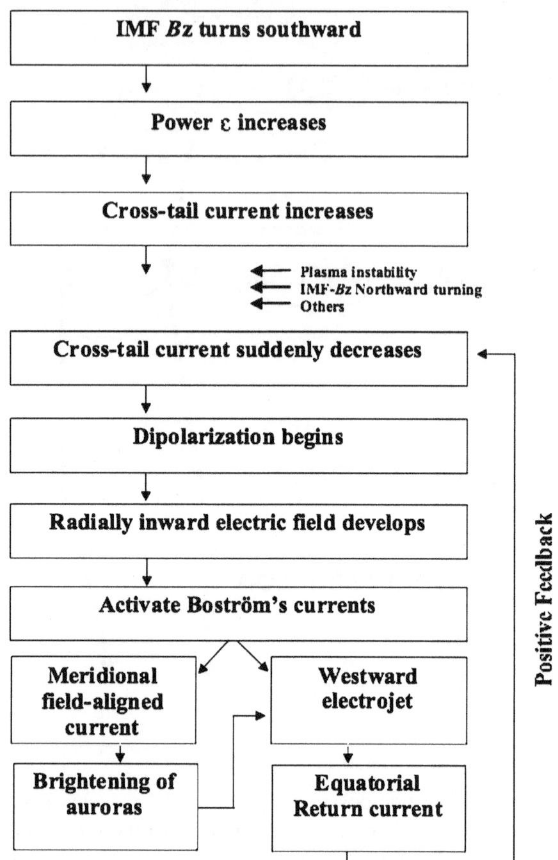

FIGURE 4.21. Block diagram showing one possible sequence of processes leading to substorm onset and the expansion phase.
Source: Akasofu, S.-I., *Space Sci. Rev.*, **113**, 1, 2004

reconnection theories of substorm onset cannot explain quantitatively the extent of the poleward expansion and its time variations, since the reconnection rate is not known. On the other hand, here is a simple explanation:

As mentioned earlier, the triangular current must be partly responsible for the dipolarization, because it produces northward flux (a positive *B*z component) inside it (Figure 4.10b). This northward flux must be added to the Earth's magnetic field, shifting the footprint of field lines poleward (Figure 4.22). It is not difficult to estimate the total flux inside the triangular current. It is assumed that the dipolarization occurs within 10 RE from the Earth; the width is taken to be 10 RE. Thus, the triangular area is about $2.1 \times 10^9\,\text{km}^2$. The magnitude of the dipolarization is taken to be 50 nT; the so-called "positive bay," which is observed in low latitudes on the ground; the field produced by the triangular

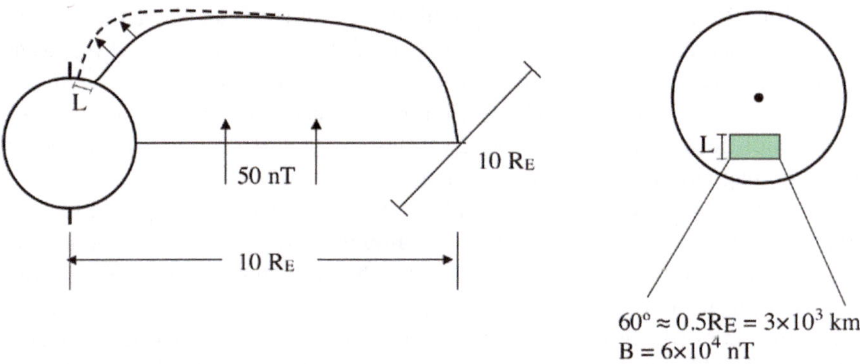

FIGURE 4.22. The geometrical relationship between the northward flux in the equatorial plane and the poleward expansion of the auroral bulge.
Source: Akasofu, S.-I.

current, not by the ionospheric current. Thus, the total flux within the triangular current is:

$$2.1 \times 10^9\,\text{km}^2 \times 50\,\text{nT} = 1.1 \times 10^{11}\,\text{nT}\,\text{km}^2$$

The flux within the triangular current moves the foot of the last closed field line. The longitudinal extent of the bulge is estimated to be about 60° or $0.5\,R_E \sim 3 \times 10^3$ km. The field intensity there is about 6×10^4 nT. The extent of the poleward-expanding bulge, L, may be estimated to be (Figure 4.22):

$$L = \frac{1.1 \times 10^{11}\,\text{nT}\,\text{km}^2}{6 \times 10^4\,\text{nT} \times 3 \times 10^3\,\text{km}} = \sim 610\,\text{km}$$

Thus, a significant part of the poleward expansion can result from the growth of the triangular current. Further, it is expected that the poleward expansion stops at about the time the westward electrojet intensity reaches its peak value (Figure 2.8c). Our estimate shows that the poleward expansion should occur in harmony with the growth of the triangular current, roughly along with the growth of the westward electrojet or the AL index.

4.4.3. Magnetic Reconnection in the Magnetotail

In order for magnetic reconnection to provide the same amount of magnetic flux $\sim 1.1 \times 10^{11}\,\text{nT}\,\text{km}^2$), the flux along the tail must be from 10 R_E to at least 100 R_E, assuming that the Bz component is about 2 nT there; the same width of 10 R_E is assumed. If so, the neutral line must move outward with a speed of 100 km/sec. Is there any systematic observation in the magnetotail to confirm such a movement?

In the past, proponents of magnetic reconnection assumed that magnetic reconnection could form at a distance of 5–6 Earth radii by calling it the near-Earth neutral line. Realizing that such a possibility is unlikely and that there is no definitive observational evidence, they extended the near-Earth neutral line to as far as 20–30 Earth radii (and still call it "near-Earth"). Now, they are trying to connect it to a distance of 5–10 Earth radii by an earthward ($\boldsymbol{E} \times \boldsymbol{B}$) plasma flow that has not been confirmed by observations; most fast observed flows occur along the magnetic field lines, not the convective flow.

I am afraid that it will not be possible to make substantial progress in understanding substorm processes so long as we cling to elusive magnetic reconnection as the primary process. Looking back through the history of magnetic reconnection, I am of the opinion that the original concept of magnetic reconnection in the solar atmosphere is not realistic (see Chapter 7). I doubted that a stable anti-parallel field condition could even be set up in the dynamical and turbulent solar atmosphere as many theorists formulate and that it could be explosively destroyed, although it is a sort of a problem theorists love to deal with, regardless if it is realistic or not. In fact, theorists assumed the initial condition of the anti-parallel field (the so-called "Harris solution"), which may not even exist in the realistic solar atmosphere and in the magnetotail, and soon found that the assumed anti-parallel field configuration is hard to destroy explosively. If it were easy to do so, the anti-parallel field configuration would not form to begin with. The only possibility would be that magnetic reconnection could be driven if two magnetic configurations were forcefully brought together to produce an anti-parallel configuration. In such a situation, magnetic energy that magnetic reconnection generates would be the same amount of energy that is needed for driving magnetic reconnection, rather than energy resulting from annihilation of the original anti-parallel field; it may simply facilitate the input energy conversion, not the generation.

As the concept of magnetic reconnection has become explosively popular, some of us have been left as an almost invisible minority. A fanatic believer in magnetic reconnection told me that I am not qualified to be a magnetospheric physicist unless I believe in a process as fundamental as explosive magnetic reconnection. On the other hand, Tony Lui and I were two of the few who very seriously attempted to find definitive indications of magnetic reconnection as the source of substorm energy in satellite data during the 1970s and 1980s. We failed to find it. A series of our papers on this subject was titled *Search for the Magnetic Neutral Line in the near-Earth Plasma Sheet*. Indeed, to date, no one has conclusively found the neutral line within about 10 Earth radii; see also a recent paper by Ge and Russell (2006). I am still not convinced that magnetic reconnection is the energy supply process for substorms; if any reconnection takes place, it may occur as a secondary process, as the return current from the westward current exceeds the cross-tail current. Magnetic reconnection will be discussed further in Chapter 7 in connection with solar activity.

Further, many of the MHD simulations of substorms are misleading in the sense that they give an impression that substorm processes are understood; what they simply show is that substorms are electromagnetic phenomena, which are describable in terms of the Maxwell equations and fluid equations. Basic physical processes cannot be revealed, so the simulation results are similar to observed phenomena. What did we learn? There are many parameters to be adjusted; therefore, realistic-looking results may be due to the wrong reasons. Indeed, there is no definite theory to determine even as fundamental as the reconnection rate.

4.5. Storm–Substorm Relationship

A geomagnetic storm tends to occur in association with a series of magnetospheric substorms. Figures 1.8a, 4.23a, and 4.23b clearly show that an intense geomagnetic storm tends to develop when intense substorms occur rapidly in succession. This is because the injection rate of ring current particles must overcome a heavy loss rate in the magnetosphere by the charge exchange process. Chapman and I suggested in 1968 that substorms are basic elements of a geomagnetic storm and that each substorm contributes to the formation of the ring current belt by injecting ring current particles. Section 2.5 described how our study of the great geomagnetic storm of February 11, 1958 led us to this conclusion.

One of the most interesting findings in this connection in recent years is that O^+ ions, instead of protons from the solar wind, are often the dominant ions in the storm-time ring current belt (Yannis Daglis, 1997). Since oxygen atoms

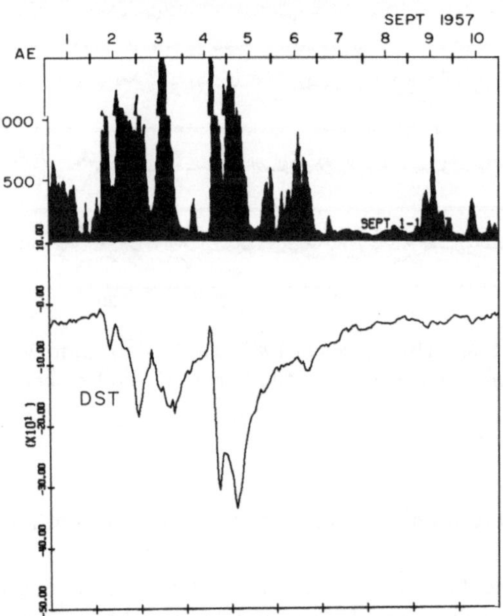

FIGURE 4.23a. The relationship between substorms (indicated by the AE index) and a storm (indicated by the Dst index).
Source: Akasofu, S.-I., *Ann. Geophys.*, **26**, 443, 1970

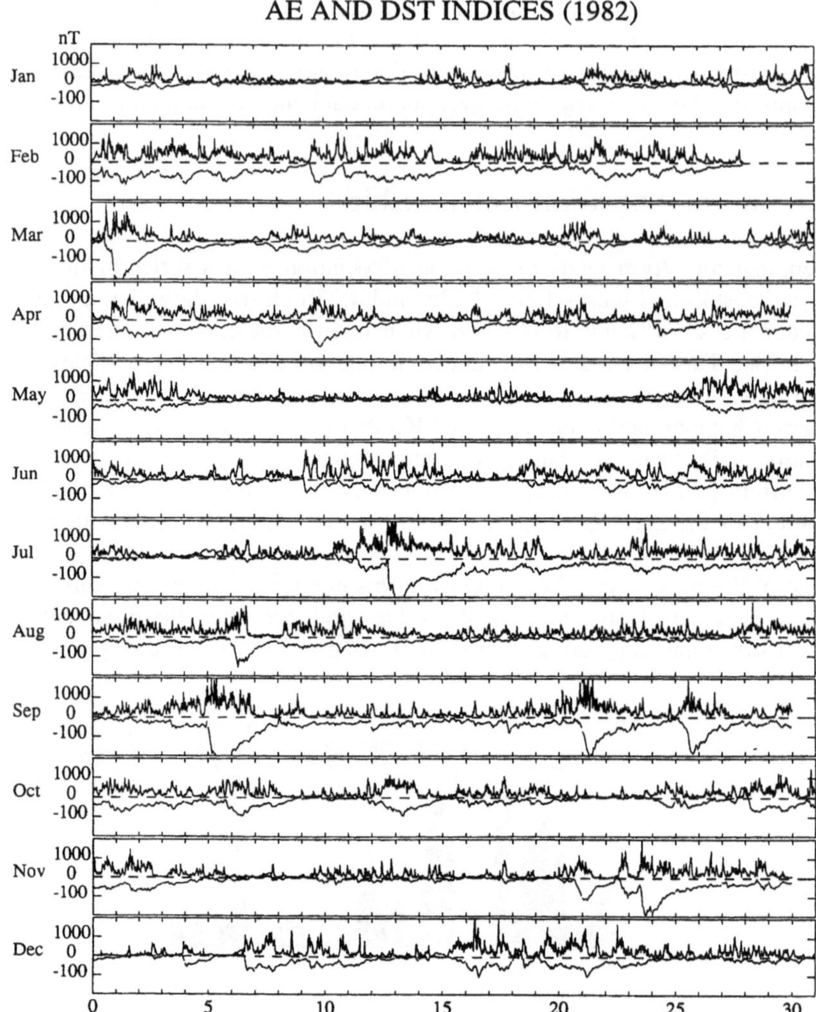

FIGURE 4.23b. AE and Dst indices in 1982. There is no storm without substorms. *Source*: Geomagnetic Data Center, Kyoto University

in the solar wind are highly ionized, say O^{+7}, ring current ions O^{+} must be of ionospheric origin.

Indeed, energetic O^{+} beams are observed to stream out from the auroral ionosphere during substorms. Hence, it is reasonable to think that the development of the ring current belt is caused by O^{+} ions from the ionosphere associated with the upward field-aligned currents during substorms. An exciting new development in this particular subject is the possibility to "visually" study the formation of the ring current belt by observing energetic neutral atoms (ENA) by High Energy Neutral Atom (HENA) imaging.

In particular, considerable progress has recently been made in studying the distribution of oxygen ions by using an imaging method, called energetic neutral atom (ENA) imaging (Figure 4.24).

Figure 4.25 shows the ion composition in the ring current belt in terms of the ratio O^+/H^+. It can be as large as 20 during an intense storm (Nosé et al., 2005). Further, it appears that they are accelerated along the geomagnetic field lines from the ionosphere into the plasma sheet and are subsequently energized and injected into the trapped region during substorms after being convected toward the Earth. Further, Figure 4.26 shows that the flux of O^+ ions increases as the aurora brightens at substorm onset (Don Michell, 2006).

It appears that a number of researchers question this simple role of substorms on the formation of the ring current belt, in spite of the fact that Daglis (1997) and others showed that the flux of O^+ ions in the ring current correlates well with the AE index. Further, a recent result showed conclusively that the flux of O^+ ions in the magnetotail increases at substorm onset, as shown in Figures 4.24, 4.25, and 4.26. The formation of the ring current belt is a two-stage process; the first is the injection of O^+ from the ionosphere into the magnetotail, and the second is the injection of O^+ from the magnetotail to the ring current belt. Thus, there can be a variety of complication before O^+ ions can find on their way to the ring current belt under a heavy loss caused by the charge exchange process. For example, unless the injection from the ionosphere to the magnetotail can occur relatively close to the ring current belt, O^+ ions may not be able to form the ring current belt after the energization; they may drift away from the magnetopause. Therefore, we need to examine why different substorms contribute differently to the ring current formation, rather than deny the storm-substorm relationship, as some suggested in the past.

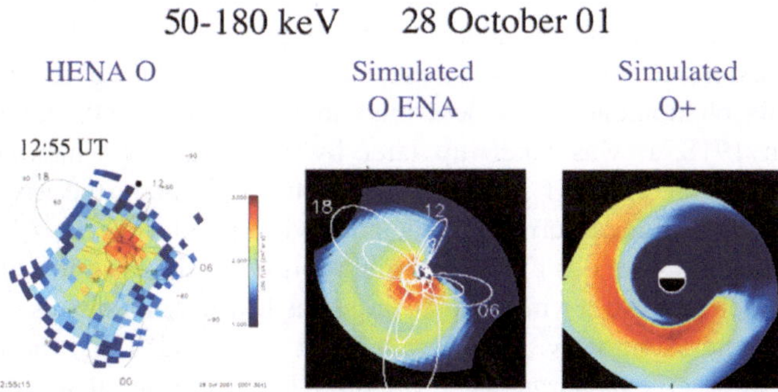

FIGURE 4.24. The energetic neutral atom (ENA) imaging of the ring current belt and its simulation, both imaging and O^+ movement.
Source: Fok, M.C., Earth-Sun System Exploration: Energy Transfer, January 16-20, 2006, Kona, Hawaii

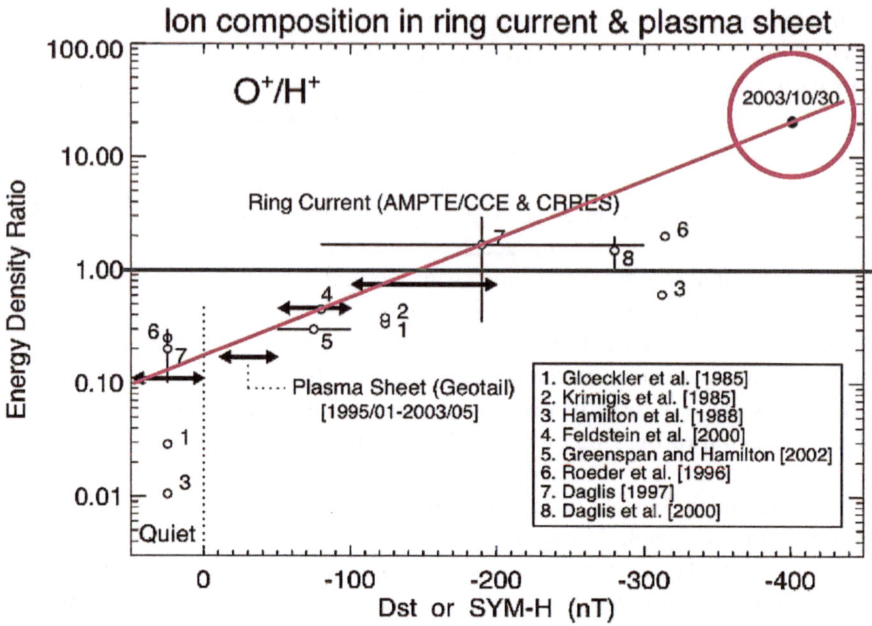

FIGURE 4.25. The ratio O^+/H^+ for different geomagnetic storms.
Source: Nosé, M., S. Taguchi, K. Hosokawa, S.P. Christon, R.W. McEntire, T.E. Moore, and M.R. Collier, *J. Geophys, Res.*, **110**, A09524, 2005

It may be added that some researchers compare the interplanetary electric field (VBs) and the Dst index, claiming that the interplanetary electric field is the cause of the main phase. This is an example of comparing an apple and an orange; Dst is a measure of the energy in the ring current belt, so it should be compared with ε. Note also that when VBs is large, ε is also large.

There is one subject that has not been mentioned in this section. The main phase of geomagnetic storms tends to develop asymmetrically. Figure 4.27 shows two examples of the asymmetric development during intense storms. Actually, this phenomenon was described in terms of the DS component by Chapman in 1918. It was taken up later by Akasofu and Chapman (1964). As mentioned in Section 3.1.3, Alfvén attempted to explain the asymmetry in terms of the field-aligned currents associated with the electrojets. Although the substorm current system can explain a small portion of the asymmetry, it cannot explain the fact that the asymmetry is larger at lower latitudes, as well as the magnitude of the asymmetry. In Figure 4.27, it can be seen that the magnitude of the asymmetry is as large as 150–200 nT. It appears that O^+ ions are preferentially injected into the ring current belt in the evening sector where the field-aligned currents are directed upward and that O^+ ions drift westward after the injection. Figure 4.24 shows an example of the ENA study of O^+ ions in the ring current belt.

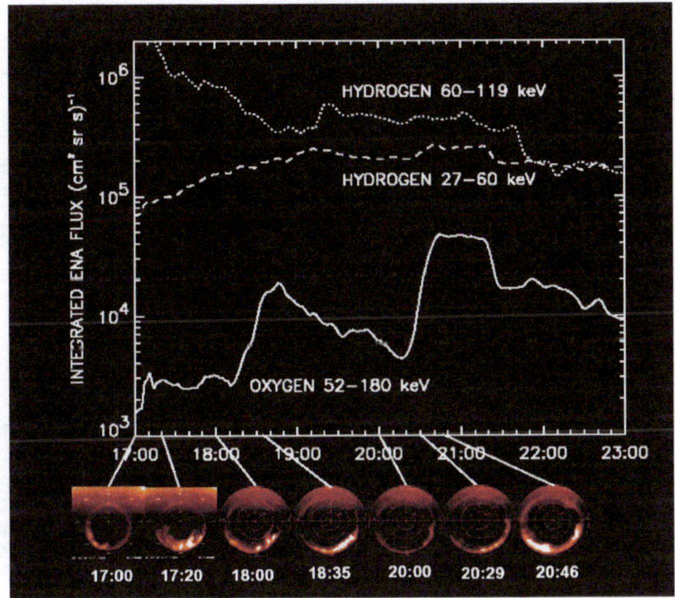

FIGURE 4.26. The O^+ flux increases as the aurora brightens.
Source: Mitchell, D.G., Earth-Sun System Exploration: Energy Transfer, January 16-20, 2006, Kona, Hawaii

4.6. Geomagnetic Indices

Many researchers attempt to use the geomagnetic indices in *quantitative* studies. Unfortunately, serious mistakes can arise as many authors attempt to find the relationship between substorms index without knowing their accuracy. The Dst and AE indices were devised in the early 1960s for individual geomagnetic storms (Akasofu and Chapman, 1961; Neil Davis and Masahisa Sugiura, 1966) and are only a very rough measure of the ring current intensity and of the electrojet activity, respectively.

If we base our quantitative study on the present AE and Dst indices, our results will be greatly limited by the accuracy of these indices. In any scientific field (or in economics) an index is a very rough measure intended to show a trend. One can make a serious mistake without knowing what the index indicates, how it is derived, and how rough it is. To begin with, *both indices are not really the physical quantities* we seek. They are not the total ring current intensity and the total electrojet current intensity. Further, the substorm current system causes a positive change in the H component in low latitudes (see Figure 4.10b). This positive change can reduce the magnitude of the Dst index. Unless this positive effect can be removed, one could conclude that substorms reduce the ring current intensity. Therefore, the present AE and Dst indices should be calibrated before the storm-substorm

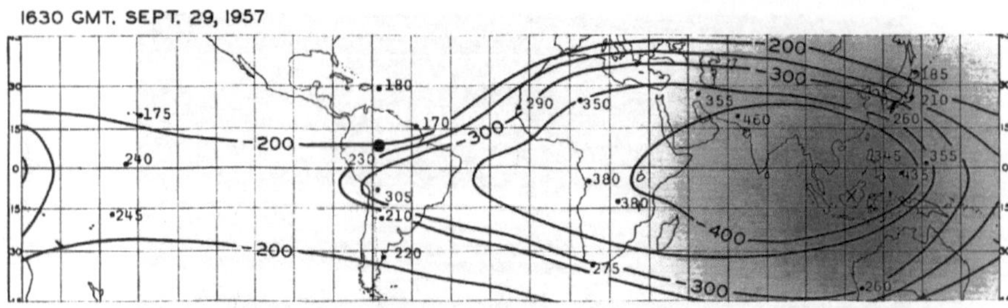

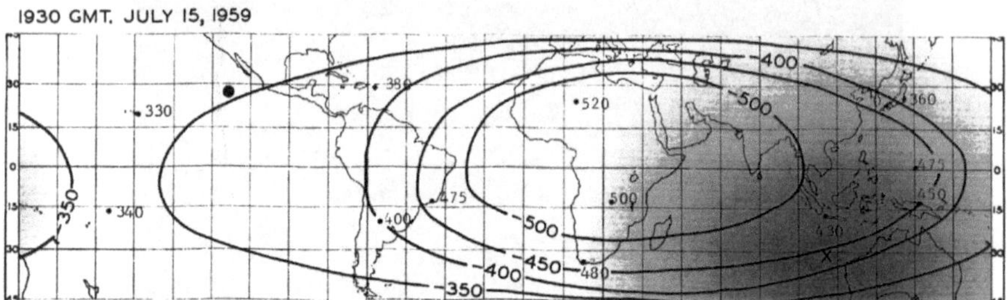

FIGURE 4.27. The distribution of the main phase field (the horizontal component) for the two intense storms of September 29, 1957 and July 15, 1959. The subsolar point is indicated by a black circle and the anti-subsolar point by a cross.
Source: Akasofu, S.-I. and S. Chapman, *Planet. Space Sci.*, **12**, 607, 1964

relationship can be studied quantitatively. Both the authors and their reviewers must learn a little more on such issues before the papers can be published.

4.7. Summary of Chapters 1, 2, and 3

It may be useful to summarize in a few paragraphs what I have learned from the IGY period to the end of the 1980s and what I described in Chapters 1, 2, and 3.

The solar wind and the magnetosphere constitute a dynamo (Figure 4.28). Its power is represented by $\varepsilon = VB^2 \sin^4(\theta/2)\ell_o^2$ (see Section 1.9). When the power exceeds 10^{18} erg/sec or so, the magnetosphere becomes primed as the north and south solenoidal currents in the magnetotail, including the cross-tail current, are intensified. During this primed time, for about 40 minutes, several processes can occur that suddenly reduce the intensified cross-tail current. A sudden reduction of ε associated with the so-called "northward turning" of the IMF *Bz* component often causes substorms after the southward turning. There also must be other processes. Then, the stretched field lines, caused by the intensified cross-tail current, start to contract. The contracting field lines cause a particular type of electric polarization, producing an earthward electric field. The electric field intensifies the Boström currents. The magnetosphere exhibits a specific response called the magnetospheric substorm as a result of the growth of the earthward

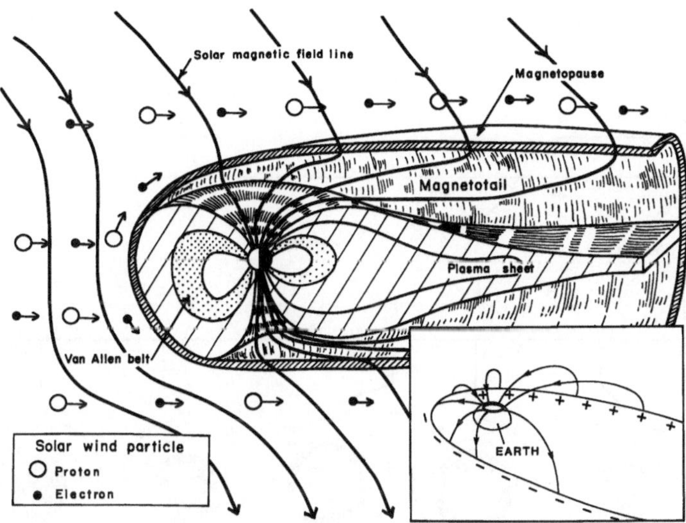

FIGURE 4.28. The solar wind-magnetosphere dynamo and its primary discharge circuit. *Source*: Akasofu, S.-I., *EOS*, **73**, 209, 1992

electric field, which drives an eastward plasma flow ($\boldsymbol{E} \times \boldsymbol{B}$). Its magnetic manifestation is the polar magnetic substorm and its auroral manifestation is the auroral substorm.

Figure 4.29 shows the resulting response of the aurora, including the auroral substorm. When the IMF Bz component is +5 nT or more, the auroral oval contact poleward. The polar cap is small. As the IMF Bz component becomes −5 nT or so (the so-called "southward turning"), the auroral oval expands. (A substorm typically begins about 40 minutes after the southward turning.) As the IMF Bz turns northward, the recovery phase begins and the polar cap becomes small. If the IMF Bz component becomes as large as +20 nT for several hours, the arc structure of the aurora disappears and an oval-shaped glow remains in the polar cap.

The discharge begins to occur between the magnetosphere and the ionosphere at substorm onset. The upward currents from the ionosphere are carried by downward flowing electrons that cause optical emissions by colliding with atoms and molecules in the ionosphere, which we identify as the aurora. The upward currents are also associated with outflow of O^+ ions. When O^+ ions are injected into the inner magnetosphere, they form the ring current belt. When intense substorms occur frequently, a large number of O^+ ions accumulate in the trapping region and its effect is mainly responsible for a large depression of the Earth's magnetic field in low latitudes. This phenomenon is prominent during the main phase of the geomagnetic storm.

It is my hope that the new generation of researchers will advance our understanding of the solar wind-magnetosphere interaction beyond this summary, if necessary by revising it completely. The fact that the paradigm of substorms has remained for a half century may indicate slow progress of this discipline. Thus, I welcome the paradigm change. However, a new one must be better than

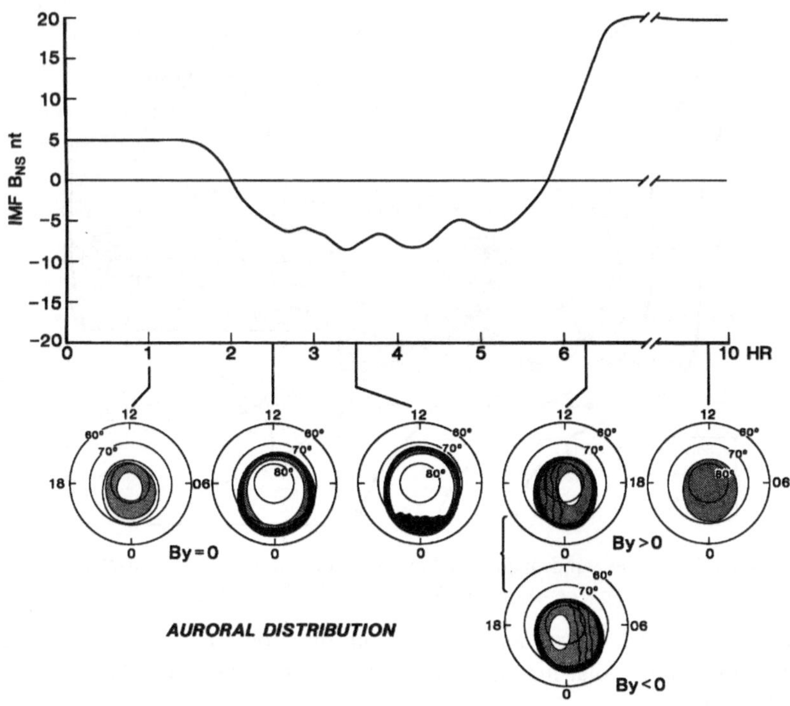

FIGURE 4.29. Schematic illustration to show how the aurora responds to a specific change of the IMF.
Source: Akasofu, S.-I., *Scientific American*, p. 90, May 1989

the present one. The electric field paradigm on the storm–substorm relationship sounded new, but had little foundation. We should avoid such a case of unfounded new findings. It simply confused the discipline for a decade or so.

4.8. Publication of Solar-Terrestrial Physics from Oxford University Press

Geomagnetism, published by Chapman and Bartels in 1940, was the classic treatise and served the development of the field of geomagnetism. I still remember the great excitement I felt when I bought it; it was a very expensive book for a poor student. In 1968 or so, Chapman and I felt that a new comprehensive treatise was needed because geomagnetism had developed into magnetospheric physics. It was a period when magnetospheric physics was developing rapidly. As a result, the manuscript had to be revised many times. I regretted that I could not complete it before Chapman's untimely death in 1970. Sir Edward Bullard gave me many valuable pieces of advice in completing it. It was finally published under the title *Solar-Terrestrial Physics*, from Oxford University Press, in 1972. I dedicated it to Katherine Chapman. I felt that Chapman would have agreed to do so. Thirty years have already passed since the publication of the book. This means that it was published before many of the present active researchers were born. Although I myself am not very inclined to read any papers published before my birth (the Chapman–Ferraro paper was published in 1931, so I had no choice but to read it!), it is my hope that the younger generation might at least flip through it to find that many unsolved problems today were present even before their birth. Our book might prevent also their rediscovery of well-established facts.

With Emiko Akasofu and Katherine and Sydney Chapman at the University of Michigan campus in 1962. Our Oxford book *Solar Terrestrial Physics* was dedicated to Katherine. *Source*: Akasofu, S.-I.

At the Geophysical Institute, we hosted many international conferences. Upper: AGU Chapman Conference "Magnetospheric Polar Cap," August 6-9, 1984. Lower: AGU Chapman Conference "The Formation of Auroral Arcs," July 21-25, 1980. *Source*: Geophysical Institute, University of Alaska

The opening ceremony of the International Conference on Substorms (ICS-2) at the University of Alaska Fairbanks. Joe Kan is introducing me as the first speaker, March 7-11, 1994.
Source: Geophysical Institute, University of Alaska

Some of the attendees of the ICS-2, March 7-11, 1994.
Source: Akasofu, S.-I.

Chapman Medal from the Royal Astronomical Society, London.
Source: Akasofu, S.-I.

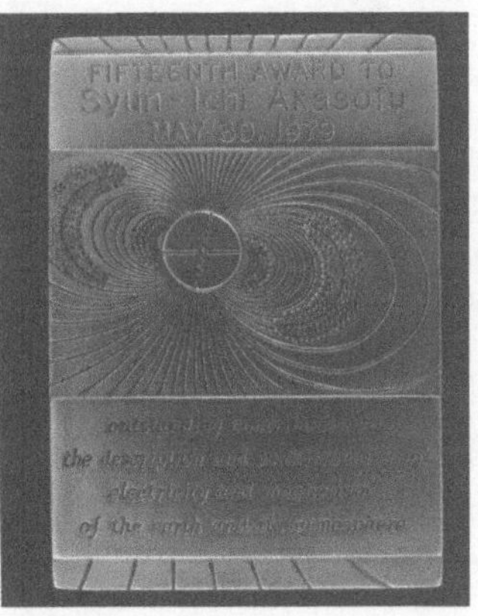

Fleming Medal from the American Geophysical Union.
Source: Akasofu, S.-I.

At the occasion of the opening of the Van Allen Hall, University of Iowa. From left: D. Venkatesan, Harold Taylor, Mary McIlwain, Carl McIlwain, Guido Pizzella, Bruce Randall, James Van Allen, Thomas Armstrong, Theodore Fritz, and Syun Akasofu.
Source: University of Alaska

With Alv Egeland in Kiruna, Sweden (1992).
Source: Akasofu, S.-I.

With Walter Orr Roberts (1979).
Source: Akasofu, S.-I.

Kiruna Geophysical Observatory (presently the Swedish Institute of Space Physics).
Source: Akasofu, S.-I.

From left: Albert Galeev, Syun-Ichi Akasofu, Renat Zinnurovich Sagdeev, and an unidentified person at the Space Research Institute, Russian Academy of Sciences. *Source*: Akasofu, S.-I.

With Newton's statue at Trinity College, Cambridge University (1988), Chapman's alma mater.
Source: Akasofu, S.-I.

With Yosuke Kamide and Tony Lui at a volcano in Hokkaido (1994).
Source: Akasofu, S.-I.

Normal-run magnetometer (the H component) operated at the College magnetic observatory from 1955 to 1992. Presented to me by J. Townshend.
Source: Akasofu, S.-I.

5
Planetary Magnetic Fields: Is the Earth's Dipole Really Off-Centered and Inclined?

5.1. Introduction

Spherical harmonic analysis has been considered the most powerful method for studying planetary magnetic fields. However, it is a mathematical tool, like the Fourier analysis method, and its results must be examined in the light of physics.

Based on the spherical harmonic analysis method, it has long been believed that the Earth's main dipole is off-centered by 0.08 Earth radii and is inclined by 11.5° with respect to the rotation axis (cf. Chapman and Bartels, 1940). These deviations have not received much attention in the past.

However, we have to face them now, because the spherical harmonic analysis method shows that the main dipole of Uranus is off-centered by 0.3 Uranus radii and is inclined by 60° with respect to the rotation axis; Neptune's dipole is off-centered by as much as 0.55 Neptune radii and is inclined by 47°. The problem is that since all the dynamo theories of planetary magnetism rely on planetary rotation, it is unlikely that they can explain easily how the main dipole of the planet can be greatly off-centered or inclined as inferred. Thus, how can we understand the greatly off-centered or inclined main dipole of Uranus and Neptune?

In this chapter, I intend to show that it is possible to learn a great deal about planetary magnetism from solar magnetic fields. (Note that both solar and planetary magnetisms rely basically on the same dynamo theory.) For this purpose, I ask the reader to consider solar magnetic fields on an imagery spherical surface of 3.5 solar radii over the Sun (2.5 solar radii away from the photosphere). This surface is called the *source surface*. In obtaining the field distribution on the source surface, it is assumed that the field in the corona can be approximated by a potential field and that the field lines reaching the source surface are perpendicular to it. Figure 5.1 shows such an example for the sunspot minimum and maximum periods. In spite of the great complexity of the photospheric magnetic field, the field at a distance of 2–3 solar radii is much simpler and is approximately dipolar.

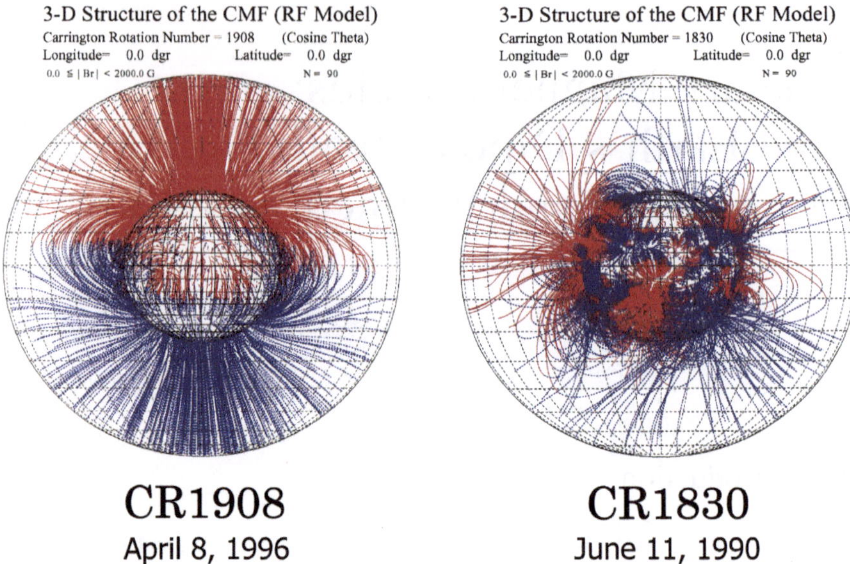

FIGURE 5.1. The magnetic field line configuration between the photosphere and the source surface; the left one in 1996 (near the sunspot minimum) and the right one in 1990 (near the sunspot maximum).
Source: Courtesy of K. Hakamada, 2005

A number of researchers have shown that the magnetic fields on the source surface can be approximated by a dipole field and that the polarity (towards/away from the Sun) of the source surface field is fairly well correlated with that of the interplanetary magnetic field (IMF) observed near the Earth.

In the upper and lower parts of the left-hand side of Figure 5.2, the magnetic equator on the source surface inferred from the photospheric magnetic field during Carrington rotation 1720 (the Carrington rotation is a sort of solar day number) is shown both in a regular rectangular map and on a source (spherical) surface. In this particular case, the first spherical harmonic term provides an inclined dipole at the center of the Sun; the equator is an inclined circle on the source surface and its rectangular projection is a sinusoidal curve (the right-hand side of Figure 5.2). In this situation, the common practice is to state that the solar dipole for Carrington rotation 1720 is inclined from the rotation axis by about 45°. Figure 5.3 shows the relationship between the neutral line and the magnetic field structure in the photosphere and the lower corona. If the top of the tallest arches of the field lines are connected, one can infer the location of the neutral line on the source surface.

However, if the dynamo theories demand that the dipole axis should be parallel or anti-parallel with respect to the rotation axis in this example, additional dipoles are required to reproduce the observed equator (the right-hand side of Figure 5.4). So long as such a combined field of the dipoles can reproduce the observed field

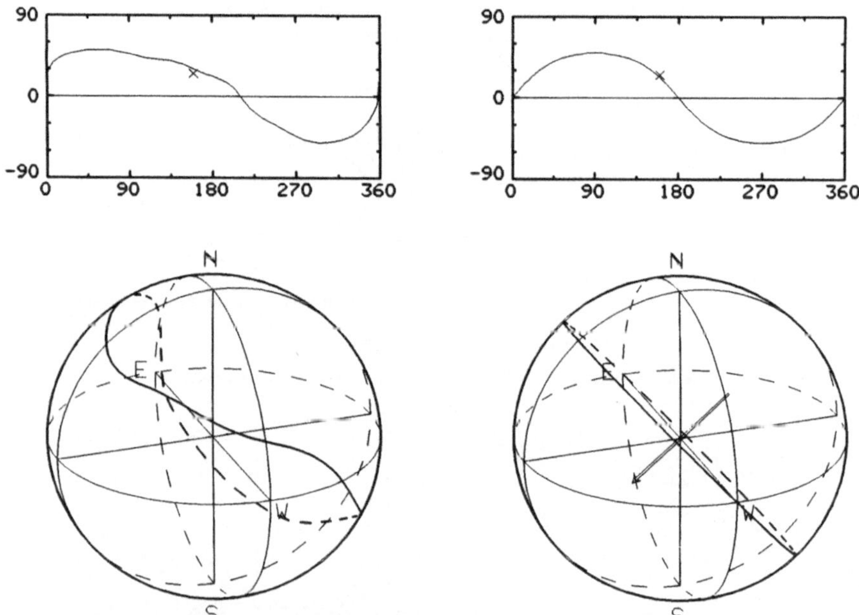

FIGURE 5.2. Left: Observed magnetic equator on the source surface. Right: Result of the spherical harmonic analysis.
Source: Akasofu, S.-I.

(expressed mathematically by the spherical harmonic analysis method), such an inference is at least a possibility. Although it is difficult to determine uniquely the characteristics of the additional dipole, our interpretation may be physically more meaningful than the spherical harmonic analysis results of the inclined dipole.

5.2. Triple Dipole Model

In this section, it is shown that three dipoles, the central dipole (parallel or anti-parallel) along the rotation axis and two hypothetical dipoles on the photosphere, can reasonably well reproduce the neutral line on the source surface. Actually, it can be shown that the two hypothetical dipoles do exist on the photosphere.

In Figure 5.4, the source surface field in the upper left diagram is modeled in the upper right diagram by a combination of an axially parallel dipole and two hypothetical photospheric dipoles. It is somewhat surprising that the two dipoles, together with the central dipole, can fairly well reproduce the neutral line. The two dipoles, thus determined, are then transferred to the photospheric magnetic field map in the lower left diagram. Each of the two dipoles is fairly well co-located with the observed large-scale dipolar field; this is a sort of blind

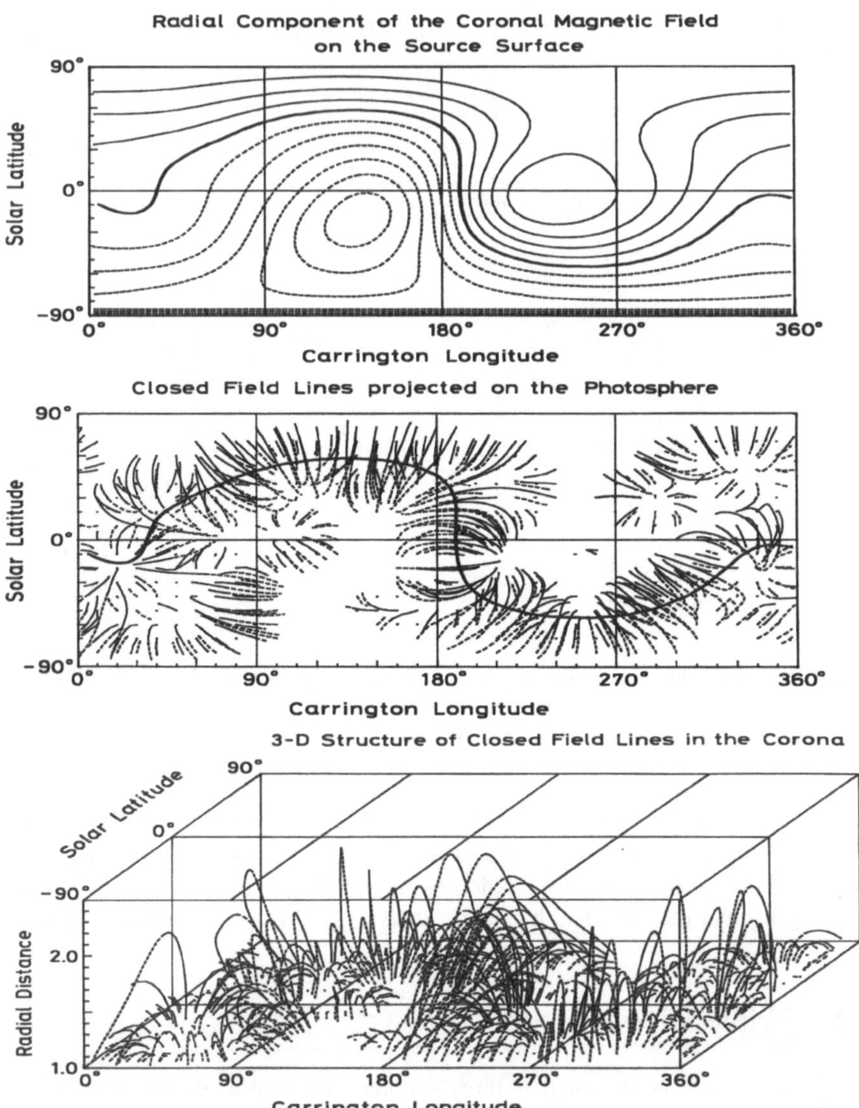

FIGURE 5.3. The relationship between the neutral line and the magnetic field structure in the photosphere and the lower corona. Top: The field distribution on the source surface; Middle: Field line configuration near the photosphere; Bottom: The 3-D structure. *Source*: Courtesy of K. Hakamada

experiment. Therefore, the two hypothetical dipoles inferred from our modeling on the source surface do actually exist. These two dipolar fields are not individual sunspot pairs, but are larger scale fields in active regions. This feature will be discussed in more detail in Sections 6.6 and 6.7.

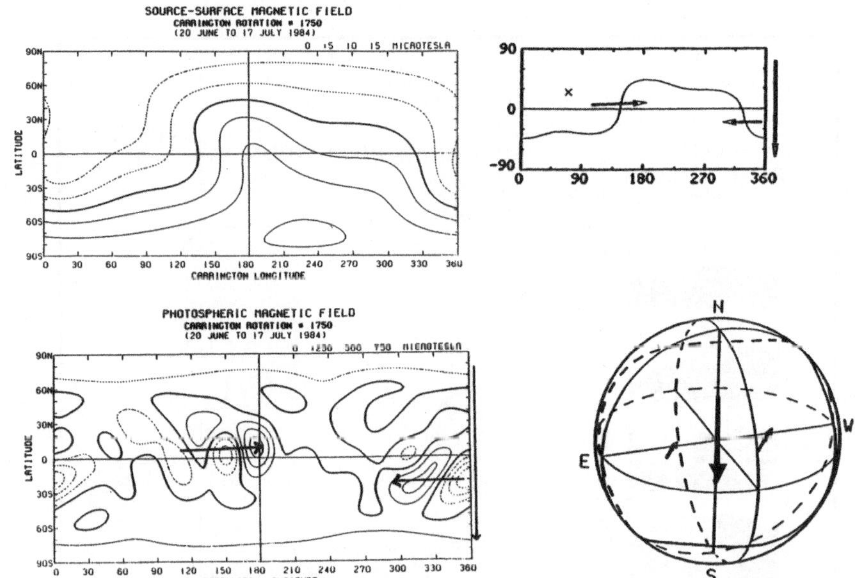

FIGURE 5.4. Upper left: Observed magnetic equator. Upper right: Observed magnetic equator and the auxiliary dipoles on the photosphere. Lower left: Photospheric field and the location of the auxiliary dipoles on the photosphere. Lower right: Spherical representation of the axial dipole and the auxiliary dipoles.
Source: Saito, Takao, Y. Kozuka, T. Oki, and S.-I. Akasofu, *J. Geophys. Res.*, **96**, 3807, 1991

It is possible, therefore, to infer that the main dipole is actually axially parallel or anti-parallel and that the inclination (with respect to the rotation axis) of the dipole on the source surface is produced by a combined effect of the axially parallel (or anti-parallel) field and the two photospheric dipoles.

5.3. Rotation of the Solar Magnetic Field on the Source Surface

Todd Hoeksema and P.H. Scherrer (1984) and Takao Saito (1987) examined sunspot cycle variations of the magnetic equator (which is usually referred to as the neutral line) on the source surface and demonstrated that the neutral line varied fairly systematically during Sunspot Cycle 21 and earlier cycles. Results by Saito et al. (1989) are shown in Figure 5.5.

The neutral line lies near the ecliptic plane at the beginning of the cycle and tilts gradually as the cycle advances, standing almost vertically (with respect to the equatorial plane) during the maximum epoch of the cycle. It is in this situation that the polarity of the interplanetary magnetic field is divided vertically.

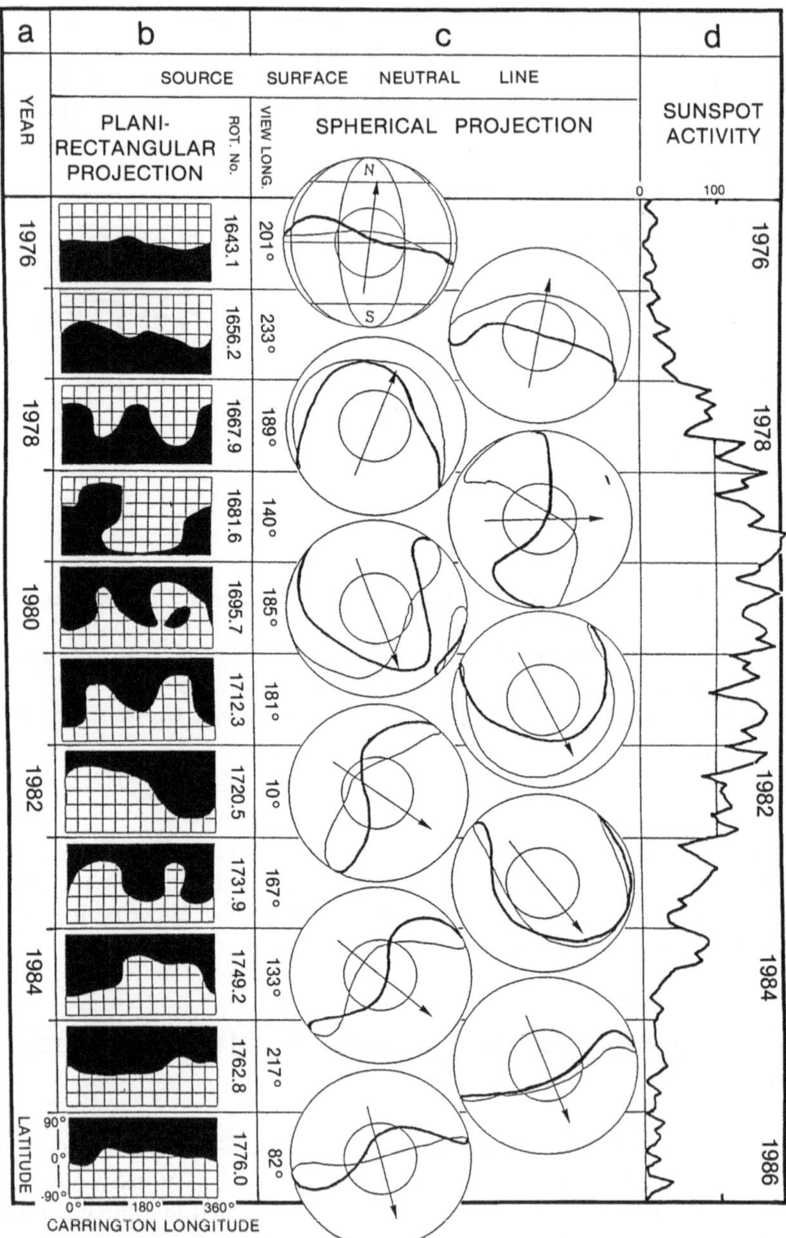

FIGURE 5.5. Rotation of the assumed central dipole and the observed magnetic equator during Sunspot Cycle 21, which peaked in about 1979–1980.
Source: Saito, Takao, T. Oki, S.-I.Akasofu, and C. Olmsted, *J. Geophys. Res.*, **94**, 5453, 1989

John Wilcox called such a situation "the sector boundary" structure. The neutral line tilts further during the declining epoch of the cycle and lies near the ecliptic plane at the end of the cycle. If one approximates the magnetic field on the source surface by a central dipole, this change can be represented by a gradual rotation of the dipole by 180°, so it changes from pointing northward to pointing southward (Figure 5.5).

On the other hand, it is important to realize that there is no indication that the main dipole of the *photospheric* magnetic field shifts in such a way to produce the rotation from one polarity to the other. Indeed, it is well known that the so-called *unipolar region* is located near each pole throughout the sunspot cycle. It is believed that the reversal of the dipolar field occurs as a result of the migration of a large-scale unipolar field (say, positive) from low latitudes to the polar region, canceling the pre-existing unipolar (negative) field there. Meanwhile, a cancellation of the opposite polarity occurs in the opposite hemisphere at about the same time. Thus, the reversal of the polarity does not involve a gradual shift of the main dipole pole from one hemisphere to the other across the equator. These observations show that one must look for other causes of the rotation of the dipole field on the source surface, namely causes that do not rely on the rotation of the main dipolar field on the photosphere. It is difficult to comprehend why researchers in planetary magnetism do not consider such important solar information. It appears that planetary scientists decide that the planetary dipole should rotate and they try to find a way for it to do so, regardless if such an effort is realistic or worthwhile.

As a solution to this puzzle, Takao Saito and I suggested that the major changes of the neutral line on the source surface during a sunspot cycle can be well represented by a combined effect of changes of an axial dipole located at the center of the Sun and of two nearly antipodal (equivalent) dipoles near the equatorial plane on the photosphere. Specifically, the observed variations of the neutral line during Sunspot Cycle 21 can be expressed by assuming that:

1. The magnetic moment of the central dipole (parallel to the rotation axis, say, directed northward) decreases as a new sunspot cycle advances and becomes null at about the sunspot maximum;
2. Subsequently, a small central dipole of the opposite polarity (directed southward) appears and its moment reaches maximum intensity near the sunspot minimum; and
3. A pair of dipoles on the photospheric surface, located at low latitudes, increases its magnetic moments from the beginning of a sunspot cycle until about the sunspot maximum and then decreases during the declining phase.

The growth and decay of the central dipole and the two dipoles during Sunspot Cycle 21 are illustrated in Figure 5.6. The top row in the figure shows the sunspot number. The second row shows the neutral line determined by the Wilcox Observatory. The next two rows show the axial and the auxiliary photospheric equivalent dipoles, respectively, for eight different epochs. The second row

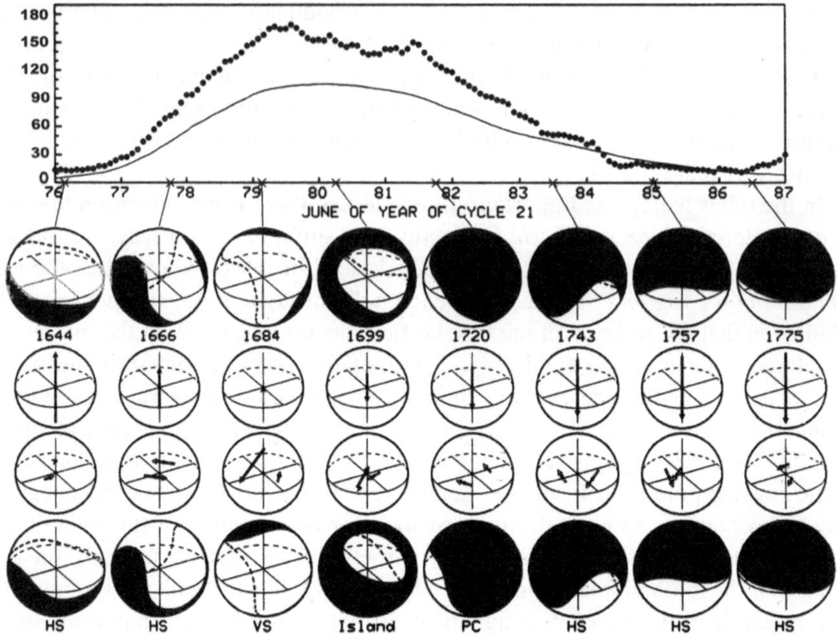

FIGURE 5.6. Sunspot number, the observed neutral line (Wilcox Observatory), the central dipole parallel or anti-parallel to the rotation axis, the auxiliary photospheric dipoles and the neutral line determined by the combination of the central dipole and auxiliary dipoles.
Source: Saito, Takao, T. Oki, S.-I. Akasofu, and C. Olmsted, *J. Geophys. Res.*, **94**, 5453, 1989

shows the observed neutral line and the last row shows the neutral lines computed based on our three-dipole model. The combined magnetic fields of these dipoles result in the neutral lines shown in the bottom row. The agreement between the observed and computed neutral lines is quite reasonable.

The apparent rotation of the dipole throughout the sunspot cycle is thus produced by a relative change of the strength of the axially parallel dipole and the photospheric dipoles, together with the reversal of the axial dipole as a result of the migration of a low-latitude unipolar field to the polar region in each hemisphere.

5.4. Large Inclination and Eccentricity of the Dipole-like Field of Uranus and Neptune

It is interesting to speculate that the photospheric surface corresponds to the surface of the core of magnetized planets and that the source surface of the Sun corresponds to the surface of the magnetized planets. An assumption is that the mantle of the planets is inactive in generating the main dipole field. It is expected that the dynamo process is the same or similar for the Sun and magnetized planets. In fact, all the dynamo theories treat them basically in the same way.

If this is the case, one should be able to assume that the main dipole of the magnetized planets is located at the center of the planets and is parallel (or anti-parallel) with respect to the rotation axis and also that additional dipolar fields, together with the main dipole field, give the result of the off-centered and inclined dipole by the spherical analysis method. This is physically a more plausible situation than what the spherical harmonic analysis can provide.

Norman Ness et al. and his colleagues (1986, 1989) suggested that the magnetic fields of Uranus and Neptune indicate that the main field can be represented, as a first approximation, by an eccentric dipole and that the dipole is greatly inclined with respect to the rotation axis, Table 5.1.

Their model is often referred to as the offset tilted dipole (OTD) model. Their results are based on the spherical harmonic analysis of the magnetic field observed along the flyby trajectory of the Voyager spacecraft.

The large inclination and eccentricity of the dipole-like field of Uranus and Neptune can be described, as a first approximation, by the combined field of an axial dipole and a single auxiliary dipole.

TABLE 5.1.

	Inclination Angle	Location from the Center	Reference	
Earth	11.5°	$0.08R_E$	Chapman and Bartels, 1940	
Uranus	~60°	$0.3\,R_U$	Ness et al., 1986	
Neptune	~47°	$0.55\,R_N$	Ness et al., 1989	
	Two-dipole Model of Uranus			
	M (Gauss R_u^3)		Location	Orientation
Axial dipole	0.143		$x = 0$	$\theta = 0°$
			$y = 0$	
			$z = 0$	
Auxiliary dipole	0.157		$x = 0$	$\theta = 90°$
			$y = 0$	$\theta = 0°$
			$z = -0.3\,R_U$	
	Two-dipole Model of Neptune			
	M (Gauss R_N^3)		Location	Orientation
Axial dipole	0.0641		$x = 0$	$\theta = 0°$
			$y = 0$	$\phi = 0°$
			$z = 0$	
Auxiliary dipole	0.0769		$x = 0.14$	$\theta = 60°$
			$y = 0.42$	$\phi = 0°$
			$z = 0.24$	

The upper left diagram of Figure 5.7 shows the OTD located near the surface of the core of Uranus, as proposed by Ness et al. (1986). The lower left diagram shows some magnetic field lines in the plane that contains the OTD. We determine the magnitude and the orientation of both an axial dipole and an auxiliary dipole in the same way as we examined the solar source surface field. The results are presented in the upper right diagram of Figure 5.7. For simplicity, we assume only one auxiliary dipole, which is located at the position calculated for the single offset dipole. The parameters for the two dipoles are given in Table 5.1.

Comparing the lower left and right diagrams, one can see that the simple two-dipole model can reproduce reasonably well the observed field, which is represented by a single off-centered dipole.

There is no doubt that one or two additional dipoles in the model can better reproduce the observed field. However, the main point of this section is to illustrate the basic idea that even a simple two-dipole model could reproduce the observed field fairly well and thus may be able to remove the great puzzle of the large inclination angle and the large eccentricity of the main field of Uranus.

The upper left diagram of Figure 5.8a shows the OTD of Neptune, as proposed by (Ness et al., 1986) (1986). The lower left diagram shows some magnetic field lines in the plane that contains the OTD. The magnitude and the orientation of an axial dipole and an auxiliary dipole giving a similar magnetic field are shown

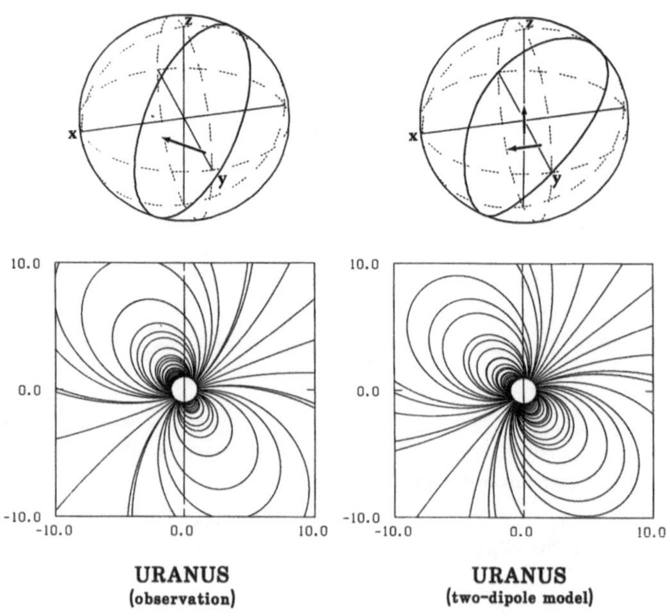

FIGURE 5.7. Left: Uranus' main dipole and the field lines based on the spherical harmonic analysis. Right: Uranus' magnetic field based on the axial and an auxiliary dipole.
Source: Akasofu, S.-I., L.-H. Lee, and T. Saito, *Planet. Space Sci.*, **39**, 1259, 1991

in the upper right diagram. The parameters for the two dipoles are given in the Table 5.1. Some magnetic field lines of the one- and two-dipole models are shown in the lower part of Figure 5.8a.

The discovery of the large inclination angle and the eccentricity of the main field of Uranus and Neptune provided a great puzzle. However, it is important to realize that the finding is based on spherical harmonic analysis of the planetary fields observed by a spacecraft flyby. Certainly, the dipole representation based on spherical harmonic analysis provides us with the unique mathematical description of the planetary magnetic field. However, the result obtained does not necessarily indicate that the field inside the planets is physically given by such an analysis. Indeed, since the dynamo process is thought to rely so strongly on the rotation of the magnetized planets, it is possible that the observed dipole field consists of the combined field of an axial dipole (parallel or anti-parallel to the rotation axis) and a few auxiliary dipoles. This possibility is physically more plausible than the mathematical representation and is further supported because our interpretation was tested in Section 5.3.

It is generally believed that the magnetic axis of neutron stars is also inclined significantly from the rotation axis (Van den Henvel, 2006; Figure 5.8b), so that the same issue may be raised with them.

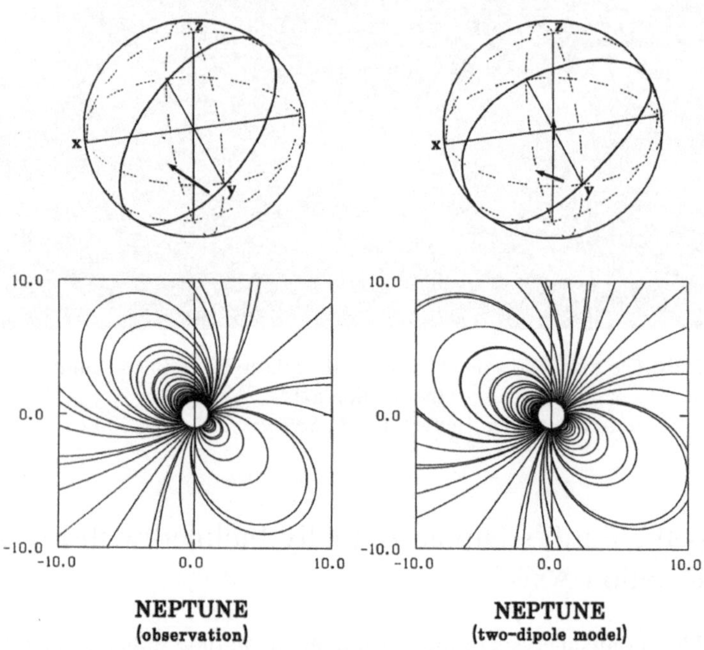

FIGURE 5.8a. Left: Neptune's main dipole and the field lines based on the spherical harmonic analysis. Right: Neptune's magnetic field based on the axial and an auxiliary dipole.

Source: Akasofu, S.-I., L.-H. Lee, and T. Saito, *Planet. Space Sci.*, **39**, 1259, 1991

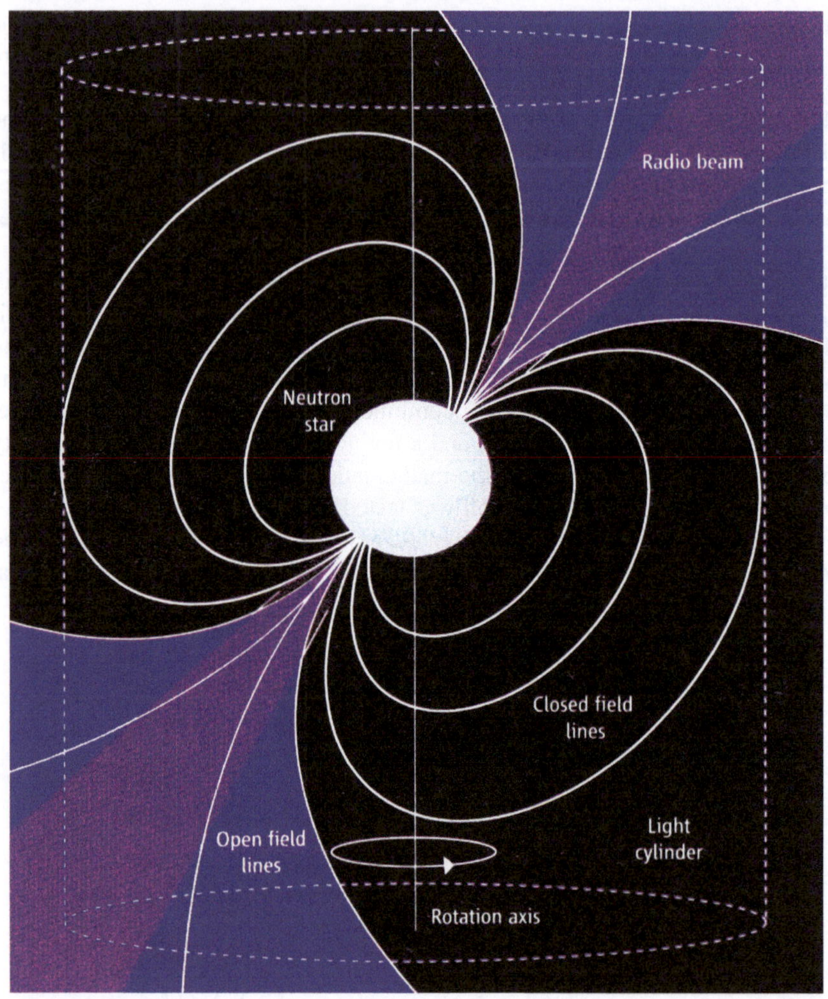

FIGURE 5.8b. The magnetic axis of pulsarsis thought to be inclined significantly form the rotation axis.
Source: Van den Henvel, E.P.J., *Science*, **312**, 539, 2006

5.5. Is the Earth's Dipole Actually Inclined with Respect to the Rotation Axis?

Spherical harmonic analysis of the Earth's magnetic field indicates that the main field can be represented, as a first approximation, by an off-centered dipole, and that the dipole axis is inclined with respect to the rotation axis by about 11.5°. Since the present dynamo theory for generation of the Earth's magnetic field relies heavily on the planet's rotation, it may be worthwhile to examine whether the Earth's magnetic field could consist of an axially anti-parallel dipole and a few dipoles on the surface of the core.

It is our finding that three dipoles near the core surface, together with the axially anti-parallel dipole, can reproduce fairly well the magnetic equator (Figure 5.9). The three dipoles are located at longitudes $\sim 105°$, $\sim 210°$, and $\sim 330°$, respectively; thus, they are located southeast of Hawaii, at the Atlantic Ocean between Africa and South America, and at the southern part of Thailand, respectively. It is suggested that the main dipole is aligned with respect to the rotation axis and that the combined effect of the three dipoles provides the tilted and off-centered main dipole.

It is unfortunate that researchers in this discipline believe so firmly that the spherical harmonic analysis method can provide the physical solution. The late Sir David Bates, editor of the *Journal of Planetary and Space Science*, accepted our paper on this subject against the recommendation of a referee. I learned from him later that he received an angry letter from the referee, who stated that he would not submit his future papers to the journal. A referee should not forget

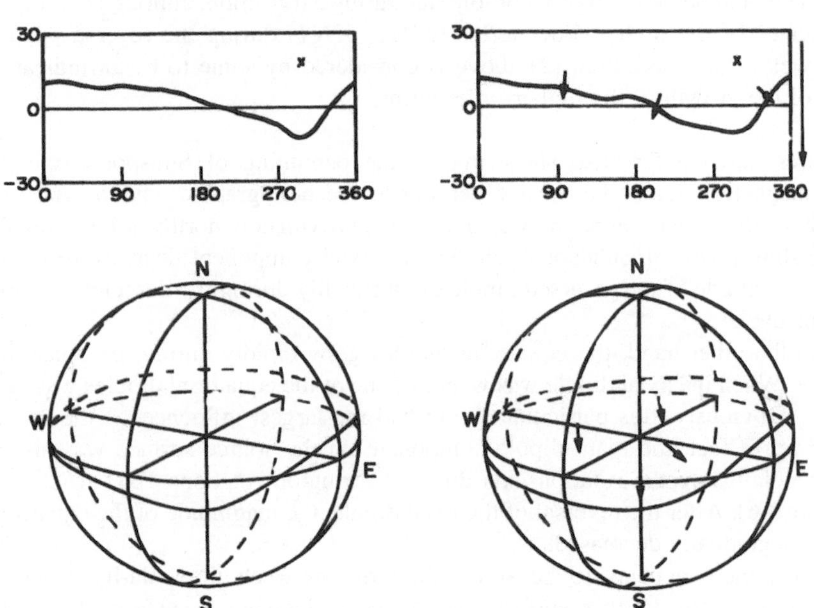

FIGURE 5.9. The upper- and lower-left presentations show the observed magnetic equator of the Earth in the standard and spherical projections, respectively. The upper- and lower-right presentations show the magnetic equator reproduced by the axially antiparallel dipole and three dipoles on the core surface, which are indicated in both presentations.

Source: Akasofu, S.-I. and Saito, Takao, *Planet. Space Sci.*, **38**, 1203, 1990

that the editor of a scientific journal makes the final decision, not the referee. This is not to say that I have an accurate solution to this difficult problem. We have to be open-minded in facing it.

5.6. Does the Main Dipole of the Geomagnetic Field Rotate during the Reversals?

After a long debate, the reality of reversals of polarity of the geomagnetic field through geologic time has been established. The phenomenon of reversals is considered by many researchers to be the rotation of the dipole axis, either from the normal to reversed (N $\rightarrow$ R) or from the reversed to normal (R $\rightarrow$ N). It is generally accepted that:

(1) The dipole pole shifts often along a restricted sector of longitude.
(2) There occurs a significant reduction of the field.
(3) The transition period is relatively short, $\sim 4500 \pm 100$ years.
(4) The frequency of reversals has increased from 0.5 to 0.15 Ma during the last 70 Ma.
(5) The field becomes highly non-dipolar during a transition, although the importance of the higher order terms (g2/g1, g3/g1) during the reversals is not well established; item (2) above is considered by some to be an indication of the growth of the higher-order terms.

Figures 5.5 and 5.6 also show that at the beginning of Sunspot Cycle 21, the magnetic equator lay nearly parallel to the heliographic equator and also the axial dipole was large and was pointing approximately northward. During the ascending phase of Sunspot Cycle 21, the axial component decreased rapidly. The magnitude of the reversed dipole grew steadily during the descending phase of the cycle.

On the other hand, the equatorial dipoles grew rapidly during the ascending phase. When the axial dipole was weakest, one of the equatorial dipoles was very large. Obviously, this particular dipole had the largest influence on the source surface. In fact, the main dipole component on the source surface was almost perpendicular to the rotation axis during Carrington rotations 1681–1685 (see Figure 5.6). After the reversal of the axial dipole, the magnitude of the equatorial dipoles gradually decreased.

Thus, there appears to be some similarity between the polarity reversals observed on the Earth's surface and on the solar source surface. There are common features on the rotation of the dipole axis of both Earth's field and the solar field. Both reversals occur during a relatively short period during the period of one polarity. A significant decrease of the main dipole field occurs and there is some indication of the growth of higher order fields. The major difference is that the solar reversals are quite regular compared with those of the geomagnetic field. In fact, there is no physical reason why one cannot assume as a first approximation that the source surface corresponds to the surface of a magnetized planet, and that the photosphere corresponds to the core surface. The most important point here is that the photospheric magnetic field does not show any indication of the rotation of the dipole axis, in spite of the fact that the dipole field on the source surface rotates; the polar regions remain as the magnetic poles.

As mentioned in Section 5.3, the reversal of the solar dipole field on the photosphere occurs as a result of the migration of a large-scale unipolar field, not by the rotation of the dipole axis. Perhaps a similar process is responsible for the reversal of the Earth's dipole field. Without the reversal of the direction of the Earth's rotation, it is very difficult to explain the reversal of the dipole field.

5.7. Heliospheric Current Sheet

As the solar wind stretches the dipolar field on the source surface, an extensive current sheet is formed, dividing the magnetic regimes into two, the northern and southern hemispheres. However, as shown in Figure 5.5 and 5.6, the magnetic equator has a complicated wavy character. As a result, the heliospheric current sheet also has a complex configuration. It may be possible to test the inferred configuration of the heliospheric current sheet near the Sun by observing the outer solar corona. Kazuyuki Hakamada, Ghee Fry, and I developed a method to construct the heliospheric current sheet near the Sun (1986). We decided to predict the shape of the outer solar corona on the basis of the magnetic equator inferred by the Wilcox Observatory (Figures 5.10a). The predicted coronal configuration is shown in Figure 5.10b. Takao Saito and his colleagues

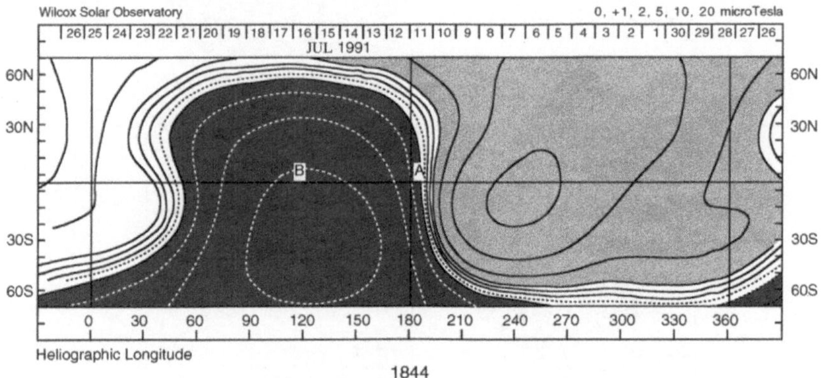

FIGURE 5.10a. Distribution of the magnetic field on the source surface for Carrington rotation 1844.

Source: Saito, Takao, S.-I. Akasofu, Y. Kozuka, T. Takahashi, and S. Numazawa, *J. Geophys., Res.*, **98**, 5639, 1993

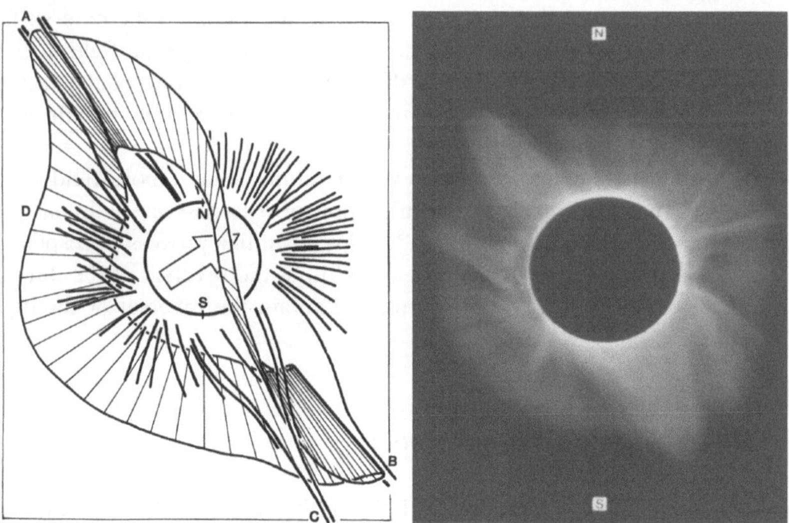

FIGURE 5.10b. Left: Predicted structure of the outer solar corona. Right: Observed outer solar corona during the 1991 solar eclipse.
Source: Saito, Takao, S.-I. Akasofu, Y. Kozuka, T. Takahashi, and S. Numazawa, *J. Geophys., Res.*, **98**, 5639, 1993

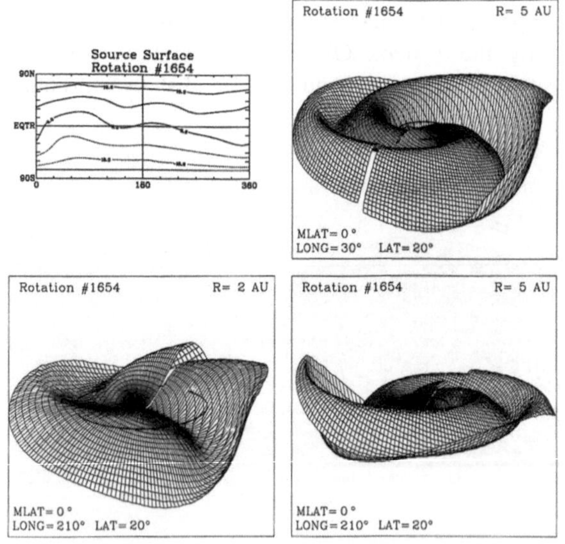

FIGURE 5.11a. The heliosphere current sheet for Carrington rotation 1654. The Earth's orbit is shown.
Source: Akasofu, S.-I. and C.D. Fry, *J. Geophys. Res.*, **91**, 13,679, 1986

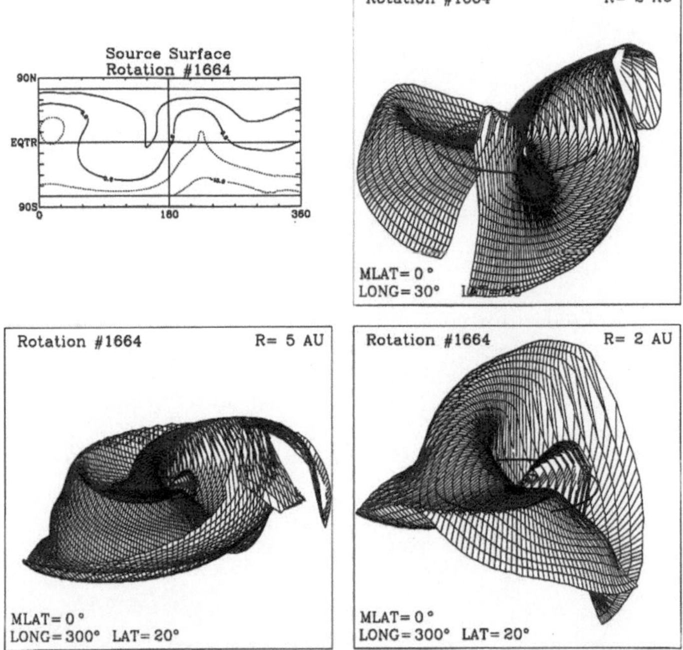

FIGURE 5.11b. The heliospheric current sheet for Carrington rotation 1664. The Earth's orbit is shown.
Source: Akasofu, S.-I. and C.D. Fry, *J. Geophys. Res.*, **91**, 13,679, 1986

successfully photographed the outer solar corona during the 1991 eclipse. The agreement between the predicted and the observed solar corona was unexpectedly good. The outer corona appears to be bright at the places where the heliospheric current sheet develops folds. As far as I am aware, this was the first time that the configuration of the outer corona was predicted so realistically.

The method we developed to infer the heliospheric current sheet can be extended to the Earth's distance and beyond for the realistic magnetic equator. One complexity is that the formation of the wavy magnetic equator does not propagate with the speed of light or infinite speed, but is carried by the solar wind particles with a speed of a few hundred kilometers per second. At the same time, the current sheet rotates with the Sun. Taking this into account, we developed a method to construct the current sheet for any given (observed) magnetic equator. Two examples are given in Figures 5.11a and 5.11b for Carrington rotation 1654 and 1664, respectively. Space probe observations indicate that the current sheet extends into the outer heliosphere. John Wilcox was the first to infer the wavy current sheet for an ideal case, but we could extend his work for realistic magnetic equators. In Section 6.1, we shall see that the azimuth angle of the interplanetary magnetic field changes its direction (from toward to away or vice versa) as the Earth crosses the current sheet (see also Figure 6.4a).

6
Recurrent Geomagnetic Disturbances and the Solar Wind

6.1. Modeling the Background Solar Wind Flow

The solar wind exhibits considerable variations even without any specific solar events, such as solar flares, CMEs, and sudden filament disappearances, which we are going to discuss in Chapter 8. Such variations often cause what is called recurrent geomagnetic disturbances as noticed first by Maunder (1905); his work was described in the Prologue. Recurrent geomagnetic disturbances occur when high-speed streams flowing from long-lasting coronal holes. Since any effects of specific solar events propagate into the existing solar wind structures and interact with them, it is important first of all to learn about the basic background solar wind and to devise a simple way to model it.

Conditions on the source surface, the imaginary spherical surface of 3.5 solar radii (Chapter 5), are important in modeling interplanetary conditions. One of the most important aspects of the source surface is the magnetic equator (or the so-called *neutral line*). The axis of the dipolar field on the source surface rotates from 0° to 180° (or from 180° to 0°) during the Sun's 11-year cycle variations (Section 5.3, Figures 5.5 and 5.6). It will be shown that we can reproduce most of the main features of the background solar wind variations during the whole sunspot cycle at the Earth or at any point to about a distance of 2 AU by assuming that the solar wind speed is minimum at the sinusoidal magnetic equator and increases toward higher latitudes. Although more realistic models can be adopted, this chapter is intended to provide the basic principle involved in this modeling.

As the Sun and its source surface rotate about every 25 days, a fixed point in space (not on the source surface) at a distance of 3.5 solar radii from the solar center scans horizontally the velocity field from solar longitude 360° to 0° in one solar rotation along a heliographic latitude line (e.g., 0° at the June and December solstices). Figure 6.1 shows an example of model distribution of the solar wind speed on the source surface and the resulting wind speed at an equatorial point (fixed in space) on the source surface. Solar wind particles leave radially from this particular point one by one with different velocities as the Sun rotates; the point depicts a sinusoidal variation of the speed of solar wind particles during one rotation. As a result, each solar rotation sends out two

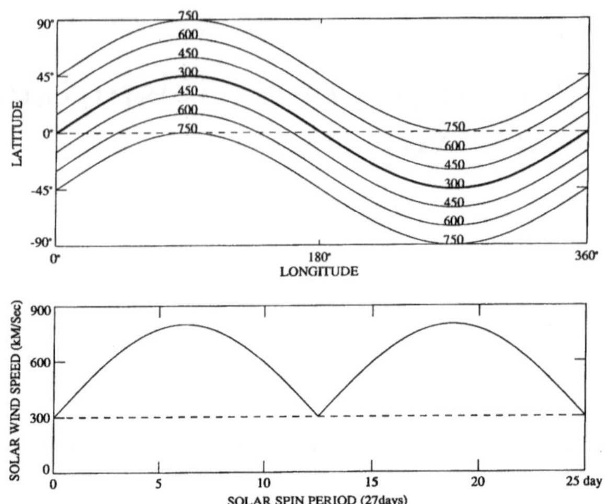

FIGURE 6.1. Upper: A sinusoidal magnetic equator and the associated solar wind speed distribution on the source surface. Lower: Solar wind speed observation at an equatorial point (fixed in space) at a distance of 3.5 solar radii.*

waves of the solar wind. Subsequent changes of the radial speed of individual particles can be modeled by adopting a kinematic method developed by Kazuyuki Hakamada (Hakamada and Akasofu, 1982). A faster flow of particles interacts with a slower flow of particles to form a shock wave and a reverse shock. Thus, by integrating the velocity as a function of time graphically, one can determine the distance traveled by individual particles (Figure 6.2).

A magnetic field line originating from the source surface can be traced by following particles leaving a particular point on the source surface (not a fixed point in space). The resulting interplanetary magnetic field structure is the familiar Parker spiral, together with the corotating interaction region produced by the formation of the shock wave structure (Figure 6.3). The computed velocity (V), density (n) and IMF magnitude B agrees reasonably well with the observed ones (Figure 6.4).

It is important to realize that such a simple scheme can reproduce reasonably well the observed 27-day variations of the solar wind observed at the Earth. The arrival of the fast wind is associated with a sharp change of the azimuth angle (PHI) of the IMF (from toward to away or away to toward), indicating the crossing of the heliospheric current sheet. It is for this reason that the corotating structure is called the sector boundary by John Wilcox.

---

* The Sources of Figures 6.1 and 6.2 are
Akasofu, S.-I. and C.F. Fry, *Planet Space Sci.*, **34**, 77, 1986
Akasofu, S.-I., Space Weather, *Geophysical Monograph*, 125, AGU Workshop, 2001
Hakamada, K. and S.-I. Akasofu, *Space Sci. Rev.*, **31**, 3, 1982

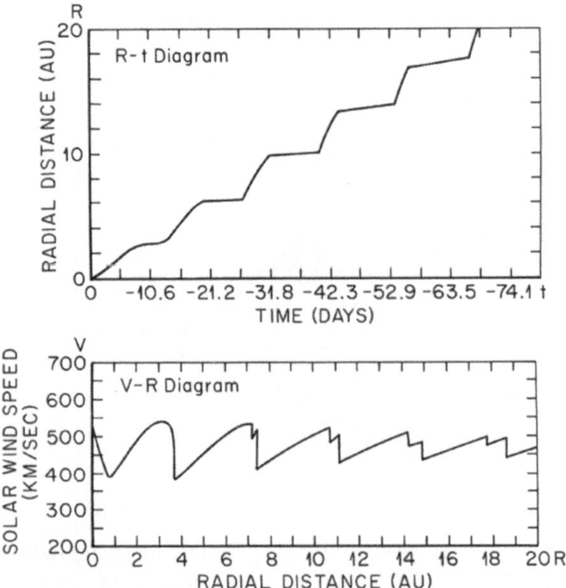

FIGURE 6.2. Upper: The distance (R) and time (t) relationship for a solar wind particle. Lower: The velocity (V) and distance (R) relationship for a solar wind particle.*

This modeling method has been improved many times and is called the Hakamada-Akasofu-Fry (HAF) model. The improved model uses the observed neutral line, instead of a sinusoidal curve.

Some solar wind physicists were upset by this modeling scheme. In general, scientists tend to ignore poor work by others. Since they were upset, it means that this modeling has something worthwhile. In fact, to begin with, before Figure 6.3 was published, there had been no quantitative or semi-quantitative pattern of the interplanetary corotating structure available except for hand-drawn sketches. Note that one requires the $R-t$ diagram in Figure 6.2 for this particular purpose, which requires, in turn, the integration of the equation of motion twice. Solar wind researchers are generally not interested in the second integration because satellites and space probes measure only the velocity. In Chapter 8, it will be shown that the HAF modeling method becomes useful when we deal with interplanetary disturbances caused by the occurrence of a series of solar flare events.

6.2. Recurrent Magnetic Storms

It was Bartels (1931), a geophysicist, who identified the so-called "M-region," which is now understood to be a source of high-speed, geoactive stream of the solar wind. In the Prologue, it was noted that "no spots" area in Maunder's ninth statement coincides with the M-region. It has long been known that during the

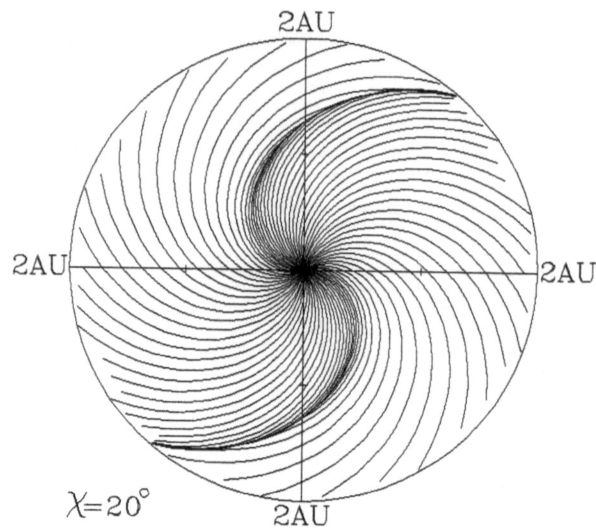

FIGURE 6.3. The spiral structure of the interplanetary magnetic field lines on the equatorial plane. The magnetic equator on the source surface is assumed to be sinusoidal, the amplitude being 20° in latitude (see top diagram of Figure 6.1).*

late declining phase of the sunspot cycle, two high-speed streams are present. In this section, we show that the two streams are the basic feature of the solar wind and are separated by roughly 180° in longitude; one extends from the northern coronal hole and the other from the southern coronal hole.

Effects of the high-speed solar wind streams on geomagnetic disturbances during the late declining phase of the sunspot cycle can be effectively studied using the magnetic index C9. Figure 6.5a shows the C9 index for the declining period, namely 1973–1975, 1983–1985, and 1993–1995 and the sunspot maximum period, namely 1948–1949, 1957–1958, and 1979–1980. The first column shows sunspot activities in terms of numbers from 0 to 9, the second column shows the year and the Bartels rotation number, the third column the beginning date of the Bartels rotation number, the fourth column the C9 index (0–9), and the last column shows the first part of the C9 index during the next Bartels rotation; note that higher C9 numbers are shown by a thicker bold and larger character. Bartels devised this particular presentation in order to study a long-term solar–terrestrial relationship. The daily C9 index for 27-days, one solar rotation period observed from the earth, is given in one row. Therefore, if large C9 index numbers appear vertically for several times, it means that geomagnetic disturbances recur approximately every 27 days, for several solar

---

* The Sources of Figures 6.3 through 6.17 are
Akasofu, S.-I., H. Watanabe, and Takao Saito, *Space Sci. Rev.*, **120**, 27, 2005
Akasofu, S.-I. and L.-H. Lee, *Planet Space Sci.*, **37**, 73, 1989 and **38**, 575, 1990
Akasofu, S.-I., K. Hakamada, and C. Fry, *Planet Space Sci.*, **31**, 1435, 1983

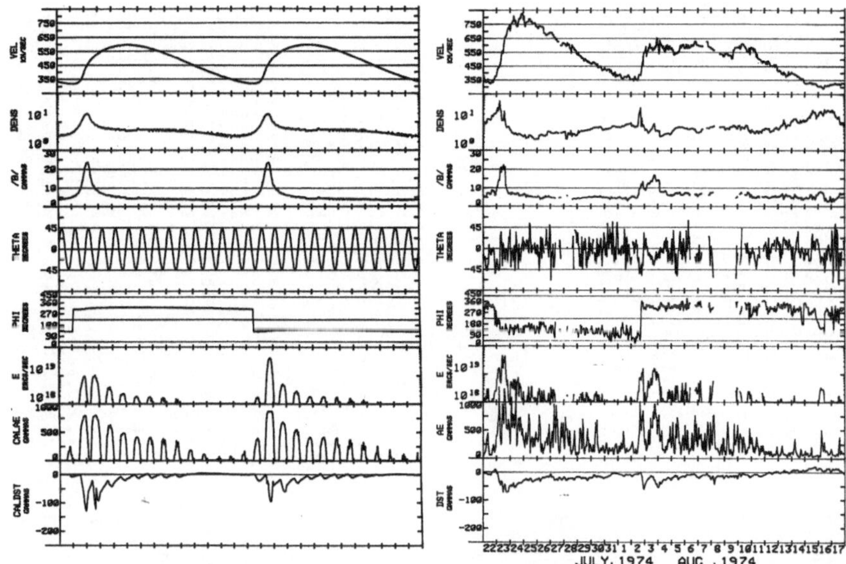

FIGURE 6.4. Left: The velocity, density, IMF magnitude, IMF latitude angle (THETA), IMF azimuth angle (PHI), ε parameter, calculated AE and Dst indices at the Earth (or any other point at a distance of 1 au) for one rotation period of the Sun (seen from the Earth), namely 27 days for the same conditions as those for Figure 6.1. The IMF THETA angle variations are assumed to have sinusoidal variations of amplitude 45°. It can be seen that they can reproduce reasonably well the observed storm activities associated with the arrival of the sector structures. Right: A typical 27-day variation of the solar wind speed V, IMF magnitude B, IMF THETA and PHI angles and the AE and Dst indices, when two large coronal holes (separated approximately by 180° in solar longitude), between July 17 and August 12, 1974. These observed variations may be compared with the simulated variations in the left.*

rotations, indicating that there is some fixed structure on the Sun, which causes magnetic disturbances. If small C9 index numbers appear vertically for several solar rotations, it means that there is some fixed structure on the Sun, which prevents a high-speed solar wind to blow, and thus geomagnetic disturbances on the earth are weak.

The C9 index is the best index suited for our particular purpose of studying the relationship between the Sun and the magnetic disturbances; for the relationship between the Kp, C9, and the other indices, see Akasofu and Chapman (1972). The C9 index has been available since 1884, but unlike the Kp index, the C9 index has not been used much by solar physicists and geophysicists; Figure 6.5b illustrates and also assures that the C9 and the familiar Kp indices are similar.

Based on the degree of magnetic disturbances measured in terms of the magnetic index during several 11-year sunspot cycles, it is generally known that magnetic disturbances tend to peak *well after, not during*, the sunspot maximum period, and that the so-called 'recurrent storms' are responsible for it.

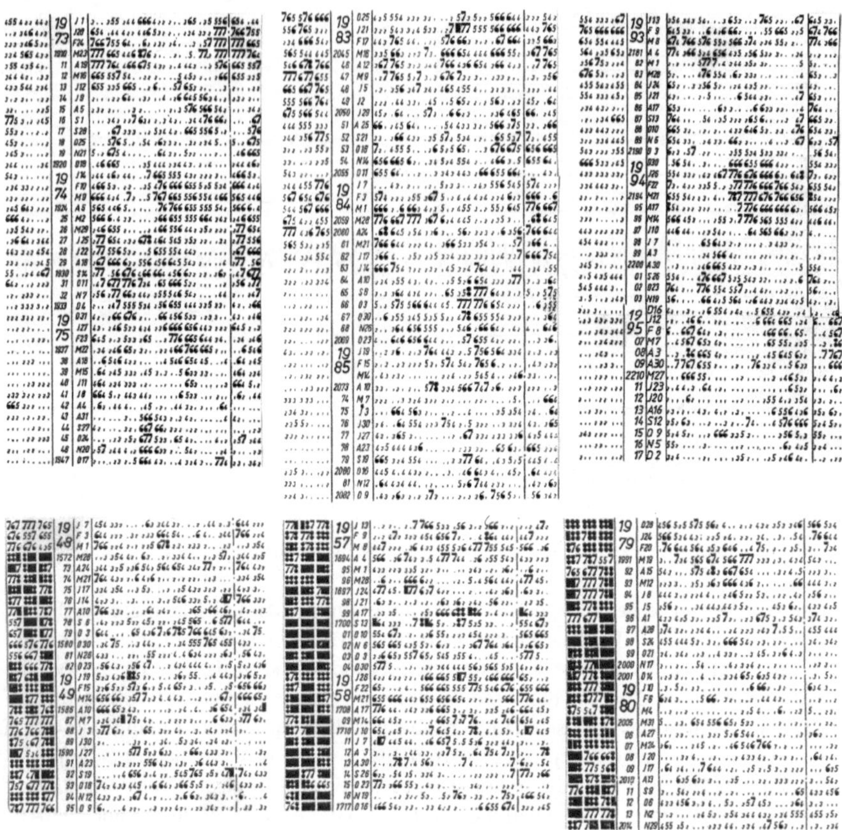

FIGURE 6.5a. (a) The C9 index and the sunspot number during the declining phase of the sunspot cycle, 1973–1975, 1983–1985, and 1993–1995. (b) The C9 index and the sunspot number during the maximum phase of the sunspot cycle, 1948–1949, 1957–1958, 1979–1980.*

Figure 6.5b, which shows the sunspot number R together with the daily C9 index (see below) and the three hourly magnetic index Kp (27-day average) for the period from 1955 to 2002, illustrates this trend clearly for several recent sunspot cycles. Unfortunately, this well established relationship is often forgotten and must be learned again during every new sunspot cycle by a newer generation of researchers. This chapter is based on joint work with Takao Saito.

Figure 6.5c shows the sunspot number and the 27-day average solar wind speed during 1965–2002, including the sunspot cycles 20–23. Contrary to the general belief that the solar wind speed is highest during the sunspot maximum period, Figures 6.5b and 6.5c clearly illustrate that the 27-day average solar wind speed tends to become higher toward the late phase of the sunspot cycle, well after the sunspot maximum period. The solar wind speed during sunspot maximum years is relatively low, even at a minimum; this was clearly seen in 1980. Indeed, the 27-day average solar wind speed is almost out of phase

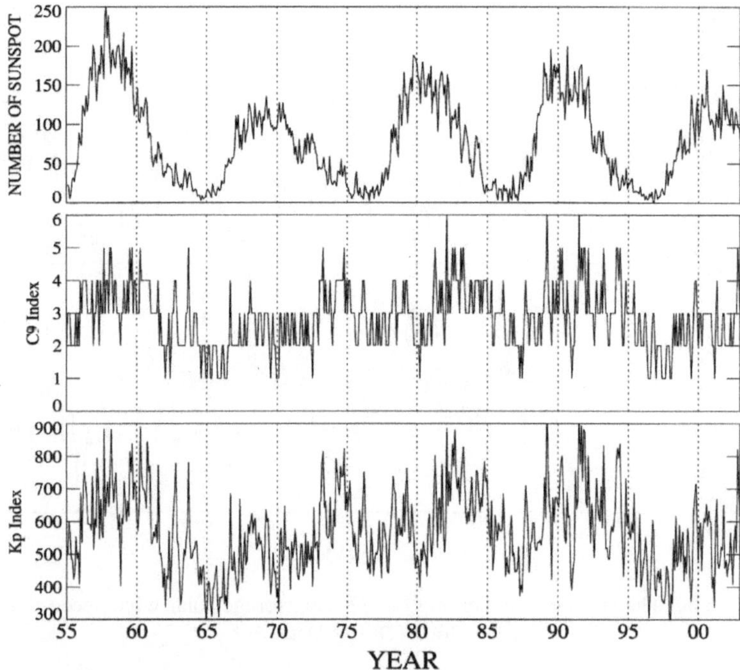

FIGURE 6.5b. From the top, the sunspot number, the C9 index (the occurrence of the number of C9 storms), and the Kp during 1955–2003.*

with the sunspot variations. This trend is also evident for other kinds of average speed. Using the C9 index, we examine the implications of Figure 6.5c in more detail in the next section. In summarizing the above facts, we have:

(1) Geomagnetic disturbances measured in terms of 27-day average, tend to peak well after, not during, the sunspot maximum period.
(2) The 27-day average solar wind speed tends to peak towards the end of a sunspot cycle..
(3) The solar wind speed during sunspot maximum years is relatively low.

These important facts have not necessarily received full attention from magnetospheric physicists and solar physicists, who tend to emphasize the importance of the sunspot maximum years. As early as 1955, Pecker and Roberts (1955) discussed the tendency of sunspots to suppress the solar wind. They noted this trend based on geomagnetic disturbances and proposed the concept of "cone of avoidance," but their findings received little attention from most solar physicists and magnetospheric physicists. This point will be discussed further in later sections.

The occurrence of high-speed solar wind streams, together with highest levels of geomagnetic disturbances, is why the late declining phase of the sunspot

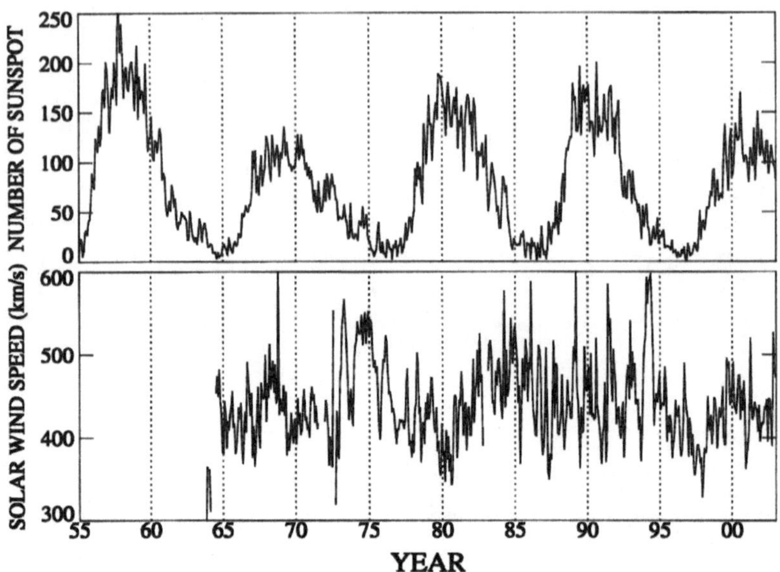

FIGURE 6.5c. The sunspot number and the 27-day average solar wind speed during 1965–2002 (OMNI Data).*

cycle is chosen here in this chapter. Here, it is shown that the Sun shows its fundamental and yet simple aspects of solar activity, which can best be studied during the declining phase.

One purpose of this chapter is to identify the reasons why the 27-day average solar wind speed becomes highest and why the magnetosphere is most disturbed during the late declining period of the sunspot cycle. Although it is widely known that the coronal hole is somehow responsible for those features, the specifics of the relationships are not generally understood. One of the purposes of this chapter is to show that such an inquiry can lead us to understand some of the most basic aspect of solar activity.

6.3. Solar Wind Speed during the Declining Sunspot Period and Associated Geomagnetic Disturbance

Figure 6.6 shows the C9 index during the declining period of the sunspot cycle. In some years, there are two storm groups, but in some other years, there was only one storm group. Note that there were only a few sunspots during highly disturbed periods. Sunspots in those years may be compared with those in 1948–49, 1957–58, and 1979–80 when the sunspot number was very high (see Figure 6.5a). In those years, magnetic disturbances were not necessarily high and even relatively low, except for occasional high values of the C9 index.

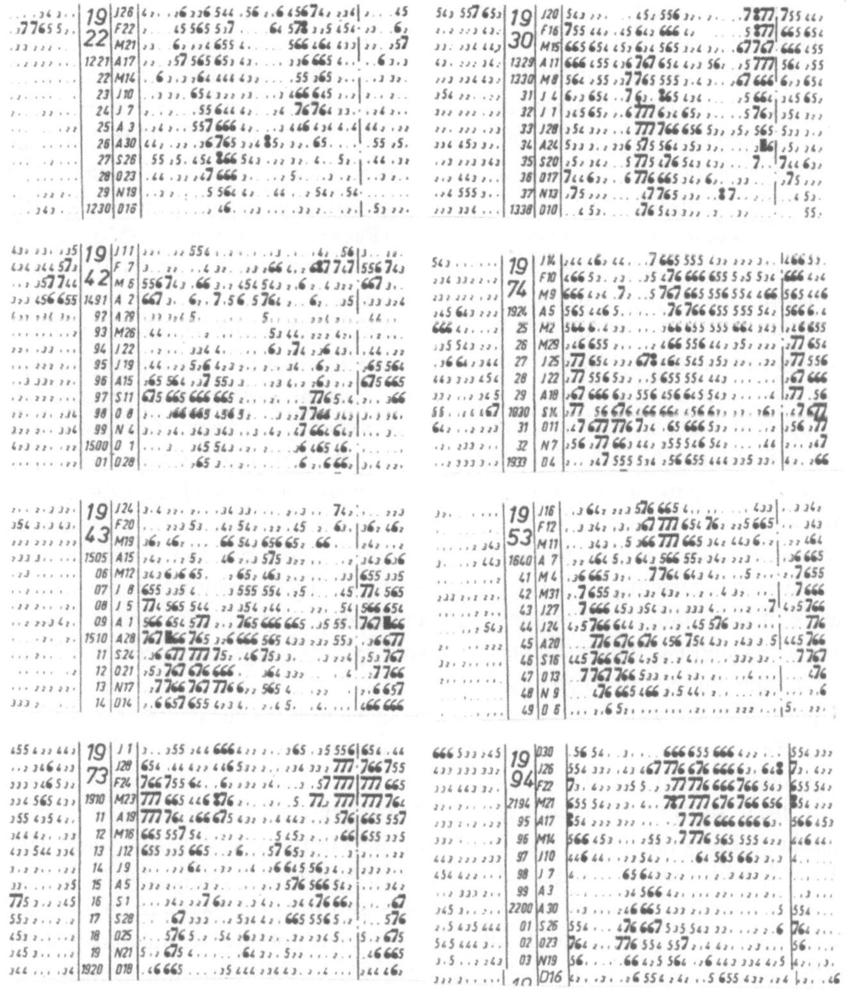

FIGURE 6.6. The C9 index during periods when both one geomagnetic storm group and two geomagnetic storm groups were present.*

Second, the high-speed years (1973–74, 1983–84, 1994–95 in Figure 6.5c) corresponded to the period of a series of magnetic storms – called *recurrent storms* – which recurred for at least several solar rotations, each of which tended to last for 7 $\sim$ 10 days, instead of a few days for solar flare-associated storms. Third, it is most important to note that those years of recurrent storms occurred well *after* the sunspot maximum years (1968 for cycle 20, 1979 for cycle 21, 1989 for cycle 22); this can be seen by simply comparing the upper and lower parts of Figure 6.5a

It may be pointed out here that contrary to a common myth, the number of sunspots is not directly related to the degree of geomagnetic disturbances. This can easily be seen in Figure 6.5b. Years of high sunspot numbers are not necessarily years of high magnetic disturbances. C9 = 8 or C9 = 9 storms were rare in 1948–49 or 1957–58 and absent in 1979–80, in spite of the fact that the sunspot number was very high. Sunspots are not generally considered to be geoactive unless flares and CMEs occur in their vicinity (Mustel, 1964). Sunspots without such activities may even "suppress" (Pecker and Roberts, 1955), "divert" (Saemundsson, 1962), or "deflect" (Sito, 1965) high-speed solar wind, because the magnetic field of sunspots forms an arch-like structure.

Another important feature illustrated by Figure 6.6 is that there was a pair of recurrent storms during each solar rotation in 1974 that occurred during sunspot cycle 20 (1973–1975). Further, these two recurrent storms were separated by roughly two weeks, indicating that the responsible high-speed streams were separated roughly by 180° in longitude. Further, a pair of recurrent storms is separated by a pair of very low C9 values. On the other hand, in 1994 (during sunspot cycle 22 in Figure 6.6) there was only one recurrent storm in each solar rotation. In order to understand about the occurrence of only one recurrent storm, note that one recurrent storm was present in the spring equinoctial months and the other in the fall equinoctial months, with a minimum of geomagnetic disturbances in June/July and December/January. Sunspot cycle 21 was intermediate between the two sunspot cycles 20 and 22.

Figure 6.7 show the daily average solar wind speed in 1981–1985 and 1990–1994. It is interesting to examine several of their features in light of the points mentioned above. The year 1981 is typical of solar wind speed variations during high sunspot numbers. There occurred intermittent impulsive changes throughout the year. The year 1981 may be contrasted with the years 1983 and 1984, which were dominated by the two high-speed peaks during about a one- month period, actually every 27/2 days and thus the two-stream situation. The sunspot cycle 22 was somewhat different and complex compared with the sunspot cycle 21. In 1990, 1991, and 1992 the two-stream situation occurred intermittingly. However, prominent and large modulations became clear only in 1993 and 1994. Figure 6.7 also shows clearly that the double stream, not a single stream, is a fundamental aspect of the solar wind during the late declining phase. Similar two-stream situations were observed by the Ulysses spacecraft (Smith et al., 2001), so that they are intrinsic to the solar wind and are not caused by any of earth's effects. Another important feature to be noted is that many high-speed stream pairs do not necessarily have the same intensity.

From the above study of the C9 index and the solar wind, it may be concluded:

(1) *The late declining phase of the sunspot cycle coincides with the period when the high-speed stream and well-developed recurrent storm groups occur. This phase coincides with the period when a clear sine wave structure of the neutral line occurs.*

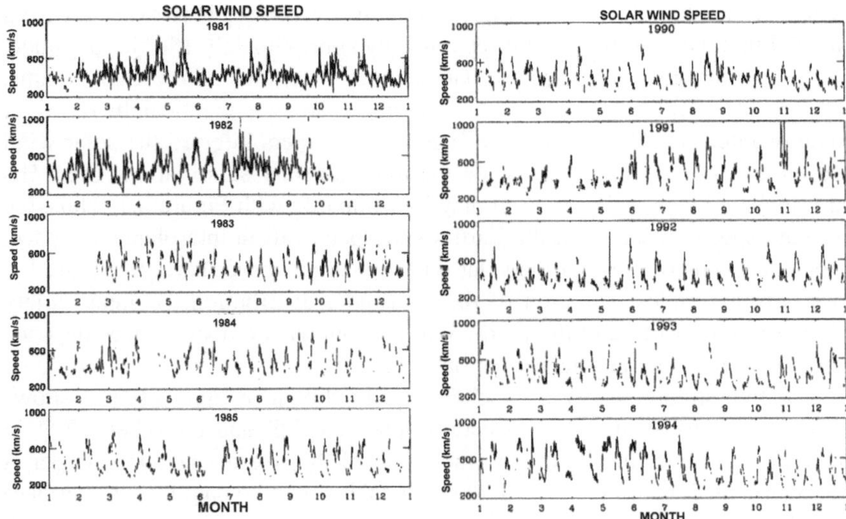

FIGURE 6.7. The daily variations of the solar wind speed from 1981 to 1985 and from 1990 to 1994*

(2) The solar wind speed is minimum at the neutral line (the magnetic equator on the source surface) and increases toward higher latitudes.

In addition, the solar wind speed as a function of magnetic latitude on the source surface has been studied by a number of workers (cf. Kojima et al., 1987; McComas, 2003), see Figure 6.1.

Although it could be said that these facts are generally understood in terms of the so-called spring and fall of geomagnetic disturbances, complexities associated with the development of coronal holes and the tilt angle of the solar rotational axis with respect to the ecliptic plane tend to obscure these facts, particularly those cases shown in Figure 6.6 (spring maximum, but no fall maximum, or fall maximum, but no spring maximum).

As is well known, the comparison between the observed 27-day variations (Figure 6.4b) and their simulations (Figure 6.4a) indicates that recurrent magnetic storms begin at the time of the arrival of the shock wave/sector boundary (the co-rotating interplanetary structure) crossing, which causes a high IMF magnitude B associated with the interplanetary co-rotating structure (note B^2 in the ε function) and 'turbulence' of the solar wind. Each sinusoidal fluctuation generates a substorm. Intense *substorms* cause a magnetic *storm* as their non-linear consequences (Section 4.5). This simulation study suggests that no other basic ingredient is needed to cause recurrent magnetic storms, except for the changes of the relative location of the earth with respect to the heliographic latitude and changes of the solar wind speed distribution on the source surface.

When the neutral line has a single sine wave structure, it is expected from Figure 6.4 that the two streams are separated by approximately 180° in longitude. In reality, it can also be expected that the source regions of the two streams, which are also separated by 180° in longitude, are not identical in terms of latitude or intensity. Hakamada (1987) and many others inferred the solar wind speed on the source surface based on the interplanetary scintillation (IPS). His figure (included here as Figure 6.8) may be compared with Figure 6.1. Thus, the solar wind speed, observed at the Earth or by spacecraft in interplanetary space, depends on its distribution of the solar wind speed on source surface. Figure 6.1 shows only a simplest situation of the sinusoidal distribution. A great variety of situations can occur and the prediction of geomagnetic disturbances depends greatly on many other secondary factors.

The double stream situation is schematically shown in Figure 6.9. It shows two *cross-sections* of the velocity pattern, one at +7.5° and the other −7.5° in heliographic latitude. In this model, the width of the streams becomes larger at higher latitudes. It takes about 7–10 days for each stream to pass by (or overtake) the earth and its magnetosphere. Thus, the magnetosphere is engulfed by both streams for 14–20 days out of 27 days (one solar rotation observed by the earth). It is for this reason that the 27-day average speed of the solar wind becomes so high during the late declining phase of the sunspot cycle.

The above situation describes the simplest case. Single recurrent storm cases in 1953 and 1994 can be explained when the Earth is located above the heliographic equator in the fall months and the southern solar wind is weak or when the Earth is located below the heliographic equator in the spring months and the northern solar wind is weak. In this way, one can explain the asymmetric C9 patterns in Figure 6.6.

Further, the solar wind distribution on the source surface is not as simple as that shown in Figure 6.1. There can exist some limited areas, from which higher

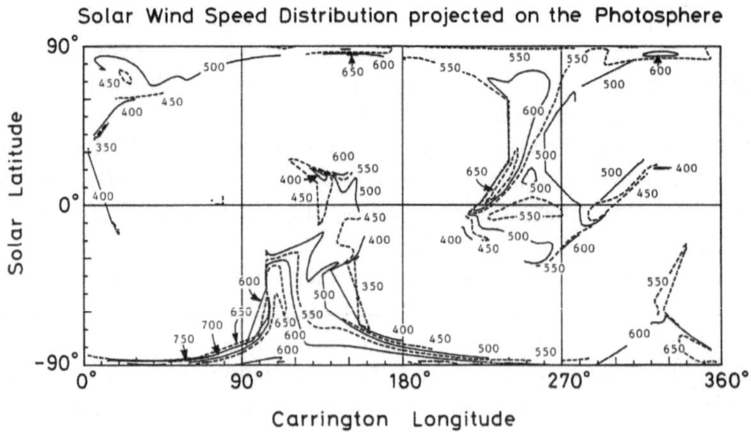

FIGURE 6.8. The solar wind speed distribution inferred from the interplanetary scintillation. The distribution is projected onto the photosphere.*

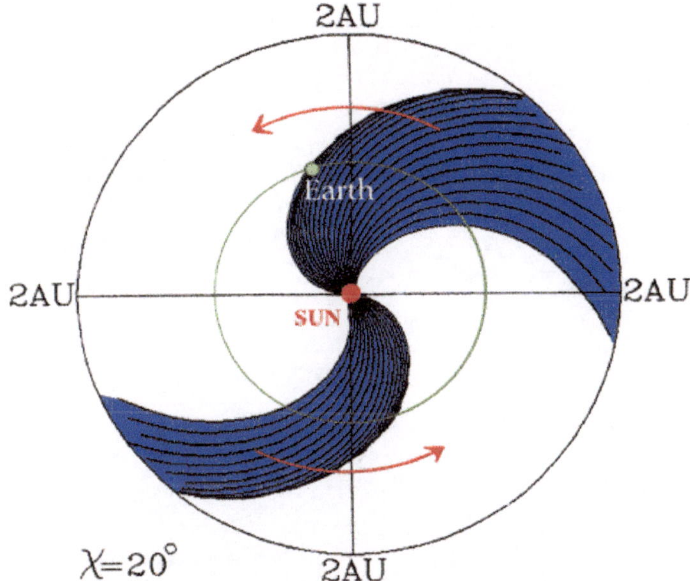

FIGURE 6.9. The schematic illustration of the two high-speed streams during the late declining phase.*

or slower solar wind can emanate. The actual 3-D structure of the streams may be between the simulated case and a case like water flow from several hoses. During the period of sunspot maximum, a large number of sunspots are expected to suppress or deflect the solar wind flow, making the situation complicated.

The suggested relationship between the coarse photospheric magnetic field structures and the source surface field in 1984 and 1994 is schematically shown in Figure 6.10. These figures show both the photospheric field between $+10^\circ$ and -10° in latitude and the vertical field structure at $+10^\circ$ to -10° in latitude, as well as the field on the source surface. These schematic sketches are based on a number of observed facts and the magnetic configurations that were discussed earlier and are thus consistent with earlier literature.

It is expected that the closed magnetic field lines, at the location of the two large-scale dipolar fields, stand high to the height of 2.5 solar radii from the photosphere and reach the source surface. It is expected that the solar wind speed becomes low several days after the central meridian passage of the two dipolar fields on the photosphere. As shown later, this is indeed the case. The magnetic field lines from individual sunspots and sunspot pairs lie low and closed (see Section 5.1, Figure 5.3). The other field lines reach the source surface, becoming open field lines. It is likely that the solar wind originates along open photospheric field lines that constitute a large bundle in the outer corona above the source surface. It is obvious that this situation can occur where the polarity of the root of these open field lines and the polarity of the source surface are the same. However, although all the field lines on the source surface are supposed to be

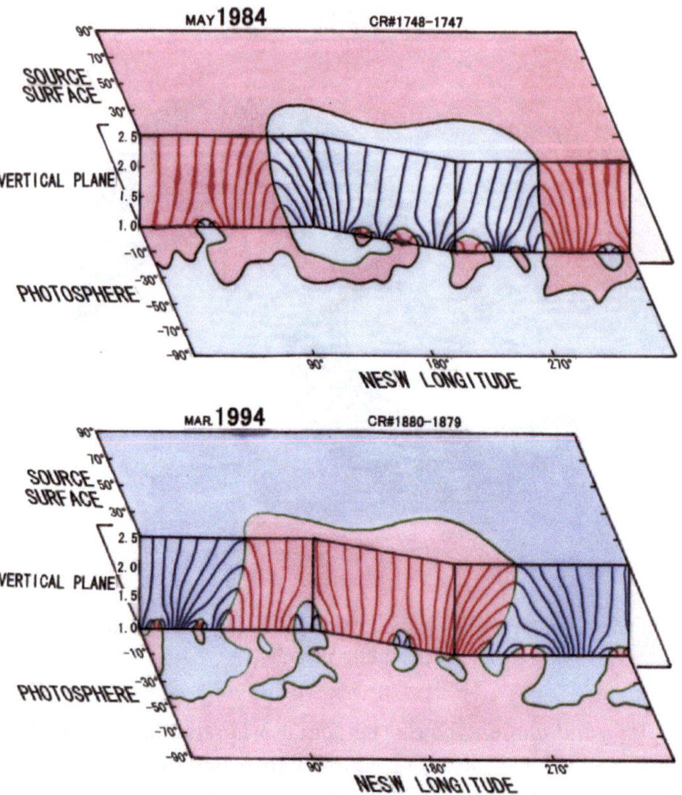

FIGURE 6.10. The inferred vertical configuration of the magnetic field lines at latitude +10° to −10° based on the photospheric field distribution (in 1984 and 1994). The field lines of individual sunspots, or sunspot pairs, are not shown.*

open field lines, the whole source surface is not necessarily the coronal hole. Coronal holes reside within the open field regions, but do not necessarily occupy the entire source surface.

6.4. NESW Coordinate System

After successfully demonstrating the importance of the neutral line on the source surface, it can be suggested that the neutral line – the magnetic equator – can be a new frame of reference in examining solar activities and geomagnetic disturbances. However, one problem we face in this attempt is that the neutral line shifts gradually or randomly in longitude in the Carrington coordinates during successive rotations. The shift can be easily seen in data from 1984, 1994, 2002, and other years as presented in Figure 6.11. Thus, we attempt to eliminate

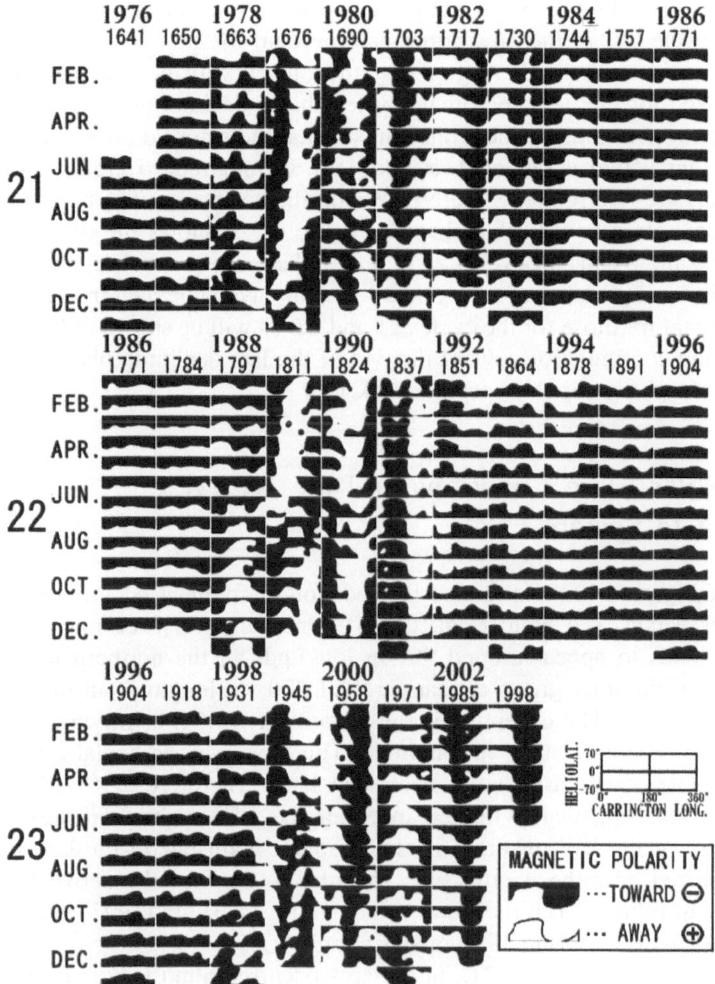

FIGURE 6.11. The polarity of the source surface field during the sunspot cycle 21, 22, and 23 (from Carrington Rotation from 1641 to 2003). The magnetic field points away from the Sun in the white area and points toward the Sun in the black area (Hoeksema, WSO Homepage).*

the shift by introducing a new heliomagnetic longitude φ for the neutral line based on the spherical harmonic analysis of the source surface magnetic field by

$$\varphi = \arctan\,(g^{11}/h^{11}) + \alpha$$

where g^{11} and h^{11} are the spherical harmonic coefficients, and;

$\alpha = 0°$ for an even cycle, or
$\alpha = 180°$ for an odd cycle.

As an example, Figure 6.12a shows the neutral line during 1993 and 1994 (sunspot cycle 22) in the new coordinate system; the new longitude does not exactly coincide with the Carrington longitude; this is why the Carrington rotation number used for the new coordinate system has a decimal number. In the new coordinate system thus defined, one can see that the sinusoidal wave (more like a rectangular wave) is centered around 180° in longitude. Hereafter, the new coordinates are referred to as the NESW coordinate system. The reason for this will become clear later in this chapter. Figure 6.12b shows the typical neutral line during three sunspot cycles 21 and 22. There are great similarities among them. (The configuration of the cycle 20 the only available data in the cycle is a little different for those for the cycles 21 and 22; as will be shown in Figure 6.15, it shows the configuration toward the end of the late declining phase when one of the photospheric dipoles became weak.)

6.5. Solar Flare Locations and the NESW Coordinate System

Note that Figure 6.12a shows also the location of frequent occurrence of solar flares; for details, see Saito et al. (2003). One can easily see that one group of flares tends to appear around 90° in longitude in the northern hemisphere (NE), while the other group appears around 270° in longitude in the southern hemisphere (SW). The exception is the Rotation 1875.54, in which a cluster of flares occurred around 180° in longitude. This case may be explained in terms of the double wave case that is described later. Therefore, 99% of the total 78 flares were statistically limited in either the NE quadrant in the northern hemisphere or the SW quadrant in the southern hemisphere. Further, most of them occurred near the neutral line. In order to confirm the above facts, we reproduce in Figure 6.12b the results obtained by Matsuura (1997), who plotted flares that occurred during 1984 in sunspot cycle 21.

During the sunspot cycle 21, most flares occurred almost directly under the neutral line, in spite of the fact that flares and the neutral line are not located on the same spherical surface. During the sunspot cycle 22, the flares in the NE quadrant is shifted a little away from the neutral line. This shift may be due to the fact that the center of the magnetic arch or archade (Figure 5.3) deviated from the radial line from the center of the sun (Saito et al., 2000).

6.6. Solar Corona and the NESW Coordinate System

Figure 6.13 shows the coronal distribution for the late declining phase during the sunspot cycle 20, 21, and 22. Note that different data sets were used for the different cycles, depending on their availability; the satellite data were not available for the cycle 20. For the sunspot cycle 20, see also Figure 6.15; because it was toward the end of the cycle, one of the photospheric dipoles was very

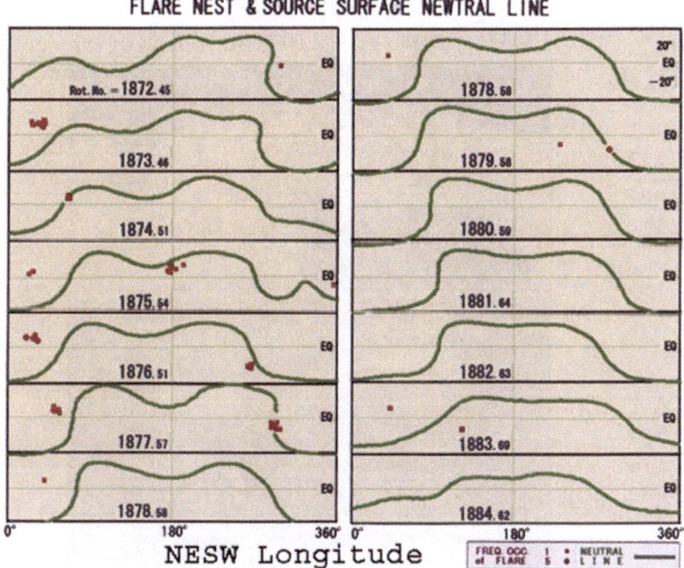

FIGURE 6.12a. Neutral Line during a successive solar rotation during the late declining phase in the NESW coordinate system.*

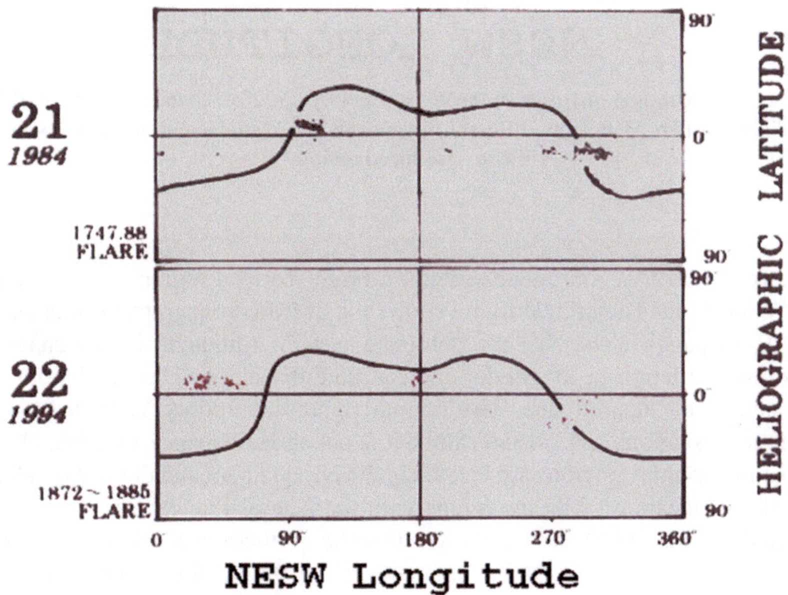

FIGURE 6.12b. Distribution of solar flare locations in the NESW coordinate system.*

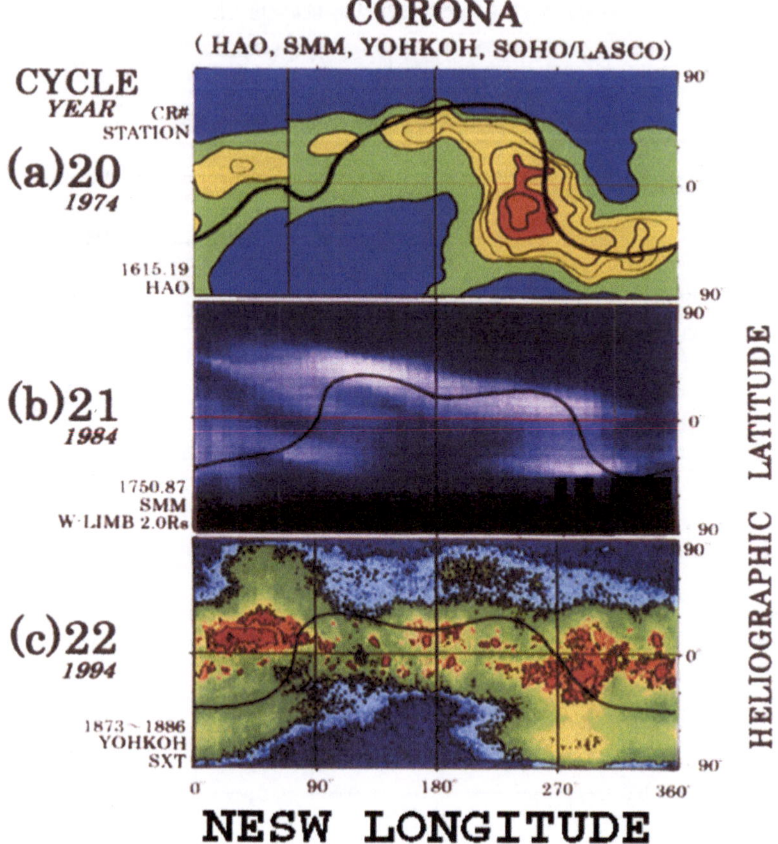

FIGURE 6.13. Coronal distribution during sunspot cycles 20, 21, and 22 in the NESW coordinate system. Data from different instruments are used. Note that the red areas is the lowest speed region.*

weak. In the NESW coordinate system, a bright coronal region was located in the SW quadrant. The middle figure shows the SMM coronagraph (Hundhausen, HAO Homepage) in the NESW coordinate system. Although this coronagraph has a low resolution in longitude, one can note that the brightness distribution is related to the neutral line. The bottom figure superposes Yohkoh data for Carrington Rotations 1873–1886 (Shibata, Yohkoh Homepage). One can clearly see that the brightest regions are located in the NE quadrant and the SW quadrant. The relative location of these regions with the neutral line can be clearly seen (Watanabe et al., 2003). It is also clear that the northern coronal hole is always located in the vicinity of 0° in longitude, while the southern one is at about 180° in this new NESW longitude, suggesting the 180° separation (or 27/2-day separation) of a pair of recurrent storms.

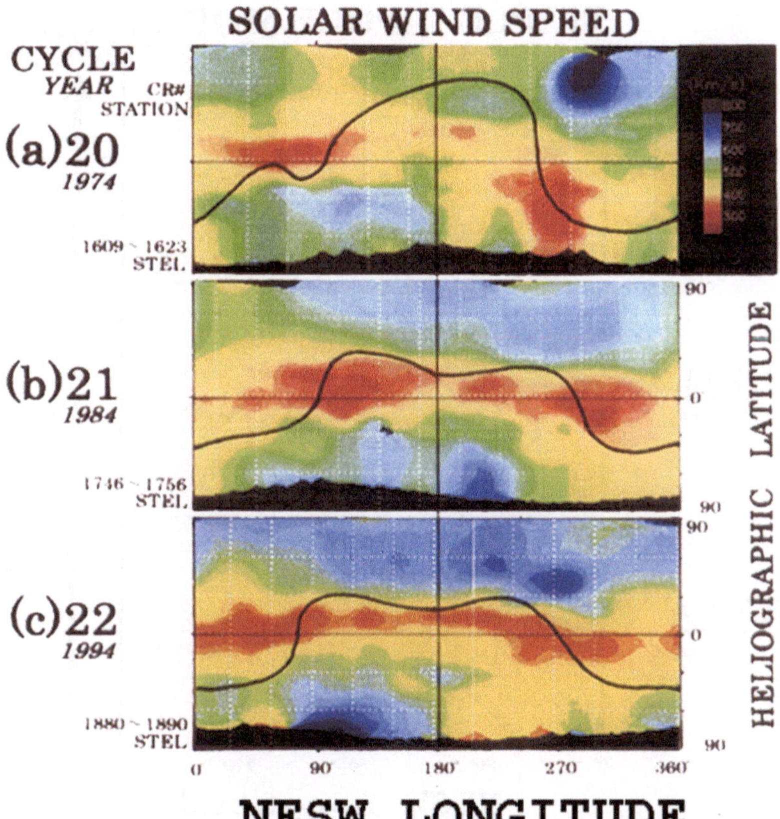

FIGURE 6.14. Distribution of the speed of the solar wind in the NESW coordinate system. Note that the red areas are the lowest speed region.*

6.7. The Solar Wind and the NESW Coordinate System

In Figure 6.14, we examine the distribution of the solar wind speed observed by the interplanetary scintillation method, IPS (Kojima, IPS Homepage). This feature is located above the source surface and the speed is projected on it. One can see that two lowest speed regions (in red) occurred, one in the NE quadrant and the other in the SW quadrant.

6.8. The Triple Dipole Model and the NESW Coordinate System

In Section 5.2, it was shown that the triple dipole model can reproduce reasonably well the configuration of the neutral line. It was also shown there that the two photospheric dipoles are present in the magnetic field map in a coarse mesh

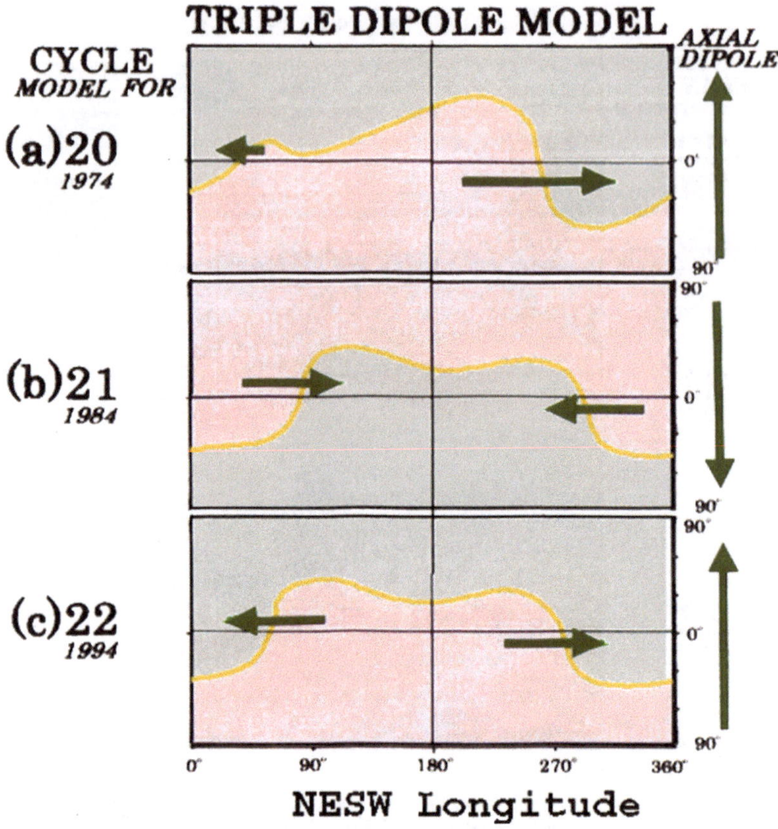

FIGURE 6.15. The location of two hypothetical photospheric dipoles that can reproduce the neutral line on the source surface (the NESW coordinate system.).*

(5° × 5°). Figure 6.15 shows the location of the two hypothetical photospheric dipoles that can approximately reproduce the configuration of the neutral line on the source surface. The location of the two hypothetical dipoles on the photosphere is determined purely on the basis of attempting to best reproduce the neutral line. The top figure in Figure 6.15 shows the neutral line for the sunspot cycle 20. Unfortunately, this is the only neutral line available for this cycle in the literature. This occurred near the end of the late declining phase, when one of the two photospheric dipoles became very weak.

In the cycles 21 and 22, the hypothetical two dipoles are located approximately where various solar activities are presented (Figures 6.12 and 6.13) and also where the lowest solar wind speed regions are present (Figure 6.14). Therefore, it is reasonable to conclude that two hypothetical dipoles are unmistakably realistic physical entities. It may be noted that, the two dipoles satisfy the Hale-Nicholson polarity law, adding credibility of the presence of the two dipoles.

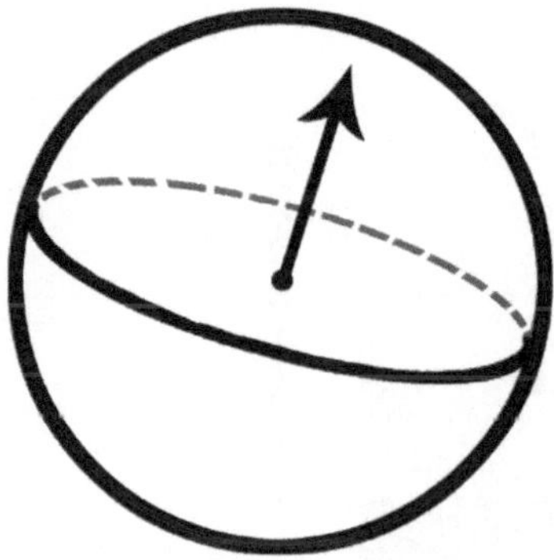

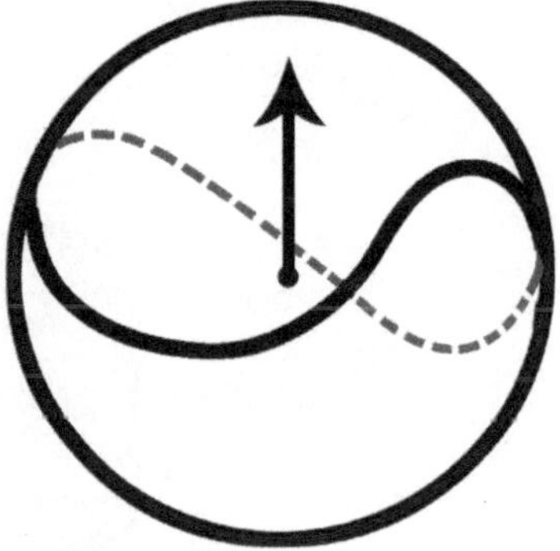

FIGURE 6.16. Schematic illustration of the equatorial plane for the single wave and double wave cases.*

6.9. Double Wave Case

The double wave of the neutral line can occur when the magnetic equatorial plane warps in a particular way as shown in Figure 6.16. In Figure 6.17, the top diagram shows a large amplitude double-wave situation, which occurred in November 1998 (see Figure 6.11). It is one of the most complicated configurations of the neutral line during the sunspot cycle. The second and third diagrams show that solar activities are not uniform and are located approximately where we can expect by considering that there are two single waves in one NESW or Carrington coordinates, as shown in the bottom diagram. There are four photospheric dipoles in the double wave case, although these are not of equal intensity (Saito et al., 2002). They follow the NESW rule and the Hale-Nicholson polarity law, so that they are not a random feature.

6.10. Summary of Chapters 5 and 6

1. It was demonstrated that certain aspects of solar activities and their consequences on the magnetosphere can be studied and understood better in terms of the magnetic field distribution on the source surface than complicated photospheric features. The late declining phase was chosen, so that the neutral line has a single sinusoidal (or rectangular) structure, and the solar wind speed and geomagnetic disturbances are highest during the sunspot cycle.
2. One of the important points we showed here is that the configuration of the neutral line on the source surface can tell us some important large-scale and perhaps some fundamental features of the photosphere, which may not be

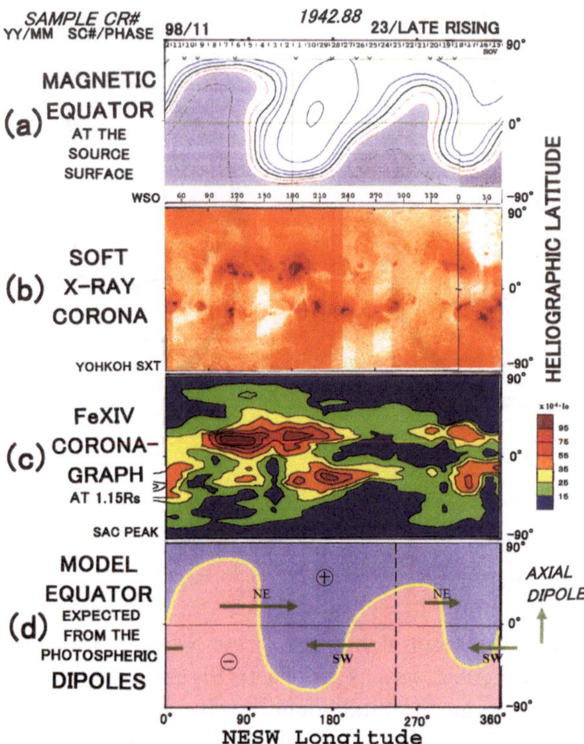

FIGURE 6.17. A large amplitude double wave case. From the top, (a) the magnetic field on the source surface, (b) soft x-ray corona (Yohkoh), (c) FeXIV coronagraph (SAC Peak) at the computed neutral line based on the photospheric fields in the NESW coordinate system.*

so obvious on the photosphere because of complexity of a variety of intense small-scale features there.

3. Demanding that there must be a field represented by an axial dipole at the center of the Sun (because the unipolar regions remain in both the northern and southern polar regions throughout the sunspot cycle), it was shown that the deviation of the neutral line from the straight line along the heliographic equator in the Carrington coordinates is caused by two hypothetical dipole fields on the photosphere. As a first approximation, two large-scale dipoles are needed to distort to the straightest neutral line to become a sinusoidal wave-like curve.
4. By introducing a new coordinate (NESW) system, it was shown that the location of the two hypothetical dipoles coincides approximately with the nest of the solar flares, the brightest region of corona and also the source of the lowest solar wind speed. Thus, the two hypothetical photospheric dipoles are realistic physical entities. Figure 6.18 summarizes all the results in this chapter.

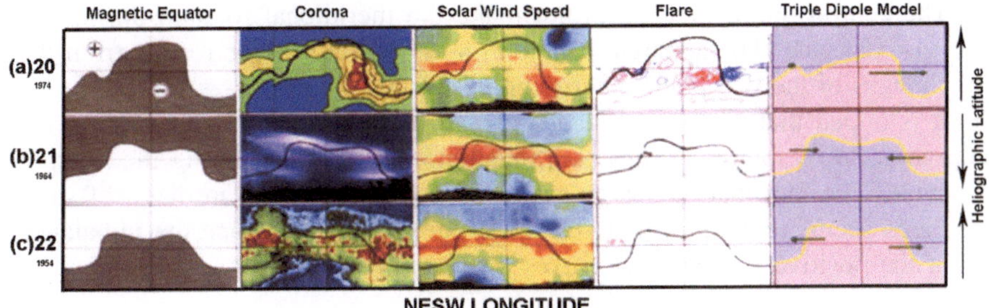

FIGURE 6.18. Summary diagram for Chapter 6.

5. The stable sinusoidal (or rectangular) wave configuration of the neutral line provides the basis of the double peak, per solar rotation, of the solar wind speed and of geomagnetic disturbances during the late declining epoch of the sunspot cycle.
6. The double peak structure, per solar rotation, in the solar wind speed and recurrent geomagnetic disturbances is a basic feature, but it is often obscured by various complexities on the Sun and the relative location of the Earth with respect to the heliographic equatorial plane.
7. Recurrent geomagnetic disturbances are related to the stable and sinusoidal (a rectangular) neutral line structure. This is why recurrent geomagnetic disturbances tend to occur during the late declining phase of the sunspot cycle.
8. The fact that the double high-speed solar wind stream is sometimes not apparent may be due to the fact that a complex distribution of sunspots during the period of sunspot maximum suppresses or deflects the high-speed streams.
9. As the intensity of the two dipoles becomes weak, the neutral line tends to align with the heliographic equator. Since the intensity of the two dipoles is, in general, not equal, one photospheric dipole, together with the central dipole, may be enough to represent the neutral line near the end of the declining phase.
10. The results, based on the single wave during the late declining phase of the sunspot cycle, may be applicable to the double wave situation and also the rest of the sunspot cycle. The classical "sector structure" may be considered as the case when the amplitude of the single wave reaches 90° in latitude.
11. Results presented in this chapter are based on the assumption that the Sun has a magnetic field that can be represented by a central, axial dipole and two photospheric dipoles. Since it is possible to explain a number of phenomena based on this assumption, such magnetic fields are expected to exist, except that the central dipole becomes very weak during the sunspot maximum phase.

12. The central dipole field does not make the meridional rotation during the sunspot cycle. The apparent rotation on the source surface results from the changes of the two photospheric dipoles.
13. It is hoped that Chapters 5 and 6 present a new hint in understanding the evolution of solar activities as a magnetic variable star during the sunspot cycle. The fact that two large-scale dipolar field regions are nests of solar flares seems to suggest that flares are the process to reduce the imbalance of the positive and negative fluxes (McIntosh, 1981).

7
Myth of the Emerging Flux Tubes: Sunspots and Solar Flares

7.1. Introduction

The present guiding concept in searching observationally and theoretically for basic processes of sunspots and solar flares is based on a hypothesis that solar activities are manifestations of interactions of intense magnetic flux tubes that emerge from beneath the photosphere, and the subsequent consequences. A typical sunspot group is shown in Figure 7.1. At present, the photosphere is considered merely a passive medium through which the magnetic flux penetrates from below. Therefore, the main theoretical efforts have so far been concentrated in examining both the emergence of *hypothetical* magnetic flux tubes for sunspots and instability processes leading to explosive annihilation of the magnetic energy carried up by the flux tubes, namely solar flares.

For these reasons, all observational/morphological features of solar activities have been discussed in terms of such theoretical implications, e.g., magnetic flux *emergence*, magnetic energy *storage*, flare *buildup*, *triggering* instability and magnetic energy *release*, instead of descriptive terms, as if the hypothesis is confirmed beyond doubt. Thus, most solar physicists working on this particular subject appear to share the paradigm of magnetic flux tubes and of magnetic field annihilation. A very large number of papers have been published with an extremely high degree of agreement on the problems to be solved within the framework provided by this particular paradigm. In these multiple papers, researchers have articulated and elaborated on the hypothesis. As in the standard paradigm, the solution is assured. Thus, if an anticipated result does not occur, this will be taken as a scientist's failure, not as the theory's failing. Gene Parker (1964) remarked:

> ... it has proved extremely difficult to progress from the general association of flares and magnetic fields to specific processes by which the field actually produces flares. At least on our scratch pads the magnetic field stubbornly refuses to dissipate on command.

Such a tendency is very unfortunate and even dangerous for the development of solar physics. This is because the identification of observational features corresponding to such hypothetical processes has not necessarily been very

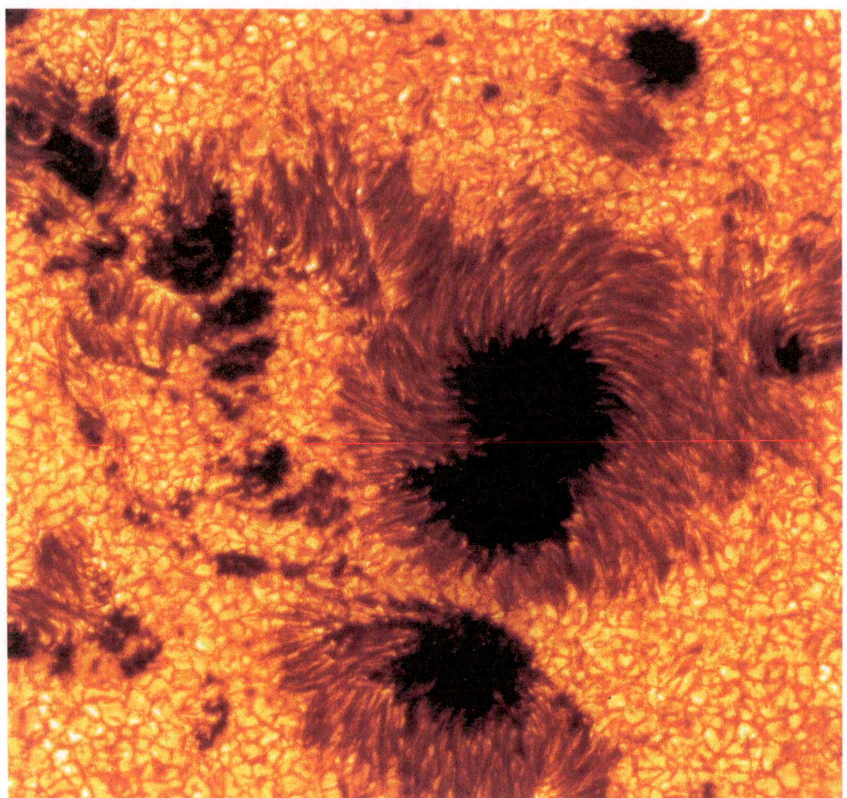

FIGURE 7.1. A typical sunspot group: Göran Scharmer, Royal Swedish Academy of Sciences, taken with the Swedish 1-meter Solar Telescope.
Source: Courtesy of Royal Swedish Academy of Sciences

definitive, at least not verified. Solar observers must be more independent of theorists for a healthy growth of the field, although this can be said in any scientific field.

It is well known that when a particular paradigm becomes dominant, it becomes very difficult to publish a paper that casts doubt on it; even kind referees could ask almost impossible tests to confirm the claims. The author attempts to overcome such a hindrance usually with great difficulty. On the other hand, observations that appear to conform to the prevailing paradigm may be accepted without much scrutiny.

7.2. Emerging Magnetic Flux Tubes

The theoretical difficulties we are facing today in understanding transient solar phenomena may not always be due to our present inability in handling theoretical problems and in sorting out observations. It is likely that the problem is that no doubt has been cast on the guiding concept of hypothetical magnetic flux tubes below the photosphere and of magnetic reconnection. It has been forgotten that the present guiding concept itself consists of a three-step hypothesis:

1. Hypothesize the presence of intense magnetic flux tubes of various sizes and orientations beneath the photosphere.
2. Hypothesize their rise through the upper boundary of the photosphere by magnetic buoyancy.
3. Hypothesize annihilation of their magnetic energy by explosive magnetic reconnection as a result of interacting with other flux tubes.

Figure 7.2 schematically illustrates these three steps. This three-step hypothesis has been held for several generations. Therefore, it has now become a *doctrine*; the occurrence of a sunspot pair is considered to be proof of the hypothesized flux tube, instead of the source of the hypothesis. A typical response from a solar physicist to my question "What is the proof for the presence of a thin magnetic flux tube?" is "A pair of sunspots." An important point to make here is that it has not been observationally and theoretically confirmed that thin flux tubes of various sizes and orientations can be created, and exist just below the photosphere. There has so far been no definitive observation of magnetic flux tube below the photosphere. Therefore, it is still only a hypothesis.

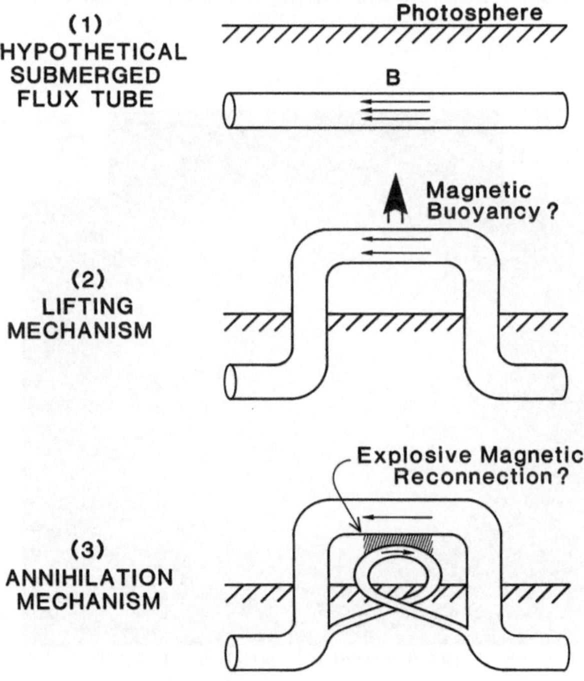

FIGURE 7.2. Illustrating the three-step hypothesis.
Source: Akasofu, S.-I., *Planet. Space Sci.*, **32**, 1469, 1984

We should have several choices of consideration: one of them is that intense magnetic flux tubes of various sizes and various orientations can be formed beneath the photosphere and form complex sunspot groups after emerging through the photosphere. In the second, large-scale, weak fields on the photosphere are basic and become seed fields for the dynamo process, which concentrates them into sunspot fields, which spread to become the large-scale fields again, becoming the seed field for sunspots. Unlike the first case, this possibility is based on magnetic field observations. Even if weak fields are residues of dead spots, the second consideration is based on the observations, as shown shortly.

Indeed, there is nothing wrong in assuming that dead spots (the large-scale, weak fields) can be recycled for new spots. Figures 7.3a and 7.3b show an example of the observed distribution of solar magnetic fields on the photosphere. Figure7.3c shows an example of the "map" of the solar magnetic fields, in which the weak, but large-scale field structures can be seen more clearly.

Figure7.3d shows a high-resolution image of the weak fields. The fields are concentrated along the boundary of cell-like structures called "supergranulation," in which the photospheric flow diverges from the center of supergranulation toward its boundary.

It is important to note that a sunspot of one polarity (say, positive) tends to form within a large-scale field of the same polarity (positive) and that a large sunspot pair tends to form in the vicinity of the boundaries of positive and

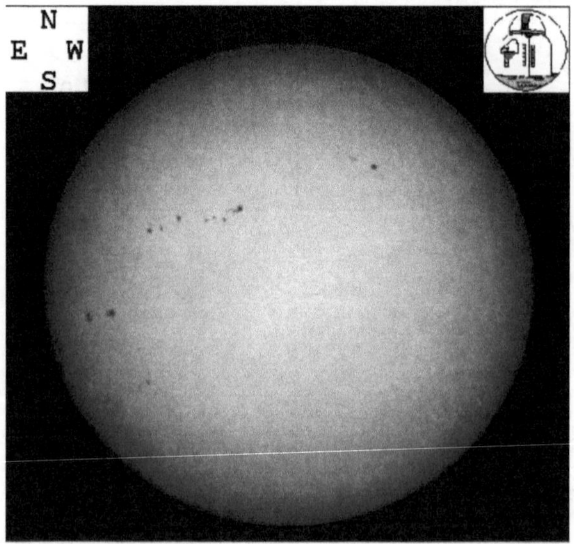

FIGURE 7.3a. Photograph of the solar disk on February 6, 2001.
Source: Courtesy of Big Bear Solar Observatory

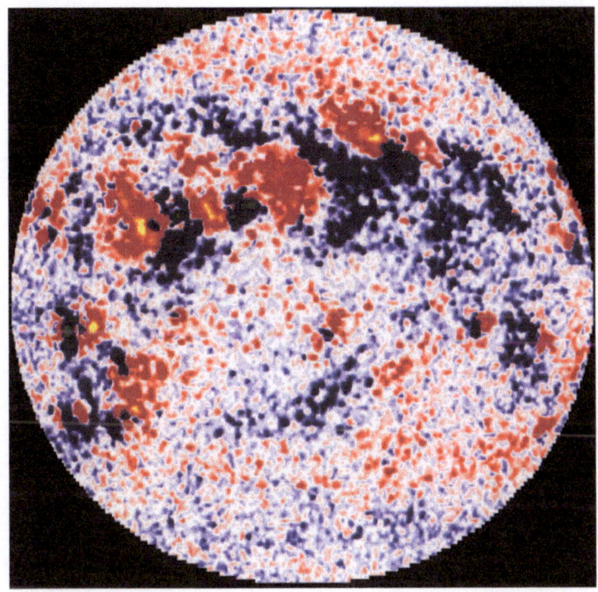

FIGURE 7.3b. Example of solar magnetograph. Red areas show where the magnetic vector points toward the Earth, while blue areas show where the vector points away from the Earth, February 6, 2001 (Figure 7.3a).
Source: Courtesy of Kitt Peak Solar Observatory

negative large-scale fields (Pat McIntosh, 1981). In Figure 7.3c, yellow dots are positive polarity sunspots in the red areas (the positive fields), and green dots are negative polarity sunspots in the blue areas (negative fields). There will not be such a relationship, if the emergence of magnetic flux tubes can occur randomly.

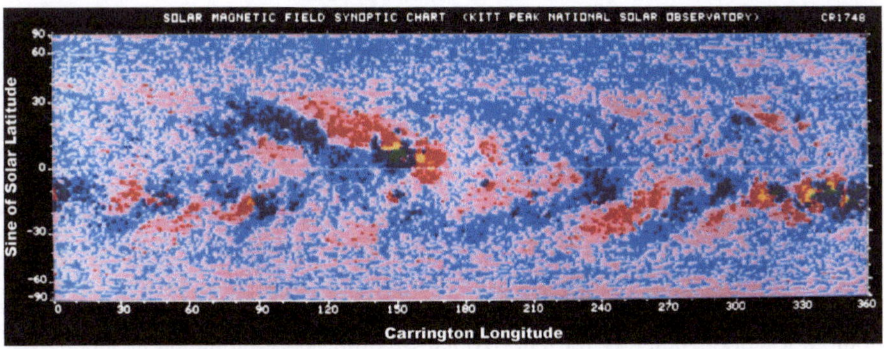

FIGURE 7.3c. An example of the "map" of the solar magnetic fields.
Source: Courtesy of K. Hakamada

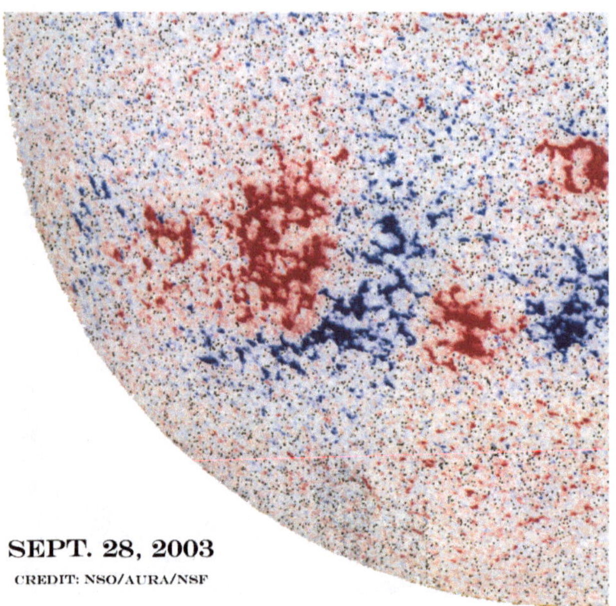

FIGURE 7.3d. Details of the weak fields. Note the net-like structure resulting from the flow around the super-granulations.
Source: Courtesy of Kitt Peak Solar Observatory

Another important point is that the magnetic flux tube hypothesis cannot explain the simplest case, single spots (Figure 7.4). Further, we shall learn in Section 8.4, that there are magnetic fluxes that cross the equator, namely the trans-equatorial flux. If we take the present magnetic flux tube concept, we have to hypothesize trans-equatorial fluxes beneath the photosphere across the equator. Thus, it is not possible for the trans-equatorial magnetic flux to be explained by the standard theory of sunspots, in which sunspot pairs in the northern hemisphere and the southern hemisphere are considered to be independent.

In this section, we consider that *the photosphere is an active medium*, rather than a passive medium through which the hypothetical flux tube merely penetrates. Specifically, we consider a process associated with vortex motions that can concentrate the *observed* weak field, forming sunspots of a variety of sizes (see Section 7.4).

One of the fascinating aspects of a vortex flow, such as a cyclone and a hurricane in the Earth's atmosphere, is that it is associated with converging flow near its base. In the photosphere, the converging flow can concentrate weak magnetic fluxes into a relatively small area and the associated dynamo process may amplify the field (Figure 7.5a). On the other hand, a concentration process would not work efficiently for a horizontal flux tube beneath the photosphere. In fact, there is hardly a paper that can demonstrate the formation of even a single thin magnetic flux tube below the photosphere.

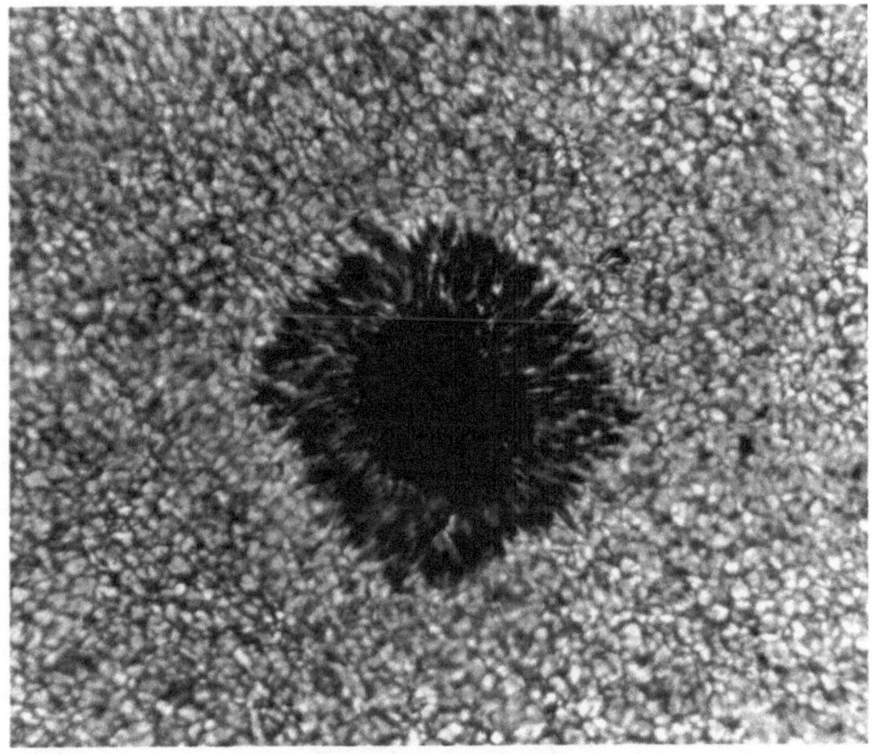

Single sunspot

FIGURE 7.4. An example of a single sunspot.
Source: Courtesy of Big Bear Solar Observatory

First of all, solar physicists must theorize the formation of magnetic flux tubes of a variety of sizes that lie horizontally and are oriented in many directions. On the other hand, a cyclonic motion has no difficulty in concentrating *vertical* magnetic fluxes in the photosphere, because the photospheric gas can escape from the top of the photosphere. In fact, such an outward flow (the Evershed flow) from the top of sunspots is a well-known feature (Figure 7.5b). Figure 7.5b shows a clear indication that a sunspot can be associated with a vortex flow, although solar physicists do not want to pay any attention to such an *observed* feature; in fact, there is no paper on this particular observation except my own (Akasofu, 1985). It is important to note that the cloud structure in a hurricane consists of both large- and small-scale features (Figure 7.5b), and all of them are associated with upward flows of air and thus converging flows near the bottom of all the clouds, the smallest one being a cumulonimbus. A similar statement may be made on the structure of sunspots in Figure 7.5b; there is a large-scale spiral structure, in which smaller scale sunspots are present. It is basically impossible

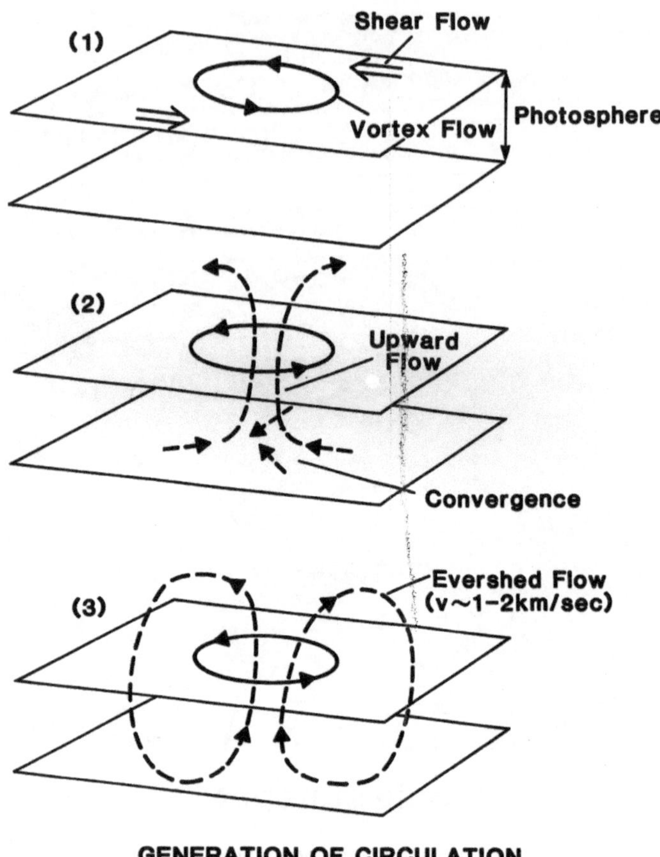

FIGURE 7.5a. Schematic representation of how a sunspot might form as a result of vortex motion in the photosphere.
Source: Akasofu, S.-I., *Planet. Space Sci.*, **33**, 275, 1985

to figure out the sunspot structure in Figure 7.5b in terms of emerging magnetic flux tubes of different sizes. Pat McIntosh (1981) examined in detail a number of sunspots and noted spiral features. It is unfortunate that solar physicists cannot think of anything other than magnetic flux tubes. Sunspots will be discussed more in Section 7.4.

In his book titled *The Solar Atmosphere*, Hal Zirin (1966) noted:

> If the theorist, who prefers things in neat packages, were presented with an ideal spherical sun, even with some magnetic field, he would never predict even sunspots, much less flares. The same is true of tornadoes or hurricanes in the terrestrial atmosphere. But faced with their existence, he must come up with some mechanism that might produce this great energy release.

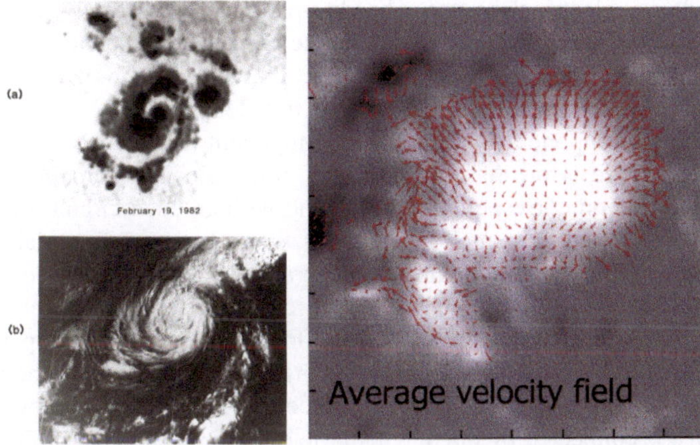

FIGURE 7.5b. Upper left: An observed sunspot group in the northern hemisphere (Kitt Peak Solar Observatory). Lower left: A photograph of a hurricane near Japan (Japan Meteorological Agency). Right: An example of the Evershed flow.
Source: Left: Akasofu, S.-I., *Planet. Space Sci.*, **33**, 275, 1985; Right: M.K. Georgulis, *Earth Transfer*, January 16-20, 2006, Kona, Hawaii

7.3. Energy Source for Transient Solar Activities

Since many solar transient phenomena are manifestations of electromagnetic processes, we must deal with the dynamo process that can supply the electric power. Thus, it is important to clarify the energy source for our process at the outset. Such a fundamental fact is often forgotten as a result of the hypothetical interactions between magnetic flux tubes. A dynamo is a machine that converts mechanical energy into electrical energy. Thus, we must identify first the mechanical energy for our dynamo. As explained in Chapter 4, the same can be said about magnetospheric substorms.

The photosphere is only a weakly ionized atmosphere, the degree of ionization being 10^{-4} to 10^{-5} in the quiet photosphere and perhaps 10^{-6} to 10^{-7} near sunspots. Since both the neutral component and the ionized component of the photosphere are expected to move with similar speeds (because of a high collision frequency), the bulk kinetic energy is mostly carried by the neutral component. Assuming an area of 10^5 km $\times$ 10^5 km (a typical radius of an umbra $\sim 10^4$ km) and depth of 10^3 km, the density of $\sim 10^{-7}$ g/cm^3 and the speed of 1000 m/sec, the total bulk kinetic energy of the photospheric flow is $\sim 10^{31}$ erg. The energy dissipation rate in most intense flares is known to be $\sim 10^{29}$ erg/sec. Thus, so long as thermal convection, the pressure gradient, and other mechanisms for the flow of the neutral component of the photosphere can be maintained, it is basically a small portion (1%) of the bulk kinetic (mechanical) energy of the neutral component that is converted into electrical energy. The bulk kinetic energy of the neutral component must be transferred to the bulk kinetic energy of the ionized component. The bulk kinetic energy of the ionized component thus transferred is converted into electrical power. The dissipation process of the power thus produced reduces the bulk kinetic energy of the ionized component and therefore its bulk speed $\boldsymbol{V}_{\mathrm{p}}$. However, the differential speed between the ionized component $\boldsymbol{V}_{\mathrm{p}}$ and the neutral component $\boldsymbol{V}_{\mathrm{n}}$ ensures the transfer of the bulk kinetic energy from the neutral component to the ionized component (Figure 7.6).

Note that if there is no dissipation and if both velocities remain the same, the resulting situation corresponds to a dynamo with an open circuit (or without load). Therefore, however small it may be, the velocity differential is most crucial for the photospheric dynamo to generate the power for the dissipative *flare circuit.* An important point is that one cannot predetermine the velocity differential based on local conditions alone. One must know the dissipation rate in the whole circuit (including the flare region) in order to estimate the velocity differential, since the dissipation can take place in a region far from the dynamo region, which is connected by the magnetic field lines. Kan et al. (1983) showed that the power of the dynamo is given by:

$$P = -\Sigma_{\mathrm{p}}\boldsymbol{E} \bullet (\boldsymbol{V}_{\mathrm{n}} - \boldsymbol{V}_{\mathrm{p}}) \times \boldsymbol{B}$$

$$\boldsymbol{E} = -\boldsymbol{V}_{\mathrm{p}} \times \boldsymbol{B}$$

Here, Σ_{p} is the Pedersen conductivity of the photosphere which, in the lower photosphere, is similar to that of sea water.

The justification of the common assumption of infinite conductivity for such a low-conductivity medium relies on a large-scale length L

$$VB/L >> B^2/4\pi\Sigma_{\mathrm{p}}L^2.$$

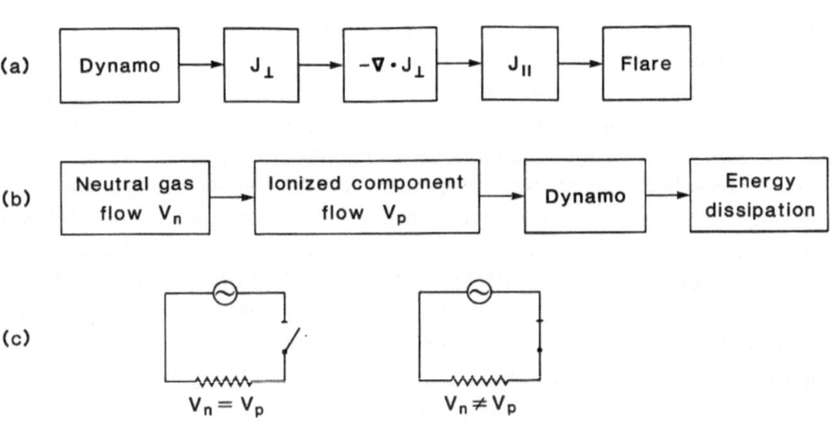

FIGURE 7.6. Diagram showing how the kinetic energy of the neutral component of the photosphere drives a solar flare.
Source: Akasofu, S.-I., *Planet. Space Sci.*, **32**, 1469, 1985

This consideration of justifying infinite conductivity of the photosphere and the assumption of $\boldsymbol{V}_{\mathrm{p}} = \boldsymbol{V}_{\mathrm{n}}$ may eliminate the possibility of understanding solar flares. In magnetospheric physics, we have found that the simplified MHD equations are hardly appropriate in dealing with the auroral potential structure. *In fact, by the initial MHD assumption of infinite conductivity along the magnetic field lines and thus* $\boldsymbol{E}_{||} = 0$, *we used to throw away the very solution of the auroral particle acceleration process* ($\boldsymbol{E}_{||} \neq 0$) even before engaging to solve the problem. The MHD method is only one of the tools in understanding magnetospheric and solar phenomena. This is often forgotten. It is a useful tool for understanding certain phenomena, but it is not the universal tool.

Hannes Alfvén was the first to insist that the frozen-in field condition ($\boldsymbol{E} + \boldsymbol{V} \times \boldsymbol{B} = 0$) should be "thawed." In his Birkeland Symposium paper titled *The Second Approach to Cosmical Electron Dynamics* (1968), he stated:

> ... One has good reasons to suspect that there often exist electric fields with components parallel to the magnetic field. The existence of such fields may invalidate the "frozen-in" picture in many cases. We may say that the first new principle is associated with a "thaw" of the frozen-in field lines.

However, in spite of the fact that he was the founder of MHD physics, some MHD theorists accused him of being a heretic. The MHD formulation became such a powerful paradigm, even its founder had to be accused as a maverick when he warned us of its limitation.

7.4. Sunspots

It so happened many years ago that Gene Parker and I ran into each other at the Logan International Airport in Boston and had a cup of tea. I asked him what was the most difficult problem he had encountered in his life. His response was simply *sunspots*. Although he may not remember the conversation, I was greatly impressed by it. I confirmed his response when we met during Van Allen Day, his 90th birthday celebration, at the University of Iowa, in the fall of 2004.

It is my belief that it is best to go back to observed facts, not a theoretical interpretation of them when we face a seemingly insoluble problem. As mentioned earlier, it has been proposed that the large-scale fields are basic in considering the formation of sunspots. As discussed in Section 7.2, a positive sunspot appears where there is a positive large-scale field and a negative sunspot appears where there is a negative large-scale field. This large-scale field pattern exists prior to the appearance of large sunspot groups (McIntosh, 1981), see Figures 7.3b and 7.3c. One of the important aspects of sunspot formation is that a pair of sunspots tends to form near a polarity reversal boundary of the large-scale fields, which is often marked by dark chromospheric filaments. Figure 7.7 shows schematically the relationship between the large-scale fields shown in

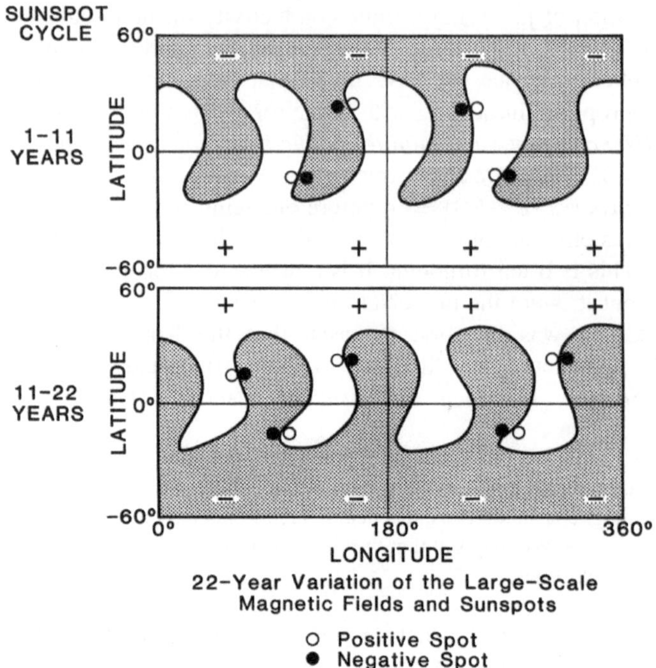

FIGURE 7.7. Relative location of the large-scale photospheric fields and sunspot pairs during the 22-year cycle variation.
Source: Akasofu, S.-I., *Planet. Space Sci.*, **32**, 1469, 1985

Figure 7.3c and the large sunspot pairs. It is of great interest that only one side (say, –/+) of the boundary in each hemisphere tends to produce sunspots and that the opposite side (say, +/–) is active in the other hemisphere. L. Svalgaard and John Wilcox (1976) referred to the active boundary as the Hale boundary.

McIntosh (1981) demonstrated that sunspots tend to form near a belt of highly sheared flows, which is far greater than the shear associated with the non-uniform rotation of the Sun. A large-scale shear flow belt forms first in high latitudes at the beginning of a new sunspot cycle and shifts equatorward, just like sunspot groups.

Thus, if a vortex motion occurs in the belt of the shear flow, a large concentration of the magnetic flux can be expected, as described in Section 7.2. Assuming that the magnetic field is nearly frozen in the ionized component, it is possible to make a rough estimate of the initial radius $r_o = r_1\ (B_o B_1)^{1/2}$, where r_1 and B_1 are the radius and the magnetic field intensity of the umbra, respectively, taking r_1 = 1000 km and B_1 = 1000G and B_o = 10G, $r_o \sim$ 10,000 km. If the formation of a very small spot can be achieved in one day, the required inward speed is about 100 m/sec.

As one vortex grows in the area of one polarity, both the inward flow and the vortex motion will be transmitted to the conjugate area along the magnetic field lines, which must exist looping across the polarity reversal boundary. As a

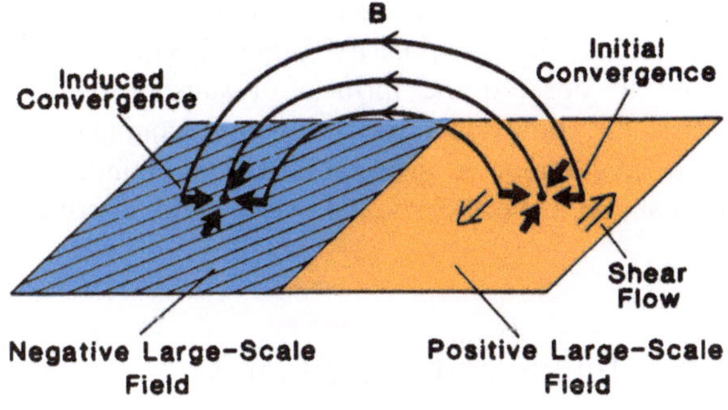

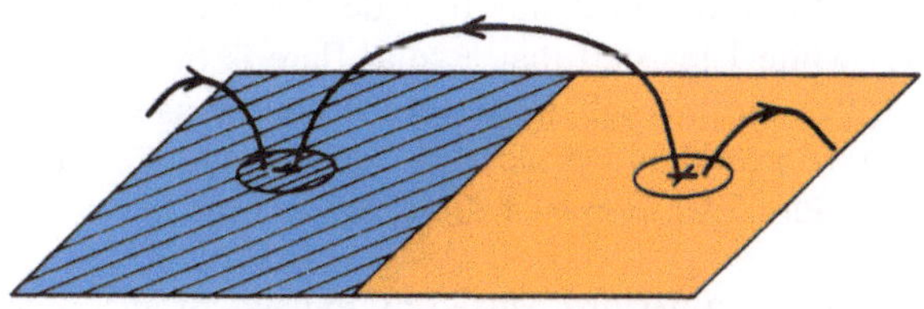

SUNSPOT PAIR FORMATION

FIGURE 7.8a. Schematic illustrations suggesting how a pair of sunspots might form across the polarity reversal boundary. The converging effect in the positive large-scale field is communicated along the magnetic field lines to the conjugate area, forming a spot of the opposite polarity.
Source: Akasofu, S.-I., *Planet. Space Sci.*, **33**, 275, 1985

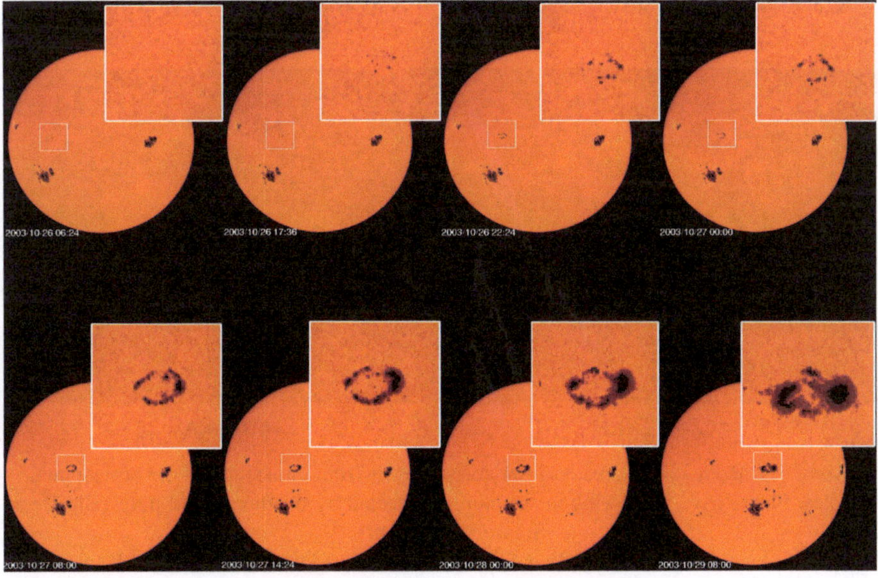

FIGURE 7.8b. Birth of a sunspot pair: NASA.
Source: http://sohowww.nascom.nasa.gov/

result, the magnetic flux will also be concentrated in the conjugate area, inducing and forming a spot of the opposite polarity (Figure 7.8a).

It is rather rare to observe the birth of a sunspot pair from the very beginning. Figure 7.8b shows that the following (conjugate) spot is not a single spot at the beginning; it grows along the boundary of the supergranulation; this can be expected by the weak field distribution shown in Figure 7.3d.

7.5. Force-free Fields and Solar Flares

In his Caltech office, Hal Zirin and I once ran both his solar flare movie and my all-sky aurora movie together side by side. He commented that the aurora is an *Earth flare*, while I insisted that a solar flare is the *Solar aurora*. In any case, a flare and the aurora result from optical emissions of the atmosphere of the Sun and Earth, respectively. In fact, there are many similarities between the two phenomena (Figure 7.9). Some of them are:

1. Both are atmospheric emissions caused by impacts of energetic electrons.
2. Both appear in a ribbon-like form (actually curtain-like).
3. Both appear at the feet of magnetic field line arches.
4. Both are associated with a great variety of electromagnetic processes, requiring a dynamo process to power them.
5. Both are associated with the field-aligned currents (see following).

Solar physicists agree in general that the *force-free* field ($\boldsymbol{J} \times \boldsymbol{B} = 0$) is vital in understanding solar activities. In a force-free field, electric currents flow along

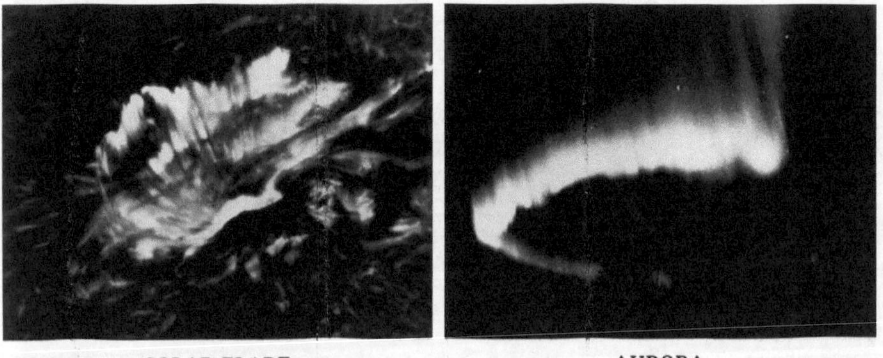

FIGURE 7.9. Both solar flares and the aurora result from emissions of the atmosphere particles. Flares have a two-ribbon structure connected by magnetic field lines (Big Bear Solar Observatory).
Source: Solar flare photo: Big Bear Solar Observatory; Aurora photograph: Kanazawa Astronomical Society

the magnetic field lines. The northern and southern auroras are connected by the geomagnetic field lines.

There have been countless papers on force-free field configurations, that simply solve $(\nabla \times \boldsymbol{B}) \times \boldsymbol{B} = 0$ under various conditions, but there have been hardly any papers on how force-free fields can be generated, and more specifically how field-aligned currents $J_{||}$ can be generated. The dynamo process associated with vortex flows is an important element in generating field-aligned currents $J_{||}$ and the resulting force-free and sheared fields near active sunspots (Figure 7.6). Solar physicists have become accustomed to considering solar activity in terms of a magnetic flux tube and thus bypassing the processes that produce the tubes. They must first consider how long and thin magnetic flux tubes of various sizes and various orientations can be generated below the photosphere. A thin magnetic tube requires solenoidal currents, and it is interesting to see what processes are hypothesized to be responsible for the solenoidal currents. In this context, Alfvén was treated as a maverick again by insisting on the need of considering electric currents for solar flare processes. However, solar physicists responded that there is no $\boldsymbol{J}$ term in their MHD equations ($\boldsymbol{J}$ is converted into $\nabla \times \boldsymbol{B}$). It is quite obvious that there is no rigid electrical circuit in the solar atmosphere. However, this does not mean we should forget the physics involved in generating $J_{||}$.

Recall that a force-free field indicates nothing but the presence of field-aligned currents $J_{||}$ in the solar atmosphere (Figure 7.6). In either solar conditions or magnetospheric conditions, the field-aligned current $J_{||}$ must be related by the dynamo process:

$$J_{||} = -\nabla \bullet \boldsymbol{J}$$

where $\boldsymbol{J}$ must be generated by the dynamo process $\mathbf{V} \times \boldsymbol{B}$, so that $\boldsymbol{J}$ must be perpendicular to $\boldsymbol{B}$. Thus, if a force-free field is crucial for solar flares, the dynamo process must be involved in generating $J_{||}$. It is unfortunate that such a dynamo process and thus the ultimate energy source for solar flares are ignored in solar physics.

It is important to note that the field-aligned currents distort the magnetic field configuration from the potential fields. In a simple case of a pair of sunspots, consider a vertical plane that contains the two spots. There is a loop of magnetic field line contained in this plane. The projection of this particular field line onto the plane horizontal to the vertical plane is a straight line that connects the two spots. The dynamo process around the two spots tends to distort the field in such a way that the projected straight line becomes an S-shaped character. This feature will be discussed further in terms of sigmoids in Section 8.4. Such a magnetic field is called a sheared magnetic field. Magnetic energy stored in the distorted portion of the magnetic field is expendable, not the potential portion of the field. The degree of the distortion provides a measure of the amount of expendable magnetic energy.

If the stored magnetic energy was expended for solar flares by magnetic reconnection, we would expect a decrease of magnetic shear; this is because the magnetic field configuration would have to relax toward a potential field during a flare. Thus, it is important to examine changes of vector magnetic fields, in particular magnetic shear associated with intense flares, on the basis of a high-resolution transverse and longitudinal magnetic field measurement (Figure 7.10). I proposed to Hal Zirin that it is important to examine how magnetic shear changes at flare onset, see Figure 7.10. Haimin Wang et al. (1993) showed the magnetic shear actually increases after several major flares, Figure 7.11 shows such an example. One can clearly see a large increase of the shear at flare onset.

It is obvious that an enhanced photospheric dynamo process provides energy for both solar flares and the sheared field. This situation is similar to magnetospheric substorms, since the size of the polar cap can increase at substorm onset, indicating that the solar wind-magnetosphere dynamo power goes to both auroral substorms and the magnetotail (Section 3.7).

Haimin Wang, Hal Zirin, and their colleagues reported their finding (Figure 7.11) for five X-class flares in their paper in *Astrophysical Journal* (1994). Toward the end of their paper they noted:

It is interesting to note that the direct driving model for solar flares proposed by Akasofu (1984) predicts that the magnetic shear should increase during solar flares. In this model, the energy source for flares is photospheric motions, which drive a dynamo process,

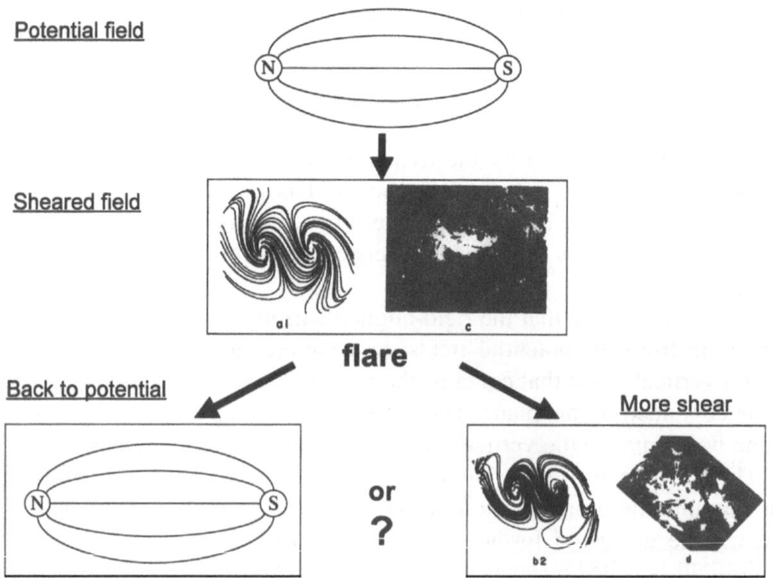

FIGURE 7.10. A potential field is distorted by the dynamo process, producing sheared magnetic field. If magnetic reconnection is the source process for solar flares, the field should relax back to the original potential field. It has actually been found that magnetic shear increases at flare onset.
Source: Akasofu, S.-I.

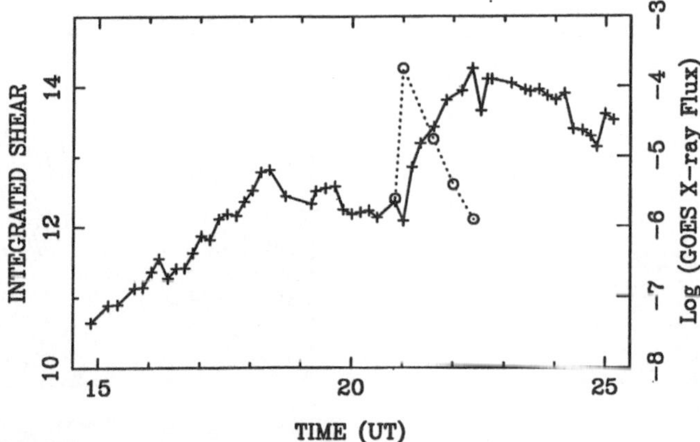

FIGURE 7.11. Changes of magnetic shear at the time of flare onset. Note a sudden increase of the shear, instead of decrease as expected from magnetic reconnection. *Source*: Wang, H., M.W. Ewell, Jr., and H. Zirin, *Astroph. J.*, **424**, 436, 1994

the energy release is due to dissipation of the field-aligned currents, and the amount of energy released as a function of time is directly related to the shear of the magnetic field. Although the model seems to explain our observations, there is a major inconsistency: the increased shear we observe persists well after the flare emission has ceased, whereas the direct driving model predicts that the shear must decrease again after the flare, otherwise the flare emission would continue.

All the later studies of this particular subject show that magnetic shear increases in most of the cases; some show no change, but there is so far no example of decrease of the shear.

Thus, it is likely that solar flares are a *local* phenomenon, where the energy dissipation rate per unit is large, but the photospheric dynamo process must occur in large-scale, so that its energy supply rate of the whole system can overcome a local dissipation.

7.6. Simplest and Most Fundamental Flares

Complex flare observations are reported in practically every issue of solar research journals. It is puzzling why solar physicists do not try to understand simplest cases. In one the simplest cases, flares appear as two parallel "ribbons" in an arcade-like magnetic field configuration, across the line, dividing the magnetic polarity. *No sunspots are present, so no magnetic flux tube is involved. No magnetic flux tube is needed in this case.*

In a model of eruptive prominences by Choe and Lee (1996) they assumed an arcade-like magnetic field configuration (Figure 7.12a). In their model, there is an anti-parallel flow of *photospheric gas* across the centerline of the arcade, namely

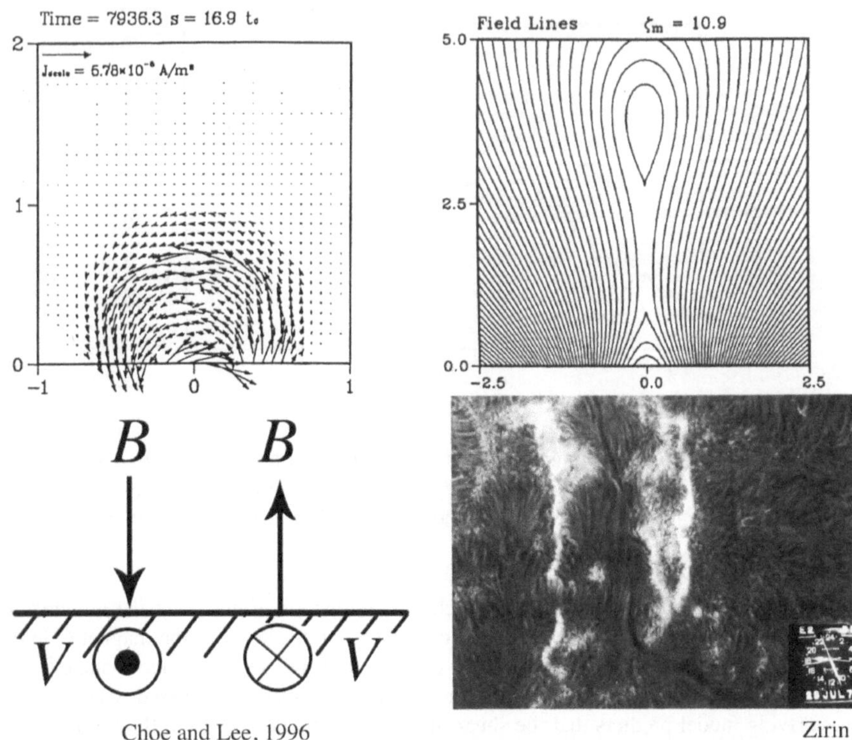

FIGURE 7.12a. A simulation of an erupting prominence and the associated electric current distribution.
Source: Courtesy of C.S. Choe and L.C. Lee (1995)

the dynamo process. The field-aligned currents flow from both sides. The arcade-like magnetic field lines will be distorted, producing a strong magnetic field component along the centerline of the arcade, along which field-aligned currents flow. Such currents tend to produce helical fields. This tendency has been observed.

It is important to note here that such a photospheric dynamo action can indeed cause flare phenomena, including magnetic reconnection *as a result*. Since this case does not involve any sunspots and magnetic flux tubes, magnetic reconnection is not the cause process in this simplest case, even if it is important for various secondary processes. Even if plasma particles are accelerated in the corona as a result, they cannot produce much Hα light. Most of them cannot descend down to the chromosphere, because they are mirrored back as the magnetic field intensity increases downward. Electric field $E_{||}$ and field-aligned currents are needed to accelerate them downward along the field lines, as we learned in magnetospheric physics.

Two important points here are: (1) a dynamo process in the photosphere can cause a complex field-aligned current system to develop, causing "shear" in the arch-like structure, which is known to be associated with flare phenomena.

(2) the upward field-aligned currents carried by downward-flowing electrons could cause an optical flare phenomenon.

In this respect, Kusano (2005) showed an interesting 3-D simulation of a flare process for a situation similar to the case studied by Choe and Lee. In one of his simulations, he drove an anti-parallel flow along a magnetic arch. As a result, the magnetic arch structure becomes "sheared." At a certain moment in this process, magnetic reconnection occurs at a coronal level and the reconnected field moves away and upward (Figure 7.12b). In this case, the magnetic arch is continuously driven and magnetic shear can continue to increase, so long as the anti-parallel flow continues.

Thus, at least in some cases, flares are not simply due to a relaxation process. This situation is similar to magnetospheric substorms. The solar wind must drive the magnetosphere to cause magnetospheric substorms, providing the power $\varepsilon = VB^2 \sin^4(\theta/2)l_o^{\,2}$, instead of a simple relaxation process in the magnetotail. In fact, the total magnetic flux in the magnetotail can increase during substorms (Chapter 4), but a substorm ends when ε starts decreasing.

Most solar physicists assume that there are magnetic flux tubes below the photosphere and that most active solar phenomena are manifestations of the tube activities, after they are brought above the photosphere by magnetic buoyancy.

It is important that the cases studied by Choe and Lee and Kusano have all the essentials needed for solar flares. It is the simplest case. Thus, it is suggested that the flux tube case may be geometrically analyzed in terms of the simplest cases in terms of basic physics involved. Indeed, the magnetic arcades tend to

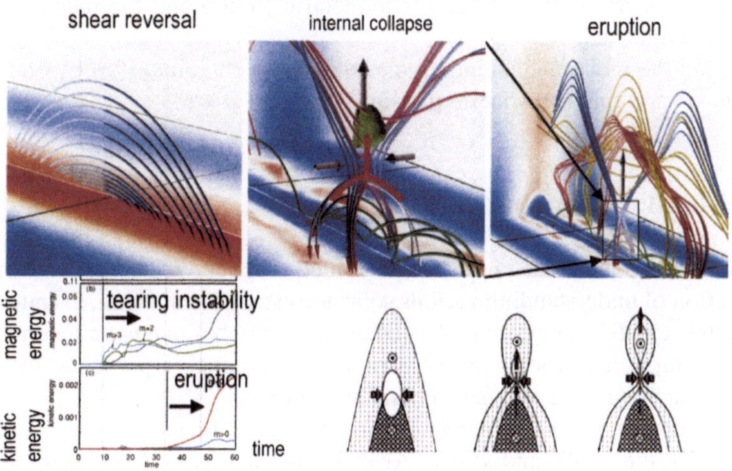

FIGURE 7.12b. A 3-D simulation of a flare process for an arch-like magnetic field configuration, together with an anti-parallel flow: K. Kusano (2005).
Source: Kusano, K, Joint International Workshop on Space Weather, April 4-6, 2005, Tokyo

have a filament along the dividing line of magnetic polarity, and the filament tends to erupt during solar flares. This case will be discussed as "disappearing filaments" in Section 8.2.

As a summary of the subject of solar flares, it is important to find simplest cases that contain the basic ingredients for flare processes. Then, the next step should be to try to analyze complicated cases in terms of the simplest cases.

1. In the models by Choe and Lee and Kusano, the driving mechanism of the flare system is explicitly expressed in terms of a *photospheric* anti-parallel flow along the centerline of an arch-like magnetic field configuration, rather than unknown processes below the photosphere. *Magnetic* flux tubes *are not needed.*
2. The photosphere is directly involved in the driving mechanism, generating $J_{||}$,not merely a passive layer through which magnetic tubes penetrate.
3. Magnetic shear results from the field-aligned currents that cannot be generated under the frozen-in-field condition ($E_{||}$=0). Thus, the dynamo process in the photosphere is crucial for magnetic shear.
4. The energy dissipation rate for the flare processes is a fraction of the energy input rate by the driving mechanism, because magnetic shear increases at flare onset.
5. From (2), it may be said that the flare process is not simply a dissipation process of magnetic energy accumulated prior to flare onset.
6. Magnetic reconnection may occur as a *result* of flare processes, rather than the driving mechanism of flares.
7. Even if magnetic reconnection occurs in the corona, it is not possible to bring down accelerated particles from the reconnection region to the chromosphere, because they are mirrored back. The electric field along the magnetic field lines $E_{||}$ is needed.
8. These are the most fundamental issues that require a joint effort by both solar physicists and magnetospheric physicists.

7.7. Magnetic Reconnection

In Section 1.9, I stated that I intuitively avoided magnetic reconnection from my consideration of understanding various solar-terrestrial phenomena, in spite of the fact that the scientific community has been emphatic that magnetic reconnection is the most important process in this particular discipline, in particular in understanding solar flares and magnetospheric substorms.

The concept of magnetic reconnection originated in an attempt to explain solar flares in the 1950s and 1960s. At that time, it was said that there was no intense flow of the photospheric gases around sunspots, so that the only source of energy that could be tapped for solar flares was magnetic energy. Actually, there are complicated flows around a large sunspot group, but they could not be observed by the Doppler method near the central part of the solar disk.

Magnetic reconnection was considered the process that releases magnetic energy from sunspots. Since solar flares appear to be explosive (in a speeded-up flare movie!), it was concluded that magnetic reconnection must be explosive. Somehow, most researchers in this field began to believe that magnetic reconnection must be an explosive process. Perhaps, most theorists learned about solar flares from speeded-up movies and have not had an opportunity to observe ones on a real-time basis; it is actually a rather slowly (boringly!) developing phenomenon.

Thus, theorists considered an idealized situation in which an anti-parallel magnetic field configuration is given as the pre-existing condition to be annihilated explosively. It did not matter for them whether a static, steady anti-parallel magnetic field configuration would exist in the turbulent solar atmosphere. It was a great surprise for the theorists to find that it is very difficult for this configuration to annihilate itself explosively. Hundreds of papers were published claiming that this difficulty could be removed. If flares are not an explosive phenomenon, such an exercise was not needed. As I mentioned earlier, it is my feeling that if a static, steady anti-parallel magnetic configuration could be produced to begin with, it would be so stable that an explosive annihilation would not be possible.

The only possible way to produce such a field configuration may be to *drive* two mutually anti-parallel fields toward each other. Thus, magnetic reconnection might occur only under a driven condition. The problem is that energy released from such a process may be exactly the same as that needed to drive the two fields together, so that no extra energy can be released from the anti-parallel field configuration itself in an *annihilation* process. This may be what happens on the dayside magnetopause, where the southward-oriented interplanetary magnetic field interacts with the northward-oriented magnetic field of the magnetosphere. In this situation, no one has claimed magnetic reconnection on the front of the magnetosphere as the magnetic energy release process.

Furthermore, since expendable magnetic energy in the vicinity of sunspot groups exists only in the non-potential component, hundreds of papers were written on force-free fields, namely on the solution of $(\nabla \times \boldsymbol{B}) \times \boldsymbol{B} = \boldsymbol{0}$. However, it appears that many researchers forget that force-free fields arise from field-aligned currents ($J_{\parallel}$) and that $J_{\parallel}$ must be produced by a dynamo process, which requires the flow of photospheric plasma across magnetic field lines.

When the magnetotail and its near-anti-parallel magnetic field configuration were discovered, many researchers thought that the magnetic energy in the magnetotail was the source of energy for magnetospheric substorms. It was said at that time that there was more than enough magnetic energy for 30 substorms. Those researchers faced the same problem as solar physicists. That is, they found that it is difficult to release energy explosively from the magnetotail. Further, I recall I asked them to consider how to stop explosive magnetic reconnection if they could succeed in initiating it. Otherwise, as mentioned earlier, the whole magnetotail will be in effect burnt up, and we know this does not happen. No one has made any study of this point!

Unfortunately, most solar and magnetospheric physicists, both theorists and experimenters, have believed that magnetic reconnection must occur in spite of such a theoretical difficulty and that the resulting X-line (the neutral line) would explain practically all substorm features, including the sudden brightening of an arc at substorm onset and the acceleration of auroral electrons. These claims have not been substantiated. This trend has considerably retarded the progress of our discipline. It took more than a decade to convince the scientific community that the magnetosphere must be driven each time for a magnetospheric substorm; magnetospheric substorms do not arise *spontaneously* from an explosive release of magnetic energy that is steadily accumulated in the magnetotail. Rather, as we learned in Section 1.9, a substorm occurs after the solar wind-magnetosphere dynamo power is substantially increased for an hour or so.

It appears that magnetic reconnection is considered to be a major process for solar physicists, which can produce all the desired phenomena associated with transient solar activities. At least some magnetospheric physicists have learned that magnetic reconnection is not the magic process. Furthermore, even if magnetic reconnection occurs as a by-product well above the photosphere, it is difficult to bring energized particles to the chromosphere and the photosphere for optical phenomena associated with solar flares. Field-aligned currents are essential in bringing down electrons.

It is understandable that scientists do not renounce easily what they have been taught and what they have based their research on, even if a new finding does not conform to what they believe in. They consider that such a finding is not credible and discredit the new finding. They begin to lose faith only when many more new findings are inconsistent with that they believe in.

It is my hope that both solar physicists and magnetospheric physicists can learn more from each other, even if solar flares and magnetospheric substorms may be found to have different causes.

We can conclude this chapter by summarizing what we have learned as follows:

(1) Both solar flares and magnetospheric substorms are manifestations of electromagnetic energy dissipation processes.
(2) Both must be driven directly by a dynamo process.
(3) Field-aligned currents are essential for both phenomena.
(4) Magnetic reconnection is not a magic process for solar flares and magnetospheric substorms. The formation of the X-line alone cannot explain the acceleration processes of plasma particles and the field-aligned currents.

8
Space Weather Research

8.1. Introduction

Our study of the solar–terrestrial relationship has developed into four major disciplines: solar physics, interplanetary physics, magnetospheric physics, and aeronomy (upper atmospheric physics). Researchers in each discipline have made considerable progress within their own field of study, although many challenging problems have still been left unsolved. Meanwhile, there has been much discussion about space weather research in recent years. The term *weather* in this context implies that the goal of *space weather* researchers should be able to *forecast* space weather and, at least, predict geomagnetic storms in terms of the two geomagnetic indices Dst and AE as a function of time after transient solar activities are observed. Geomagnetic storms are one of the manifestations of magnetospheric storms, resulting from electromagnetic disturbances around the magnetosphere, which are caused by a variety of solar wind disturbances. Another manifestation is auroral storms.

This chapter deals with space weather. Although it is closely related to solar physics, interplanetary physics, magnetospheric physics, and aeronomy, space weather research is to study how various types of transient solar activities propagate in interplanetary space and cause magnetospheric storms. Readers interested in each of the four subject areas should refer to monographs of the individual subjects. Recurrent geomagnetic disturbances were already described in Chapter 6.

To be successful in this particular effort of forecasting, space weather researchers needs to establish a new discipline that synthesizes and integrates the four major disciplines. The progress in space weather research is much more than future progress in each of the four disciplines alone. This chapter describes a research scheme needed for the success of predicting geomagnetic storms by presenting an example of this synthesis process.

However, many researchers are naturally interested in one or two subjects in solar-terrestrial physics, but very few have been interested in integrating all the subjects. The point to be made here is that it is not possible to make significant progress in predicting geomagnetic storms unless space weather researchers work together on all the above issues. In the 1970s, Kazuyuki Hakamada, Ghee Fry, and I initiated the integration of the four disciplines for space weather research.

It is encouraging to see that a number of efforts are now being made along this line, particularly in terms of the MHD modeling.

The reason I initiated space weather research in the 1970s, well before this particular research had become so popular among space physicists, was that Colonel Lee Snyder was sent from the U.S. Air Force to study under me for his graduate research. After his graduate work, he became one of the personnel who was responsible for an over-the-horizon (OTH) radar in Maine for Russian bomber surveillance; the problem was that the OTH radar could not detect the bombers when an intense auroral storm was in progress, so it became necessary to predict the occurrence of auroral storms in order to avoid a surprise attack. For these reasons, Hakamada, Fry (also sent by the U.S. Air Force to the University of Alaska), and I developed an initial space weather prediction scheme that is often referred to as the Hakamada-Akasofu-Fry (HAF) model; its improved version is called the HAFv2 model; in a sense, it was a product of the Cold War. During this project period, I was strongly convinced that it would not be possible to succeed in predicting geomagnetic/auroral storms without integrating the four fields. There are at present many missing links among the four disciplines that are needed in order to succeed in space weather prediction.

In this chapter, we examine three particular types of solar activity, which result in a phenomenon called "Coronal Mass Ejections, (CMEs)." Figure 8.1 shows an example of CME. The three known causes of solar activity related CMEs are: (1) disappearance of dark filaments (DBs) seen on the solar disk, (2) sigmoid eruption, and (3) eruptions of trans-equatorial loops. They are thought to produce CMEs, which are considered to produce a particular magnetic structure called "magnetic flux rope" which is observed in interplanetary space. However, these relationships are still a matter of intense debate.

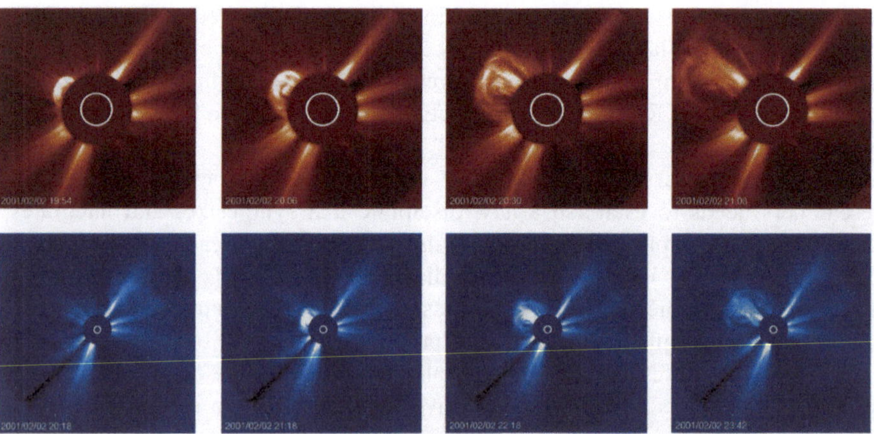

FIGURE 8.1. A typical example of expanding CME after a solar flare. The circles indicate the size of the sun.
Source: NASA/ESA SOHO Project

The relationship between prominence /filaments /sigmoids /archades /helmet streamers and CMEs belongs to the realm of solar physics. There are a very large number of papers on these subjects, and it is well beyond the scope of this book to review them, except to note that it appears to be difficult to find a one-to-one relationship among them. Some CMEs are associated with flares and sigmoids, but some others are not. Disappearing filaments (DBs) are often associated with CMEs, but not always. It is crucial to find the minimum number of the common phenomena related to this subject; see Chapter 9.

On the other hand, since the eruption of sigmoids has been discussed in terms of magnetic reconnection (Chapter 7), it may produce plasma clouds, magnetically detached from the sun. Such a structure is defined here as a plasmoid in order to distinguish it from flux ropes.

8.2. Disappearing Filaments (DBs) and their Magnetic Field Structure

As mentioned earlier, there are three types of solar activities that are related to CMEs. The first type is associated with DBs. Filaments appear as dark thin structures against the bright solar disk; they are often located in relatively high latitudes and cause two-ribbon flares of the type discussed in terms of anti-parallel photospheric flow in Section 7.5. Disappearing filaments leave behind archades (and helmet streamers). Figures 8.2a–8.2c show examples of the DBs. It may be noted that filaments in the vicinity of sigmoids may also disappear at

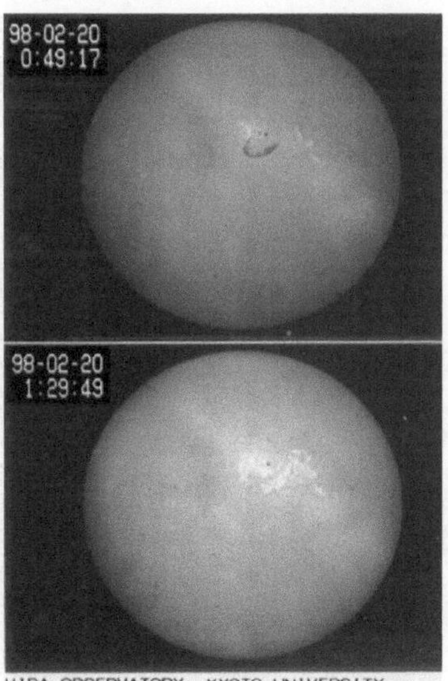

FIGURE 8.2a. A disappearing filament (DB) on the solar disk.
Source: Kyoto University Solar Observatory

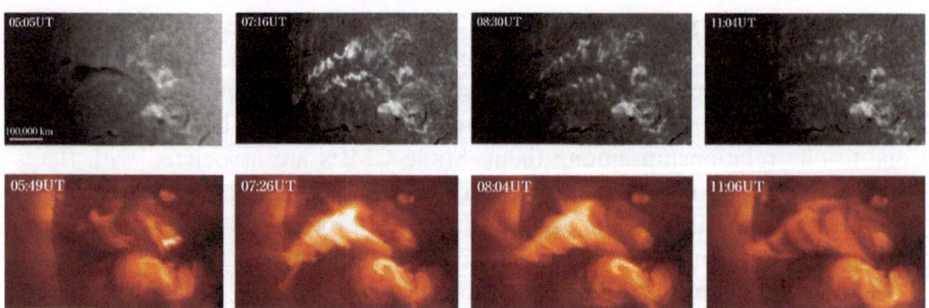

FIGURE 8.2b. A disappearing filament (DB) and the associated flare and arch-like structure.
Source: Shibata, K. and M. Oyama, *Photographic Atlas of the Sun* (in Japanese), Shokabo, 2004

the time of their eruption. A study of DBs may provide some basic clues on the causes of solar flares and CMEs as the simplest case (Section 7.6).

Hildner and Marubashi (1986) found an indication that DBs are related to magnetic flux ropes observed in interplanetary space. Bothmer and R. Schwenn (1994) examined the magnetic structure of the filaments and the observed the magnetic field at the location of the Helios 1 and 2 space probes (at a distance of ~0.5 AU from the Sun), and carefully identified the polarity coincidence. They found that the helical structure observed around both filaments and flux ropes is similar and thus concluded that the flux ropes are expanding

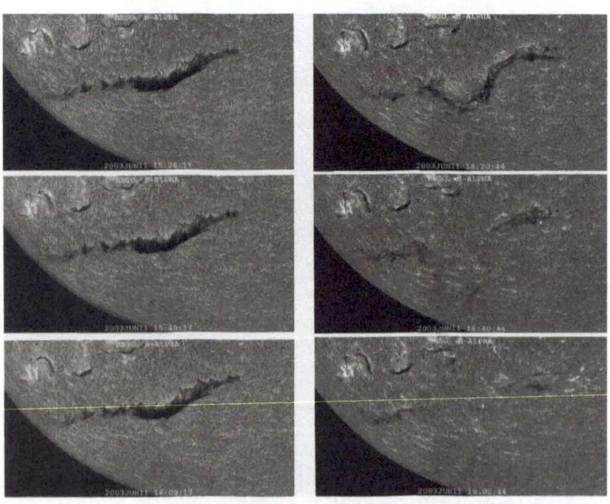

FIGURE 8.2c. An example of disappearing filament (DB).
Source: Yurchyshyn, V., Earth-Sun System Exploration: Energy Transfer, January 16-20, 2006, Kona, Hawaii

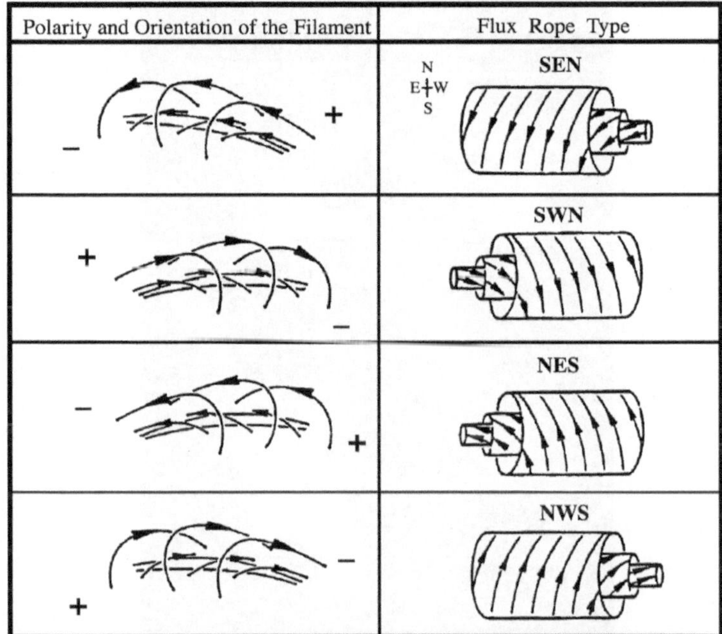

FIGURE 8.3. The magnetic configuration of filaments and the flux ropes. *Source*: Bothmer, V. and R. Schwenn, *Space Sci. Rev.*, **70**, 215, 1994

filaments (Figure 8.3). Note that the electric current configuration in Figure 7.12a can produce magnetic fields along the direction of the centerline, illustrated in Figure 8.3.

The relationship between DBs and CMEs has been extensively studied, but it is not the purpose here to review the extensive literature on this subject. Here, we only show two examples of this study. Figure 8.4a shows an example of their association. Figure 8.4b suggests that a filament and an erupted flux rope are located at the bottom of a CME.

8.3. Sigmoids and Magnetic Flux Ropes

The second type of activity related to CMEs is the eruption of sigmoids, which result from complex motions of sunspots and the sheared magnetic field structure as described in Chapter 7. Figure 8.5 shows an example of sigmoids. It is thought that the eruption of sigmoids produces the magnetic flux ropes. It may well be that a filament or filament-like structure is embedded in the sigmoid, which is ejected at the time of the eruption. Unfortunately, satellite observations of the magnetic fields in the disturbed solar wind alone cannot tell us conclusively if

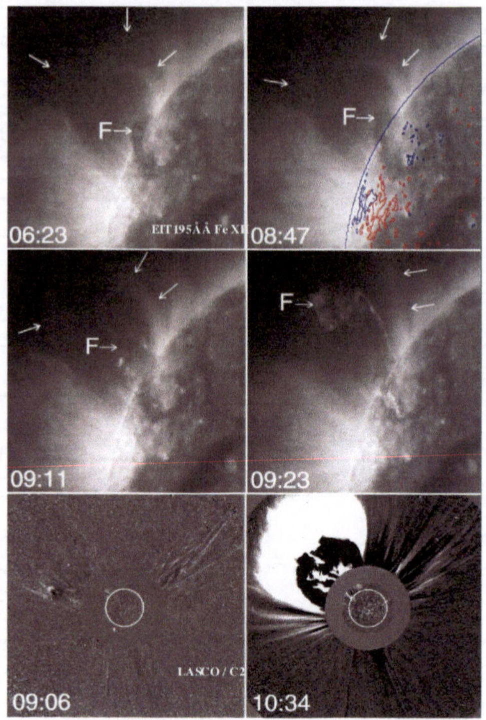

FIGURE 8.4a. An example of onset of a filament eruption and the resulting CME. *Source*: Moore, R.L., A.C. Sterling, and G.C. Marshall, Space Flight Center Report, 2005, M.K. Georgoulis, Huntsville, Alabama

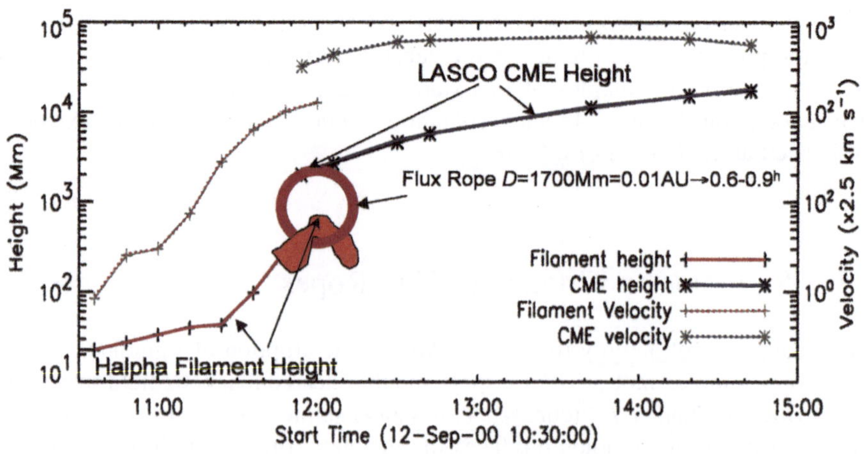

FIGURE 8.4b. Time variations of the velocity and height of both filament and CME. The filament is located at the bottom of an erupted flux rope.
Source: Qiu, J., H. Wang, C.Z. Cheng, and D.E. Gary, *Ap. J.*, **604**, 900, 2004

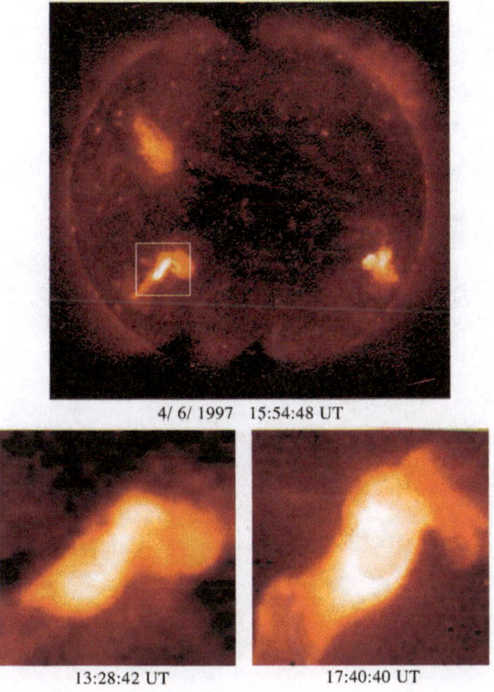

FIGURE 8.5. An example of sigmoid.
Source: Shibata, K., and M. Oyama, *Photographic Atlas of the Sun* (in Japanese), Shokabo, 2004

the observed fields are a cross-section of the flux rope (magnetically rooted in the photosphere) or a plasmoid (magnetically isolated from the photosphere).

There are so far only a few case studies of the relationship between the north-south component of magnetic fields of the sigmoids and the IMF *Bz*. Here, we examine two cases that were studied by Yurchyshyn et al. (2001), in which the north-south component of sigmoids was consistent with the polarity of the IMF *Bz*. Figures 8.6a and 8.6b and Figures 8.7a and 8.7b show their results. In Figures 8.6a and 8.6b, it can be seen that a flare occurred along the line dividing the magnetic polarity (the northern side was positive and the southern side was negative), so that the connecting field lines had a large southward component. The corresponding IMF change had first a southward and then a northward change, suggesting the passage of a flux rope-like structure slightly below the Earth. An intense geomagnetic storm developed during the passage of the IMF southward component. Magnetic flux ropes will be discussed in Section 8.6.

On the other hand, Figures 8.7a and 8.7b show more or less an opposite case. In both cases, the IMF was not simply an expansion of the photospheric fields; the presence of flux ropes imply helical magnetic field structures and field-aligned currents (Section 8.6).

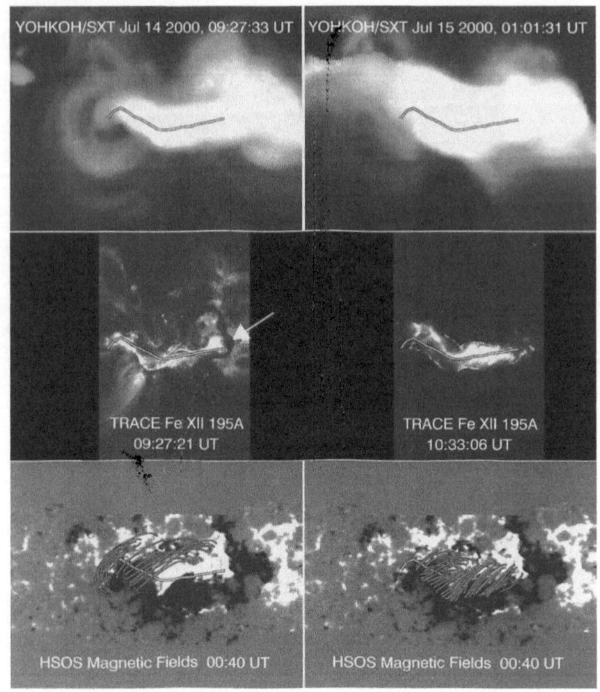

FIGURE 8.6a. An intense solar flare on July 14, 2000.
Source: Yurchyshyn, V., H. Wang, P.R. Goode, and Y. Dang, *Ap.J.*, **563**, 381, 2001

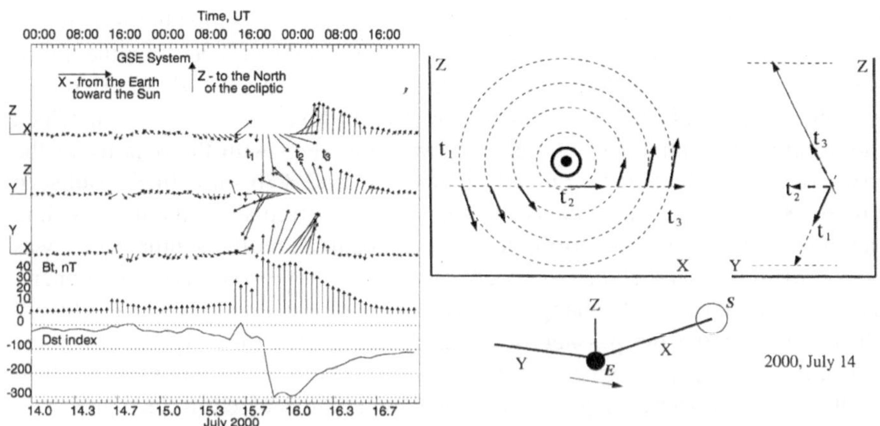

FIGURE 8.6b. The IMF and Dst changes after the July 14, 2000, event on the sun. The loop field observed after the July 14, 2000, event on the sun.
Source: Yurchyshyn, V., H. Wang, P.R. Goode, and Y. Dang, *Ap.J.*, **563**, 381, 2001

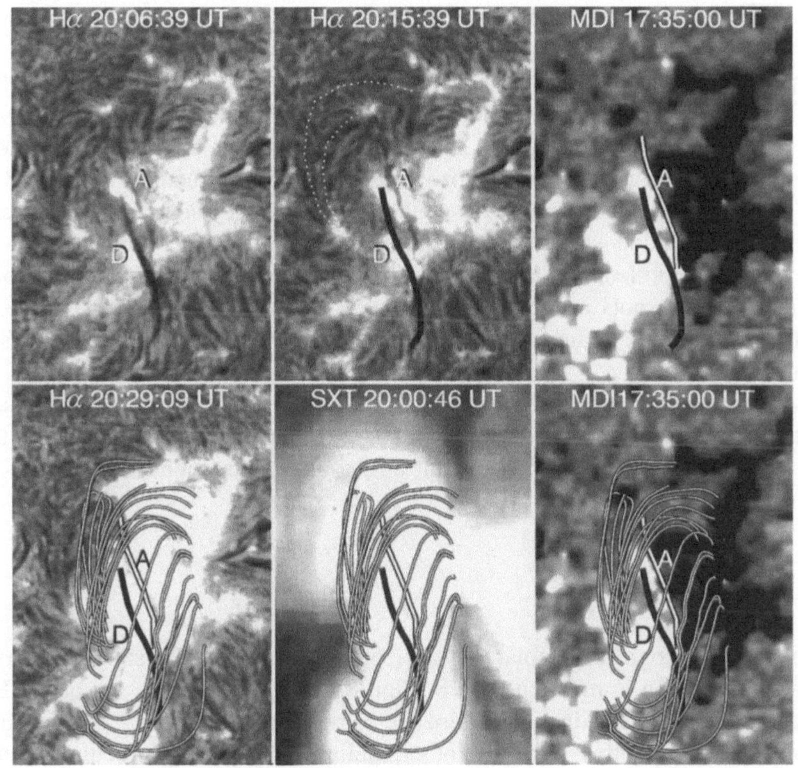

FIGURE 8.7a. An intense solar flare on February 17, 2000.
Source: Yurchyshyn, V., H. Wang, P.R. Goode, and Y. Dang, *Ap.J.*, **563**, 381, 2001

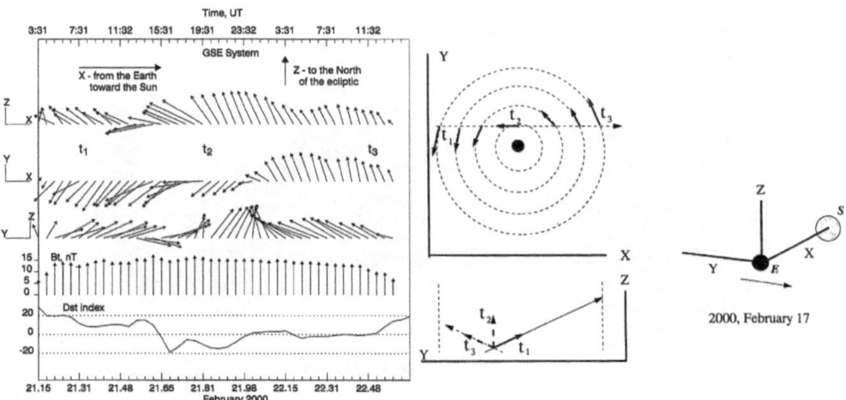

FIGURE 8.7b. The IMF and Dst changes after the February 17, 2000, flare. The loop field observed after the February 17, 2000, flare.
Source: Yurchyshyn, V., H. Wang, P.R. Goode, and Y. Dang, *Ap.J.*, **563**, 381, 2001

8.4. Trans-equatorial Loops

Sunspot group fields in one hemisphere do not necessarily close in that hemisphere. In fact, magnetic field lines from a sunspot group field in one hemisphere are often connected to a sunspot group field in the other hemisphere by loop field lines across the equator in the north-south direction (Figure 8.8a). They are here referred to as trans-equatorial loops (Pevtsov, 2000). We can confirm such a loop structure by examining them near the limb as the sun rotates. As mentioned in Chapter 7, the presence of the trans-equatorial loop indicates that sunspots in the two hemispheres are not independent, suggesting that the traditional explanation of the formation of sunspots is not correct; see Chapter 7.

As it will be shown later, it appears that some of these loops often expand and are expected to reach a distance of 1 AU (Figure 8.8b). In such cases, the loops are expected to have a significant north-south component.

8.5. Halo CMEs

An important progress in solar physics, at least in terms of solar–terrestrial relationship, which is mentioned in Section 8.1, is the identification of CMEs, particularly those that advance toward the Earth; they are called "halo CMEs" (Figure 8.9). They often have a faint expanding annular ring structure around the sun; some halo CMEs move away from the Earth after being launched from the backside of the sun. There has been much discussion on the nature of halo CMEs at the present time. The main issue is whether halo CME is a shockwave-like feature generated by the advancing filament or the filament itself.

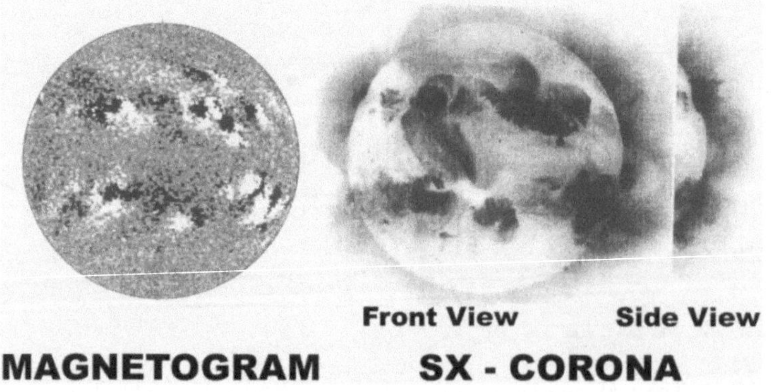

FIGURE 8.8a. An example of the trans-equatorial meridional loop.
Source: Saito, Takao

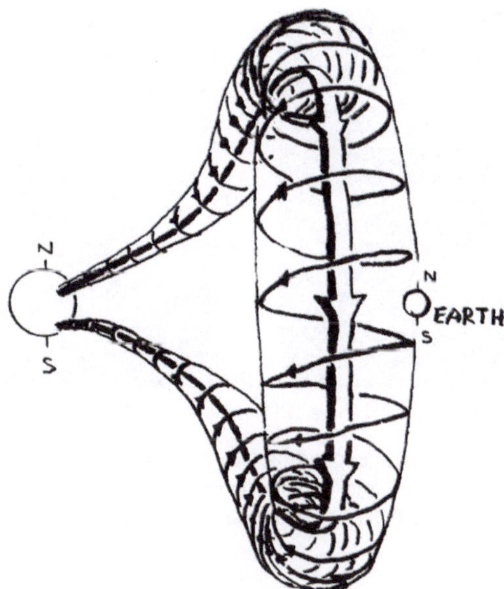

FIGURE 8.8b. Expected structure of the expanding trans-equatorial loop: Takao Saito.
Source: Saito, Takao

Chen and Krall (2003) demonstrated that an expanding flux rope can appear like a halo CME (Figure 8.10a). On the other hand, CMEs originating near the limb often appear like an expanding thin shell. It is clear that more observations and modelings are needed to find the nature of halo CMEs. At least, it may be said that they are not a massive, unstructured plasma cloud. Figure 8.10b shows a modeling of CMEs, including the magnetic field structure in and around a CME.

FIGURE 8.9. An example of "halo CMEs." It is a very faint annular ring that expands away from the sun.
Source: NASA/ESA SOHO Project

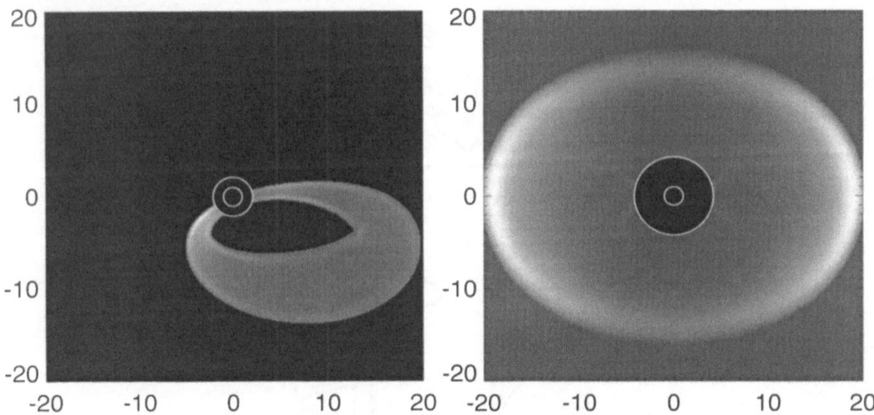

FIGURE 8.10a. Simulation of the expanding flux rope toward the earth (diagram on the left) and its view from the earth.
Source: Chen, J. and J. Krall, *J. Geophys. Res.*, **108**, A11, 1410, 2003

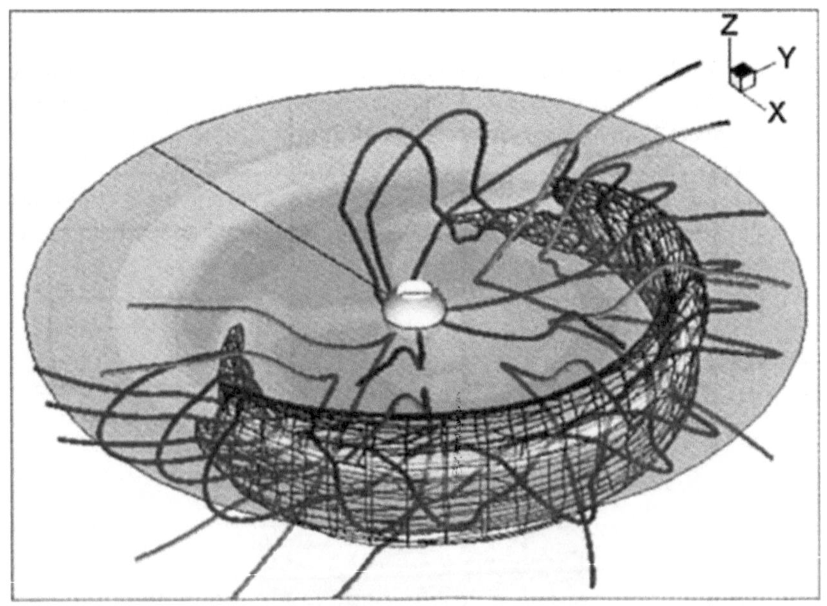

FIGURE 8.10b. A model of CME eruption in three dimensions. Meshed surface is an isosurface of density and the solid grey isosurface marks the location of the heliospheric current sheet.
Source: Riley, P., J.A. Linker, Z. Mikis, and D. Odstricil, *IEEE Trans. Plasma Sci.*, **32**, 1415, 2004

8.6. Flux Ropes Observed in Interplanetary Space

Burlaga et al. (1981), Burlaga and Klein (1982), Marubashi (1986), Burlaga (1988, 1995), and many others have shown that some passing magnetic structures in the disturbed solar wind can be approximated by a cylindrical force-free field ($\boldsymbol{J} \times \boldsymbol{B} = 0$), Figure 8.11. The helical structure they found can most easily be identified by characteristic changes of the north-south component of the IMF, when B, Bx, By, and Bz (or theta) are plotted as a function of time (Figure 8.12). During the passage of such a flux rope, the changes begin with a rather sudden increase of the northward (or southward) component, which is followed by a gradual decrease (or increase); then, the vector turns southward (or northward) and recovers. The first change (northward or southward) is called the leading component. On the other hand, the By component is relatively steady. In some other cases, the changes of By and Bz are interchanged. In the flux rope, the solar wind density tends to be low (Vandas and Geranios, 2001), while the electron temperature Te is higher than the proton temperature Tp (Osherovich and Burlaga, 1997); McAllister et al. (2001) studied one event in detail. Blanco-Cano and Bravo (2001) examined flux ropes in terms of helium enhancement, perhaps indicating that they are driver gases of shock waves, since it has long been considered that the helium enhancement is a sign of the driver gas (Feynman, 1964). If the flux rope is an expanding filament, it may be possible to detect neutral hydrogen atoms, since the filament is not fully ionized (cf. Akasofu, 1975); the neutral hydrogen emissions are observed

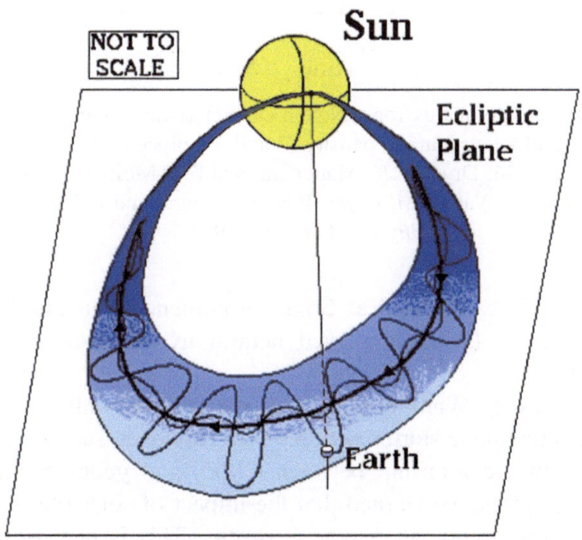

FIGURE 8.11. A schematic diagram of flux rope.
Source: Webb, D.F., R.P. Lepping, L.F. Burlaga, S.E. DeForrest, D.E. Larson, S.F. Martin, S.P. Plunkett, and D.M. Rust, *J. Geophys. Res.*, **105**, 27251, 2000

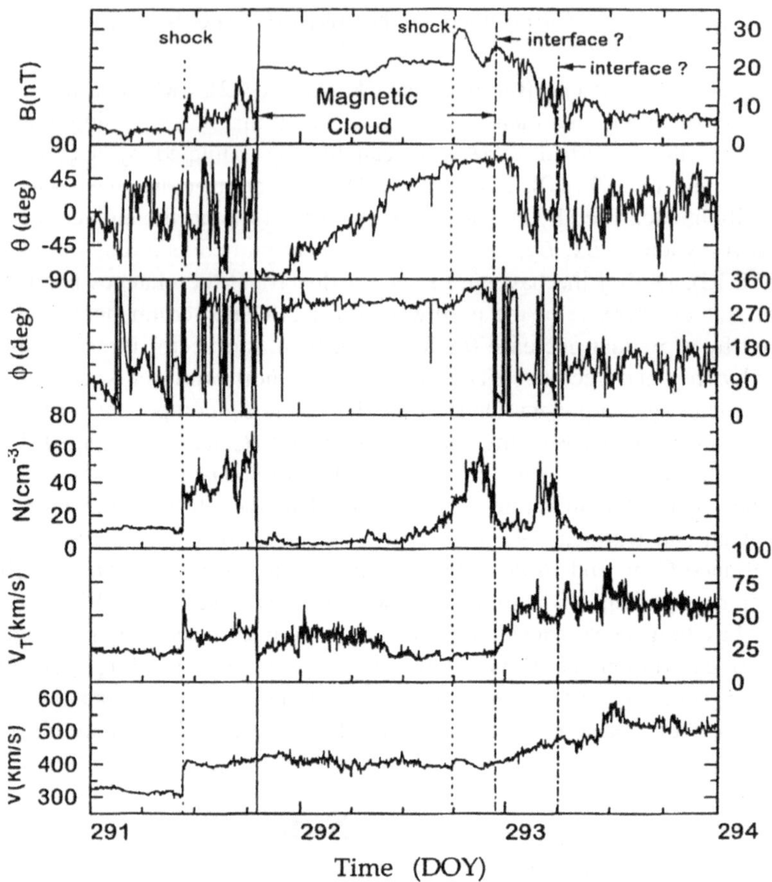

FIGURE 8.12. An example of flux rope: Note a characteristic change of the angle theta, a steady phi angle, and a low number density N and the preceding shockwave: Z. Smith, S. Watari, M. Dryer, P.K. Manohara, and P.S. McIntosh (1997).
Source: Smith Z., S. Watari, M. Dryer, P.K. Manohara, and P.S. McIntosh, *Solar Physics*, **171**, 177, 1997

when the filaments are observed as bright prominences outside the solar disk. In fact, Collier et al. (2001) observed neutral hydrogen atoms in the solar wind (Section 1.7).

A statistical study by Wang et al. (2002) showed that only 45% of 132 halo CMEs caused geomagnetic storms with $Kp > 5$. However, such statistical studies do not disprove the relationship between CMEs and geomagnetic storms. In magnetospheric physics, we learned that the impact of solar plasma ejecta does not necessarily cause major geomagnetic storms. This is because geomagnetic storms require the transfer of kinetic energy of the solar plasma ejecta to the magnetosphere; the north-south component of the IMF Bz and ε play a major role in determining the intensity of geomagnetic storms (Section 1.9).

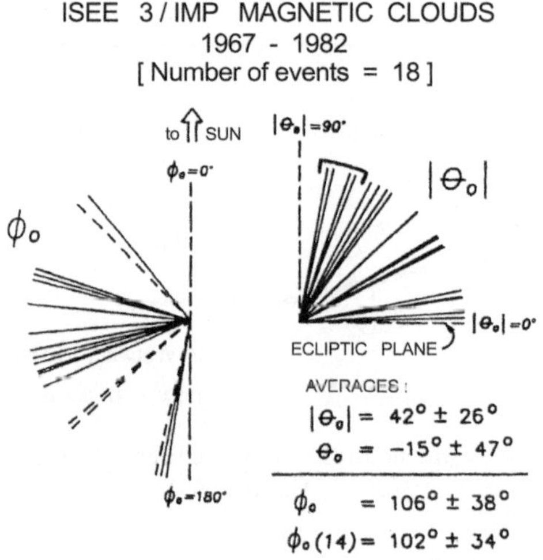

FIGURE 8.13. The orientation of the flux ropes in interplanetary space.
Source: Lepping, R.P., J.A. Jones, and L.F. Burlaga, *J. Geophys. Res.*, **95**, 11957, 1990

Many studies of flux ropes consider that flux ropes lie near the ecliptic plane, originating from a rather small, active area that is located in the vicinity of a sunspot group in one hemisphere (Figure 8.11). In Section 8.4, we discussed the possibility of magnetic loops that bridge sunspot groups in the northern and southern hemispheres across the equator. Figure 8.13 shows the observed orientation of the flux ropes, which were examined by Lepping et al. (1990). It is likely that highly inclined flux ropes result from the expanding trans-equatorial loop.

It is interesting to estimate roughly the total current that flows in the flux rope. Let us consider that the cylinder passes by the Earth in 12 hours with a speed of 700 km/sec. Thus, the radius of the cylinder can be estimated to be 1.5×10^{10} m. At the surface of the cylinder, B is about 10^{-8} web/m^2(= 10 nT). The estimated total current is about 10^9 amperes (Figure 8.14). Thus, a force-free cylindrical structure may require about 10^9 amperes of current along the flux rope. This rough estimate agrees well with the estimate, 1.13×10^9 A, made by Leamon et al. (2004), although this value is probably much less than the currents that are present in the vicinity of solar flares.

8.7. Parameterizing Solar Events

In this section, a simple modeling effort called the HAF model is presented. In the HAF scheme, a solar event is represented by a high-speed source area on the source surface, which is superposed on the background structure described in Chapter 6. The source is represented by a circular area (or an elliptical area);

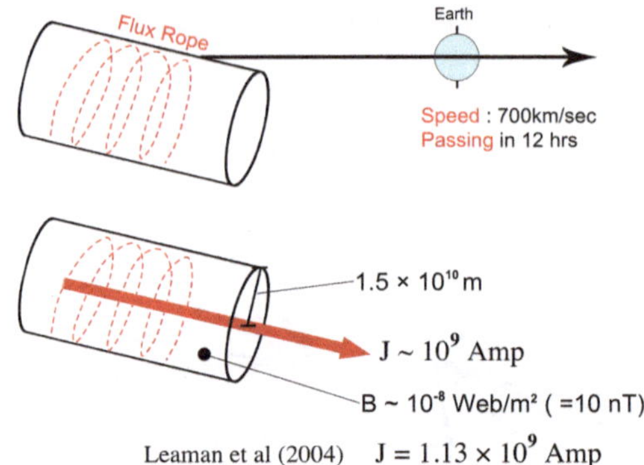

FIGURE 8.14. An estimate of the electric current along the flux rope. *Source*: Akasofu, S.-I.

the speed is highest at the center and has a Gaussian distribution. The speed at the center varies in time in a characteristic way, which is parameterized by

$$V = V_F t e^{-t/\tau_F}$$

Thus, a solar event is parameterized by the maximum speed V_F at the center at the peak of the event, the area size σ_F and the time variations τ_F, together with longitude λ_F and latitude φ_F and the start time T_F of the event. Figure 8.15a shows graphically the adopted parameters on the source surface for a hypothetical event, $\lambda_F = 180°$, $\varphi_F = 0°$, $V_F = 500\,\text{km/sec}$, $\sigma_F = 30°$, and $\tau_F = 12\,\text{hours}$. This particular set of λ_F and φ_F assumes that the event takes place on the center of the disk of the Sun, and at 12 UT on December 8. Therefore, in this particular case, the center of the Sun, and the location of the solar event and the Earth, are on the same solar radial line at the onset of the solar event.

The parameterization is obviously crucial for any modeling effort, including a MHD method. For example, a new HAF model (called "HAFv2 model") requires the information on the Type II radio emission, from which the initial speed of the shock wave is determined. Solar, interplanetary, and magnetospheric physicists must work together to find out what is the most suitable phenomenon to be used in determining the initial speed. Such issues have not been discussed much in space weather research; solar physicists are only interested in what happens on the Sun, while magnetospheric physicists consider their problems only when solar disturbances reach near the magnetosphere.

Figure 8.15b shows the propagation of the resulting shock wave in the equatorial plane at 0, 6, 12, and 18 UT, on December 10, at 36, 42, 48, and 54 hours, respectively, after the hypothetical event on December 8 (Figure 8.15a). The Earth's location is indicated by a star. Since the event is assumed to occur on the center of the disk, the center of the shock wave is propagating approximately along the Sun–Earth line. Figure 8.15c shows an example of three successive shock waves observed by three space probes and the results based on the HAF model. The HAF model can handle not only one, but also a series of flare events.

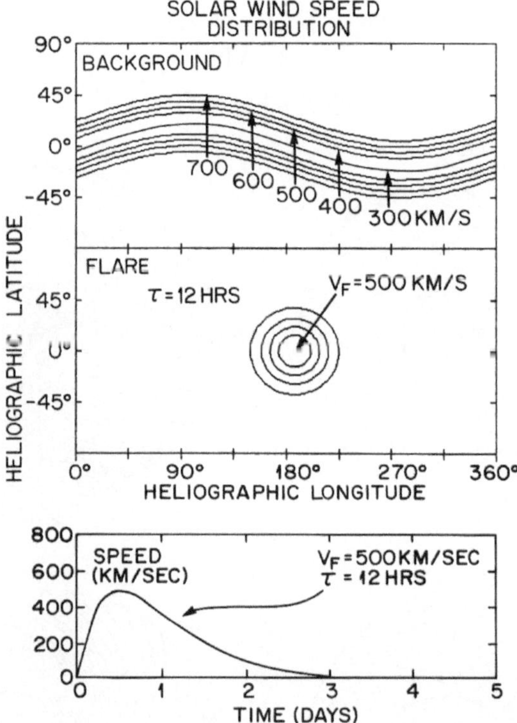

FIGURE 8.15a. From top: The background flow speed of the solar wind on the source surface (the solar longitude-latitude map); the solar wind speed is minimum (300 km/sec) at the magnetic equator and increases toward higher latitudes. The middle diagram represents a solar event; a higher speed area is added to the background flow; the speed reaches $V_F = 500\,\text{km/sec}$ after flare onset and decreases slowly ($\tau_F = 12\,\text{hr}$). The bottom diagram shows how the speed at the center of the circular area in the middle diagram varies in time (expressed by the parameter $\tau_F = 12$ hours in this particular case). *Source*: Akasofu, S.-I. and C.F. Fry, *Planet Space Sci.,* **34**, 77, 1986

During an active period of the Sun, a number of shock waves generated by flares and the co-rotating structure interact in a very complicated way. Figure 8.15c shows that the HAF scheme is useful at least as a guide to interpret solar wind variations during such a period, since it does not require an elaborate computational effort. In the next section, we show an example of the HAF modeling during one of the most active periods of the Sun.

8.8. An Example of the HAF Modeling: March 2001 Case

A large sunspot group appeared near the eastern limb of the Sun during the last week of March 2001 and reached the central meridian on about March 28. Almost 20 solar transient activities occurred from March 28 to April 18. Some

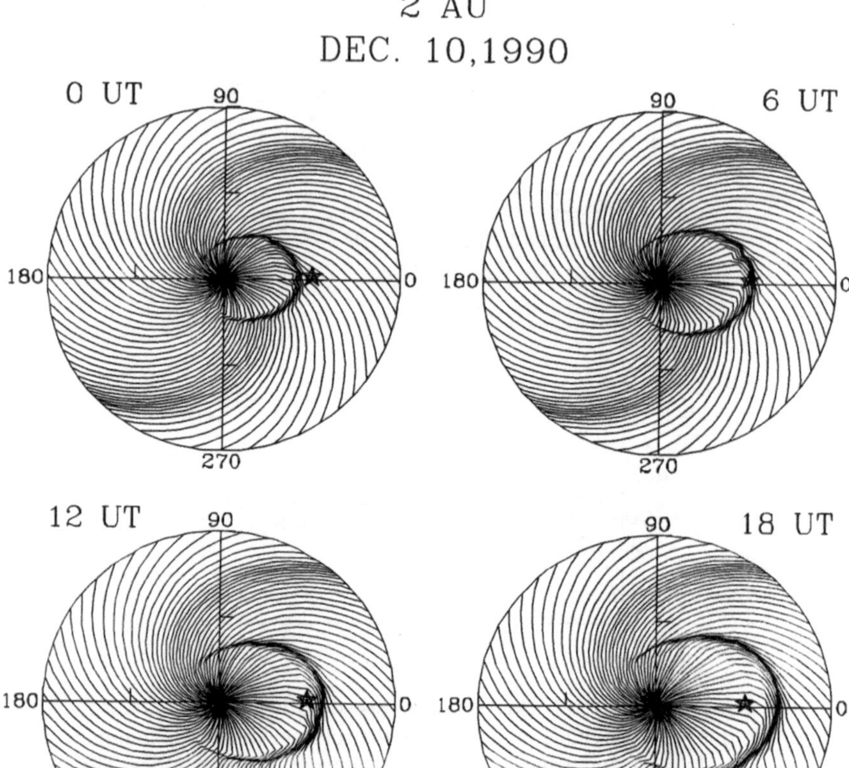

FIGURE 8.15b. The propagating shock wave generated by the growth of the high-speed flow (the middle diagram in Figure 8.15a) on the equatorial plane (the circular area of radius 2 AU). The Sun is located at the center. The location of the Earth is indicated by a star. The shock wave is represented by the IMF field lines distorted by the propagating shock wave.
Source: Akasofu, S.-I. and C.F. Fry, *Planet Space Sci.,* **34**, 77, 1986

of them caused intense interplanetary shock waves. An advanced version of the HAF model was used to simulate the propagation of the shock waves.

One of the largest transient events occurred in the vicinity of a large sunspot group (Figures 8.16a and 8.16b, at 1004 UT, on March 29). It was an X1.7 flare at longitude 12° West and latitude of 16° North on the solar disk. In Figure 8.17a, the simulated patterns of the interplanetary condition to a distance of 2 AU are

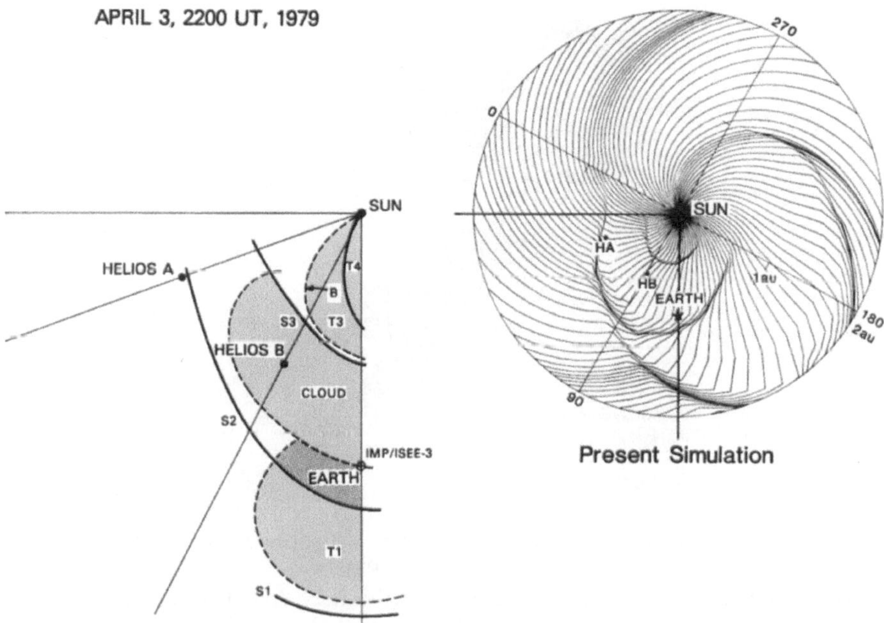

FIGURE 8.15c. Simulation (hind-casting) of the August 1979 events (right), studied by L.F. Burlaga, K.W. Behannon, and L.W. Klein (1981) who inferred the geometry of three shock waves S1, S2, and S3 (left); the geometry of the shock waves in one quarter area of interplanetary space was inferred on the basis of the observations at HELIOS A, B, and the Earth-bound satellites (IMP/ISEE-2).
Source: Akasofu, S.-I., *EOS*, **77**, 225, 1996

shown for the series of events between March 29 and April 22, 2001. In each circular pattern, the familiar spiral pattern and its deformation by the propagation of the shock waves are evident. The red lines show the field vector pointing away from the Sun, while the blue lines show those pointing toward the Sun.

It can be seen that a few shock waves merged together before reaching the Earth. This case shows the importance of the background solar wind flow pattern prior to the onset of solar activity; in the advanced HAF model, the real time neutral line can be used for this purpose. The combined shock arrived at the Earth at 0021 UT, on March 31, 2001, and caused one of the most intense geomagnetic storms since the last decade (minimum Dst $= -358\,\text{nT}$). The following patterns show the interplanetary conditions at the arrival time of the other shock waves.

The results are compared with the observations at the libration point. Figure 8.17b shows the comparison between the observation by the ACE satellite and the simulation results at L1. The initial speed is inferred from Type II the radio emissions.

In this simulation effort, the initial speed appears to be the most uncertain factor in the parameterization. The initial parameterization is crucial in studying

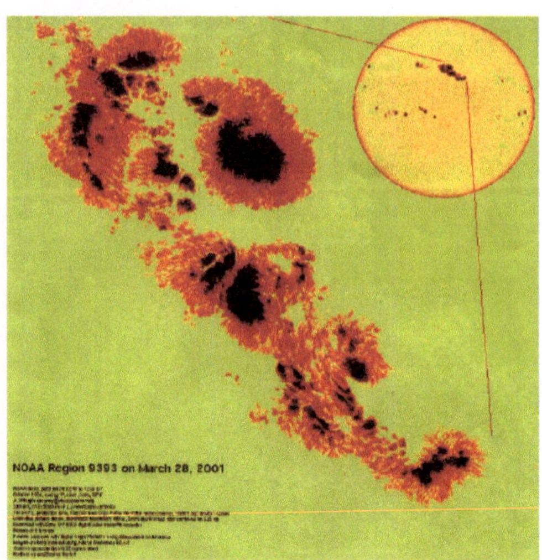

FIGURE 8.16a. One of the largest sunspot groups that appeared Sunspot Cycle 23 (NOAA Region 9393).
Source: Courtesy of D. Rust, Applied Physics Laboratory, John Hopkins University, 2001

the propagation of transient solar events, regardless of the simulation models adopted. Although the disagreement between the observation and the simulation results may depend on many factors, it appears that adjusting the initial speed alone can make a significant improvement on the agreement. Figure 8.17c shows the Dst index for the same period.

In this modeling and prediction effort, it is obvious that there is an inevitable uncertainty in identifying the relationship between particular solar events and particular interplanetary and magnetospheric events, until a direct observation of

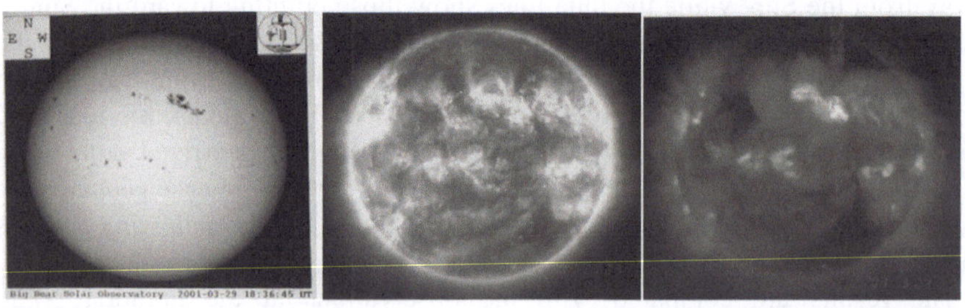

FIGURE 8.16b. Visible image on March 29, 2001: Big Bear Observatory; SOHO Image on March 29, 2001: SOHO project; and the YOHKOH Image on March 29, 2001, YOHKOH project.
Source: Left: Big Bear Observatory; Middle: SOHO image, NASA/ESA SOHO project; Right: Yohkoh project

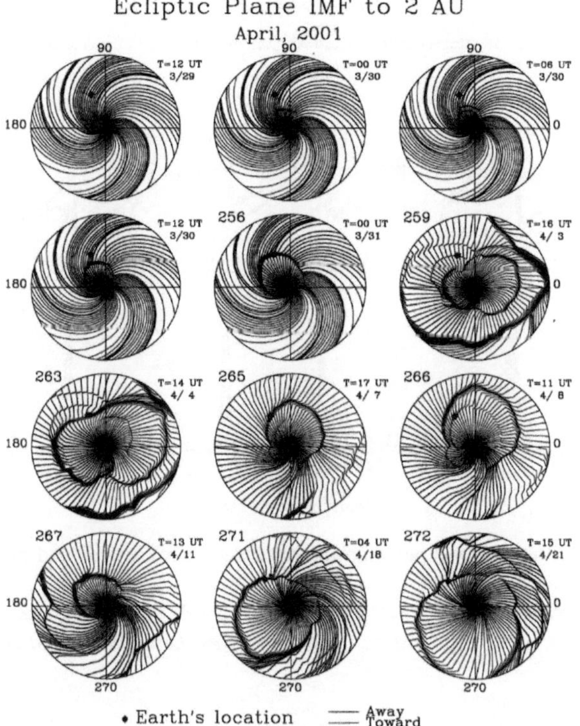

FIGURE 8.17a. A series of shock waves that propagated in interplanetary space during the period of March 29 and April 21, 2001. The location of the Earth is indicated by a star mark. The first five patterns show the propagation of the shock wave generated by the March 29 flare.

Source: Sun, W., M. Dryer, C.D. Fry, C.S. Deehr, Z. Smith, S.-I. Akasofu, M.D. Kartalev, and K.G. Grigorov, *Geophys. Res. Lett.*, **29**, No. 8, 2002

the shock waves/magnetic flux ropes/plasmoid between the Sun and the Earth becomes possible. Until then, we must make our best effort to identify them (see Section 8.10).

In the above, it is shown how the HAF modeling could handle a chain of solar transient events from the end of March to the middle of April in 2001. In addition to the prediction of geomagnetic storms, it is important to predict the propagation of high-energy particles from flare regions, particularly for manned space operations. Figure 8.18a shows an example in which the SOHO imaging device was bombarded by high-energy particles soon after the flare activity. Because of the spiral structure of the IMF lines, flares in the western hemisphere tend to bring high-energy particles (20–50 MeV) toward the Earth than flares in the eastern hemisphere. Note that because of their speed, they arrive at the front of the magnetosphere soon after optical flares are observed. Thus, unlike

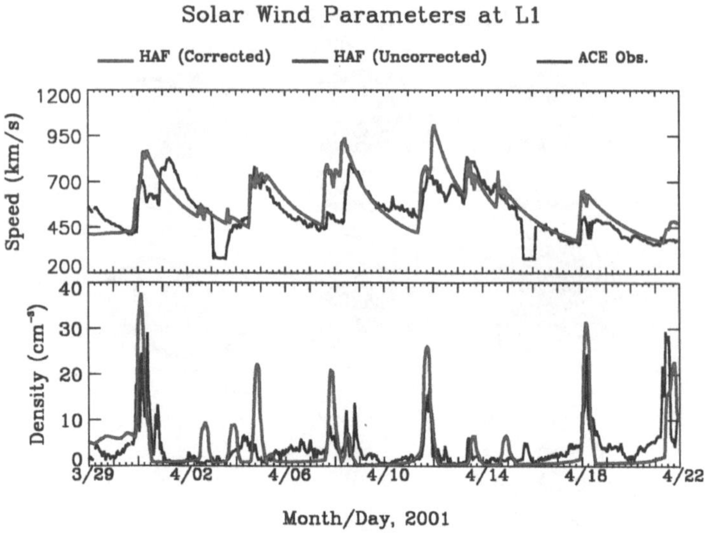

Solar Wind Parameters at L1
HAF (Corrected)
HAF (Uncorrected)
ACE Obs.
Speed (km/s)
1200
950
700
450
200
Density (cm⁻³)
40
30
20
10
0
3/29
4/02
4/06
4/10
4/14
4/18
4/22
Month/Day, 2001

FIGURE 8.17b. Comparisons between the observed shock waves and the simulation results.
Source: Sun, W., M. Dryer, C.D. Fry, C.S. Deehr, Z. Smith, S.-I. Akasofu, M.D. Kartalev, and K.G. Grigorov, *Geophys. Res. Lett.,* **29**, No. 8, 2002

geomagnetic storms, the prediction of the arrival of these high-energy particles depends on the prediction of solar flares itself and the propagation path of them. The latter depends on how accurately we can determine the background solar wind flow and IMF field lines. Figure 8.18b shows an example of the trajectories of these particles after they enter the magnetosphere.

Arriving at the magnetopause, the trajectory of high-energy protons is greatly influenced by the magnetic field of the magnetosphere. The modeling of the IMF

100
0
-100
-200
-300
-400
Dst index (nT)
3/29
4/02
4/06
4/10
4/14
4/18
4/22
MONTH/DAY, 2001

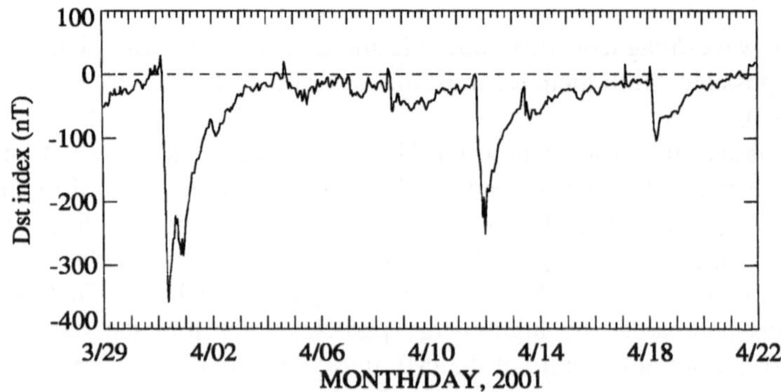

FIGURE 8.17c. A series of geomagnetic storms associated with April 2001 events.
Source: Courtesy of Geomagnetism Data Center, Kyoto University

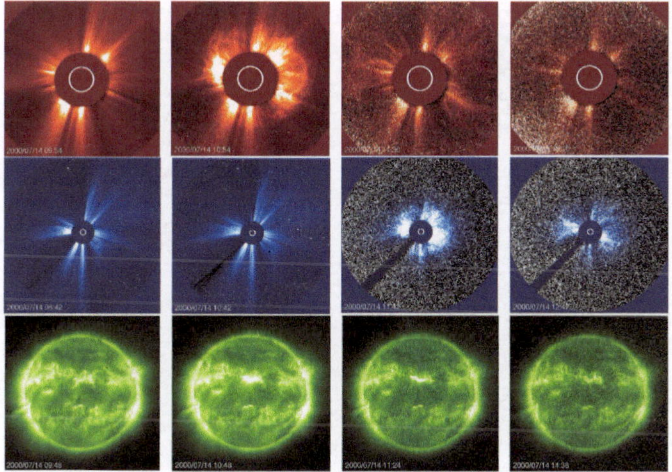

FIGURE 8.18a. An example of the arrival in the magnetosphere of high-energy particles at the libration point (SOHO) on July 14, 2000.
Source: NASA/ESA SOHO project

lines and the trajectory calculation in the magnetosphere allows us to predict their arrival at any point in the magnetosphere, such as the location of geostationary satellites and space stations, or for that matter any location.

8.9. Flux Rope Modeling

The HAF model has enabled a reasonable reproduction of the speed V(t), density N(t), and IMF B(t) magnitude, but not the IMF *theta* (= $90° - \theta$), the IMF Bz component or ε. It is quite obvious that the modeling effort is not useful in predicting geomagnetic storms, unless it can predict the *theta*, θ, IMF Bz component or ε. It can be shown here that introducing the flux rope with a helical structure (Figure 8.19a), we can reproduce reasonably well the observation. Therefore, it is crucial to find the initial condition of the flux rope, including its diameter and the orientation, as well as the sign of the helicity of the field lines and the magnitude B.

In order for the model to be useful for any loop, the geometry of the expanding loop is designed for any location on the solar disk, as shown in Figures 8.19a and 8.19b. It consists of a series of cylinder elements. The center of each cylindrical element is assumed to be aligned along a local dipolar field line that originates at the location of a transient event on the solar surface, for example a sunspot pair. Here, dm denotes the maximum radial distance of the dipole field line and Rm is the radius of the maximum cylinder element. The radius of each cylinder element linearly increases from 0 at the foot to Rm at the maximum distance dm. We assume that a force-free magnetic field configuration with

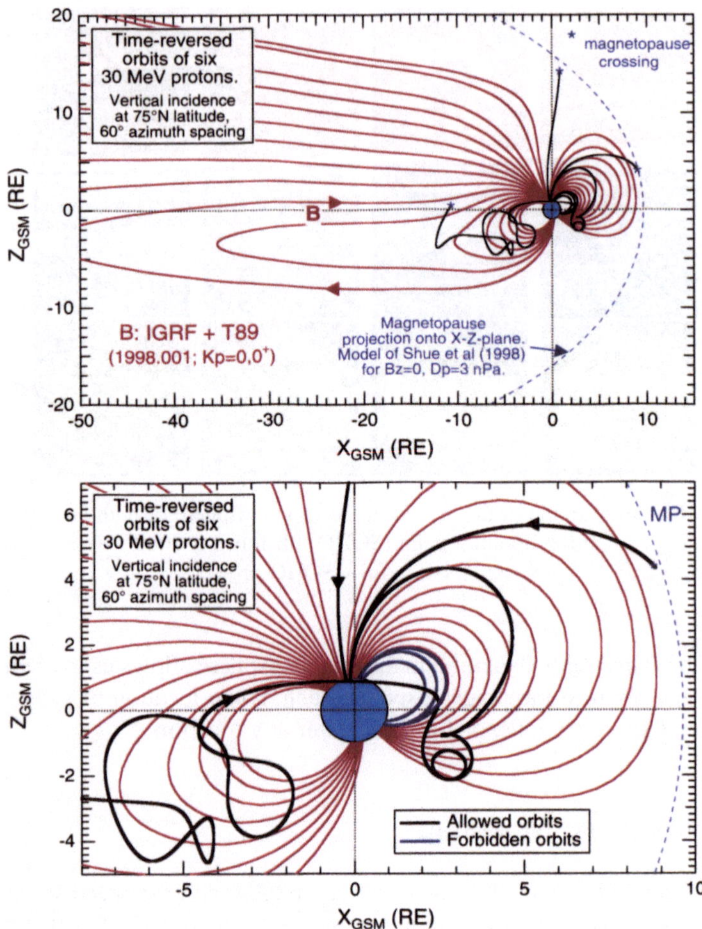

FIGURE 8.18b. An example of three trajectories of 30 MeV protons in the magnetosphere.
Source: Courtesy of R.B. Decker, Applied Physics Laboratory, John Hopkins University

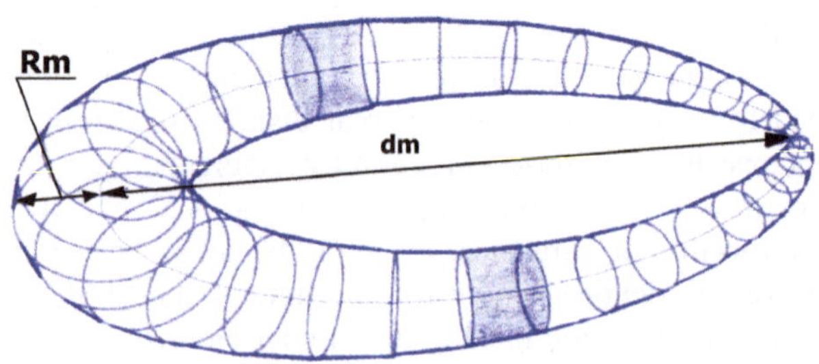

FIGURE 8.19a. The magnetic loop model.
Source: Saito, Takao, and W. Sun

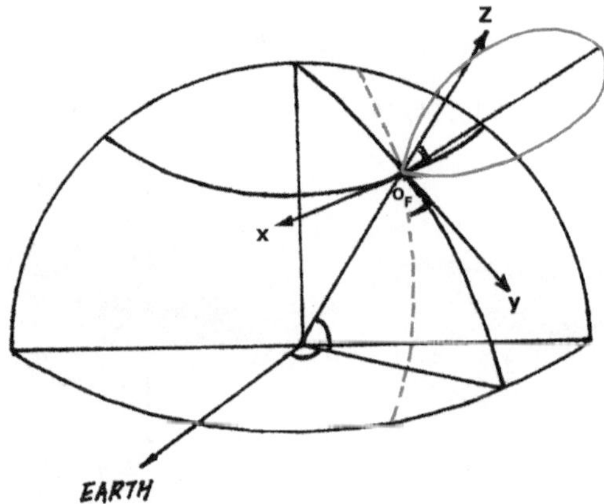

FIGURE 8.19b. Geometry of the expanding loop.
Source: Saito, Takao, and W. Sun

parameters Bm (axial magnetic field magnitude), Hm (sign of the helicity of field lines for a force-free field model), and α (the pitch of the helicity). This forms the expanding helical magnetic flux rope (MFR) model.

Figure 8.19b shows a sketch of a part of the solar surface. The point O_F indicates the location of a transient activity center, and θ_F and φ_F are solar latitude and longitude of the transient activity. A dipole field line represents an initial loop at the location of the transient activity. In the local XYZ coordinate, with the origin at O_F, φm is the angle between the plane of the dipole field line and the meridional plane of the transient activity, and θm is the angle between dm and the Z axis in the Y-Z plane. The quantities dm, Rm, θm, and φm thus determine the shape, location, and direction of a loop.

8.9.1. *Horizontal Loops*

As an example of horizontal loops (usually referred to as the magnetic flux ropes), the event of May 14, 1997 is simulated. The MFR model is combined with the HAFv2 model of the background solar wind (Fry et al., 2003 and 2004; Dryer et al., 2001, 2004; Sun et al., 2002a, 2002b) and is shown in Figure 8.19c. Note that this combination is not self-consistently made.

The parameters of the loop are obtained by using the observations from the satellite WIND (these are listed in the table below). In this model, Rm increases

TABLE 8.1. Parameters used for the stimulation of the two events

Solar Event	θ_m (°)	ϕ_m(°)	$B_m(nT)$	$H_m(\pm)$	$R_m(AU)$	α
May 17, 1997	90	21	20	+1	0.12	2.0
October 18, 1998	0	0	20	–1	0.12	2.0

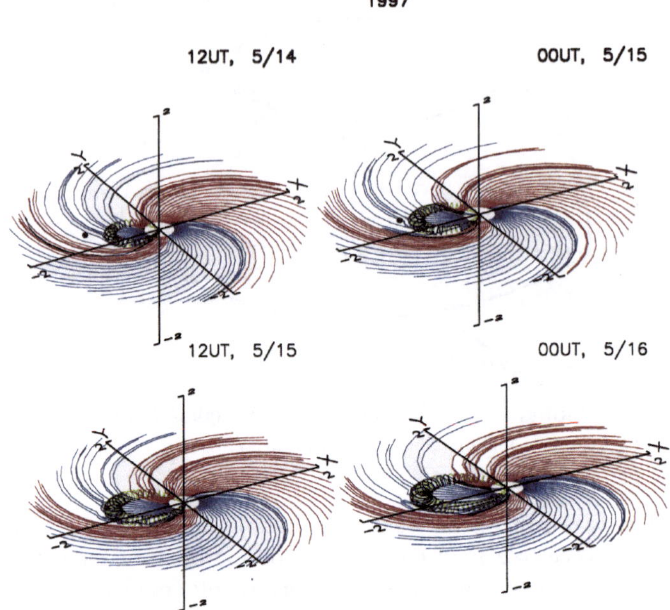

FIGURE 8.19c. Sketch of the expanding flux rope on May 14–16, 1997. The dot indicates the location of the earth.
Source: Saito, Takao, W. Sun, C.S. Deehr and S.-I. Akasofu

from 0 at the solar surface to 0.098 AU at the distance of 1 AU, and the increase of dm follows the shock wave determined by the HAFv2 model. Figure 8.19c shows the expanding magnetic flux rope thus modeled, and Figure 8.19d shows the simulated IMF changes obtained by the combination of the HAFv2 model and the MFR model as described, together with the WIND observation.

The results show that the simulated IMF agrees fairly well with the WIND observations during the passage of the magnetic cloud. As expected, the angle *Theta* agrees well with the observed changes.

8.9.2. Vertical Loops

In Section 8.4, it was mentioned that magnetic field lines from a sunspot group field in one hemisphere are often connected to a sunspot group field by loop field lines across the equator. The trans-equatorial field lines must have large north-south components. Thus, they can be a source of the IMF *Bz* components, if there is some process to expand them to a distance of 1 AU. Here, one case of trans-equatorial loops is examined. For the case of the trans-equatorial loop on October 18, 1998, we assume a loop with a helical field simulated by this force-free field model originated at the center of the solar disk. In the lower part of Figure 8.19e, a set of the standard solar and geomagnetic data associated

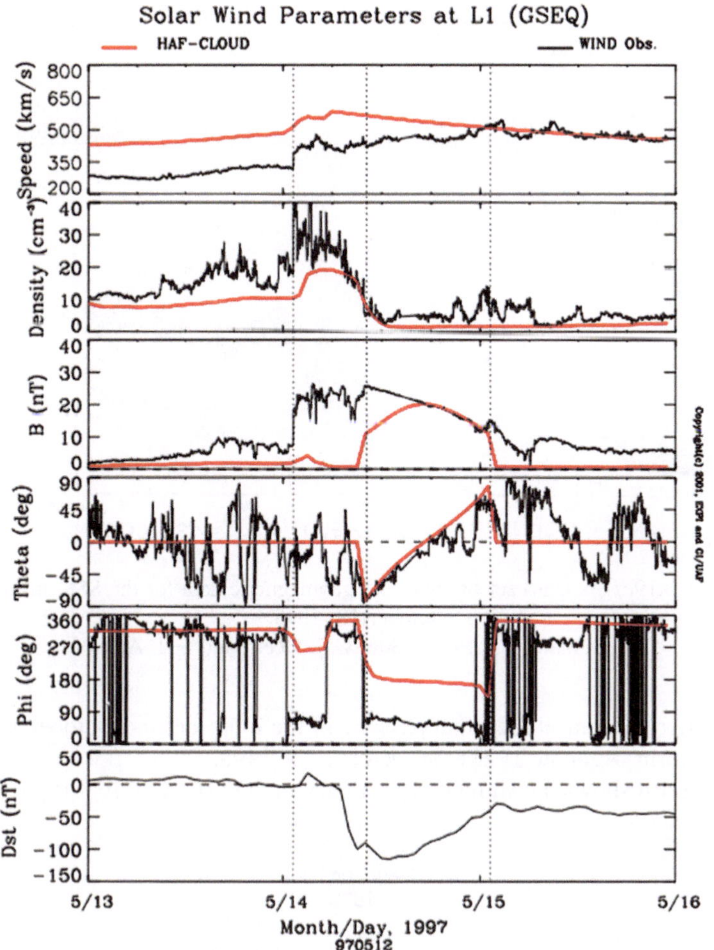

FIGURE 8.19d. The observed and simulated changes of the solar wind for the May 1997 storm.

Source: Saito, Takao, W. Sun, C.S. Deehr and S.-I. Akasofu

with the SSC storm of October 18, 1998 is shown. From the left, it shows the Hα image (October 14), the sunspot data (October 15), the solar magnetogram (October 15), and two Yohkoh images (October 15 and October 21). The October 21 image is provided to show when the trans-equatorial loop on October 15 is viewed near the western limb (the last one is a sketch of the October 21 image).

Since the trans-equatorial loops were rather faint, they were hand-traced from the Yohkoh image on the sunspot map. In the upper left, a part of the source surface map, with the neutral line, is shown; the trans-equatorial loops are projected on the map. The upper right diagram is the so-called "Kp musical

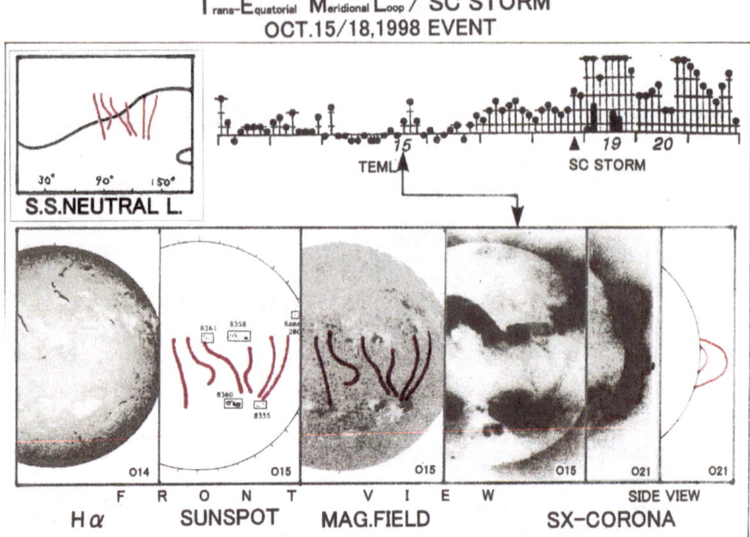

FIGURE 8.19e. Standard set of solar and geomagnetic data for the SSC storms of October 18, 1998.
Source: Saito, Takao, W. Sun, C.S. Deehr and S.-I. Akasofu

chart." After the central meridian passage of the loops around October 15, 1998, the SSC storm began at 21 UT on October 18, 1998.

Figure 8.19f shows the simulated geometry of the IMF. Figure 8.19g shows the

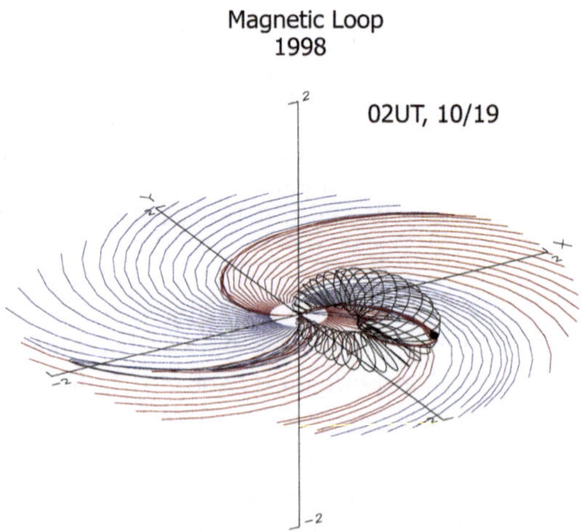

FIGURE 8.19f. The IMF field lines on the equatorial plane (red away and blue toward), together with a helical magnetic field structure at 02 UT on October 19, 1998.
Source: Saito, Takao, W. Sun, C.S. Deehr and S.-I. Akasofu

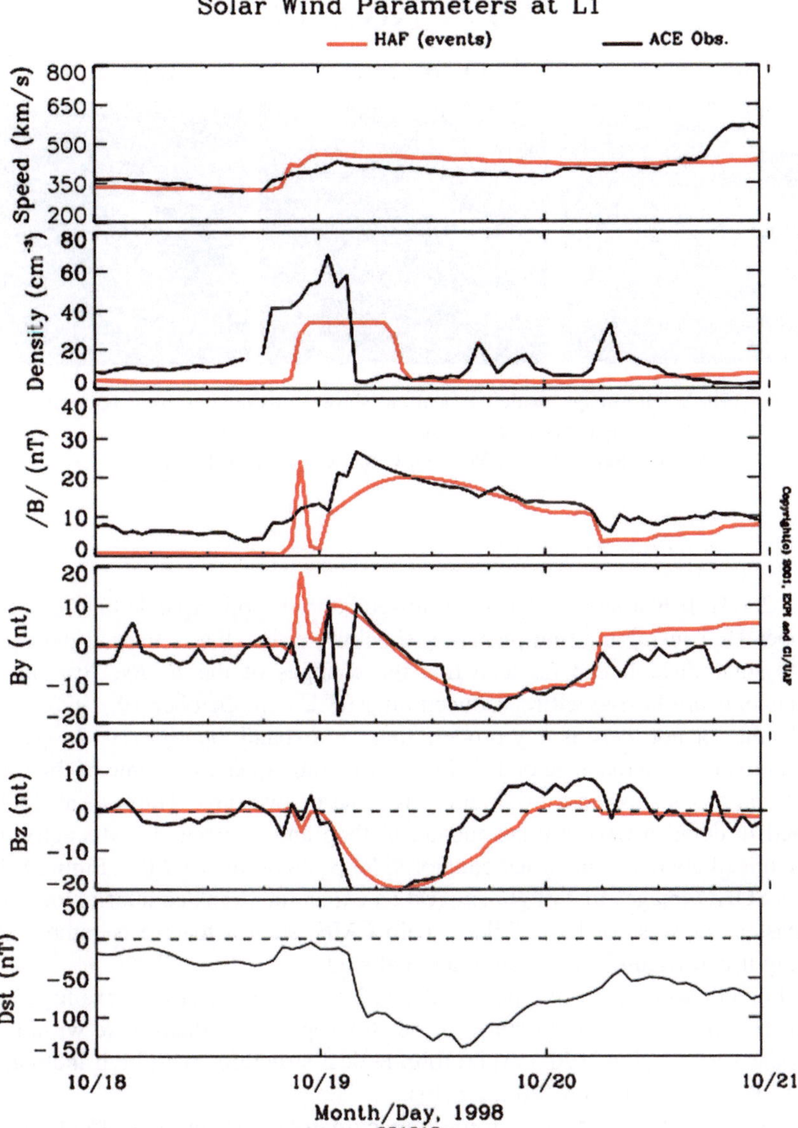

FIGURE 8.19g. The observed solar wind parameters (black) and the simulated ones (red): from the top, *V*, *n*, *B*, *B*x, *B*y, and *B*z, together with the Dst index for the SSC storm of October 18, 1998.

Source: Saito, Takao, W. Sun, C.S. Deehr and S.-I. Akasofu

observed standard solar wind parameters (black line), the speed (*V*), density (n), *B* magnitude (*B*, *B*x, *B*y, and *B*z), together with the Dst index. Added are the simulated parameters (red line) based on the HAF model (Fry et al., 2003; Sun

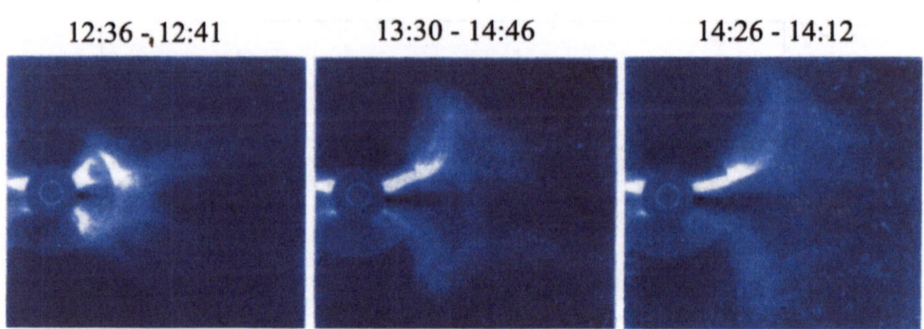

FIGURE 8.19h. SOHO image, which appears to show an expanding trans-equatorial loop near the western limb on November 6, 1997.
Source: Saito, Takao, W. Sun, C.S. Deehr and S.-I. Akasofu

et al., 2003). It is assumed that the trans-equatorial loop expanded at 00 UT on October 15, causing the interplanetary shockwave, just like a flare at the center of the solar disk. It can be seen that the changes of the B, Bx, By, and Bz components are fairly well reproduced after 03 UT on October 19, 1998.

Whether or not there is any process that can expand the observed loops near the sun to at least a distance of 1 AU is an important question. Some of the trans-equatorial loops showed a brightening for one or two days. Harra et al. (2003) studied in detail a flaring trans-equatorial loop and associated CME activities. Thus, it is likely that some trans-equatorial loops do produce CMEs. Figure 8.19h shows SOHO images that might suggest an expanding trans-equatorial loop near the eastern or western limb. Like a halo CME, such a feature is rather faint, making it is difficult to perform a detailed study.

In this connection, it may be noted that Lepping et al. (1990) examined the inclination of the helical structure for 18 flux ropes; 8 of them were within 30° from the vertical plane (90°). Therefore, helical structures with high inclination angles are not uncommon (Figure 8.13).

As mentioned earlier, in a force-free cylindrical or toroidal structure, the field lines have a helical structure. A number of authors tried to interpret the observations in terms of the constant alpha force-free structure, $\boldsymbol{J} = \alpha \boldsymbol{B}$ or $\nabla \times \boldsymbol{B} = \alpha \boldsymbol{B}$ (Burlaga, 1988; Lepping et al., 1990). In some other studies, researchers found that the flux rope consists of several layers of different pitches that cannot easily be explained in terms of the constant alpha. Romashets and Vandas (2003) considered a constant alpha force-free field in a toroidal "magnetic cloud" and found that a large-scale toroidal case is similar to a cylindrical case. Hidalgo et al. (2002) considered a helical structure, but a non-force-free one, finding that the current $\boldsymbol{J}_\perp$, which is perpendicular to the flux, is present.

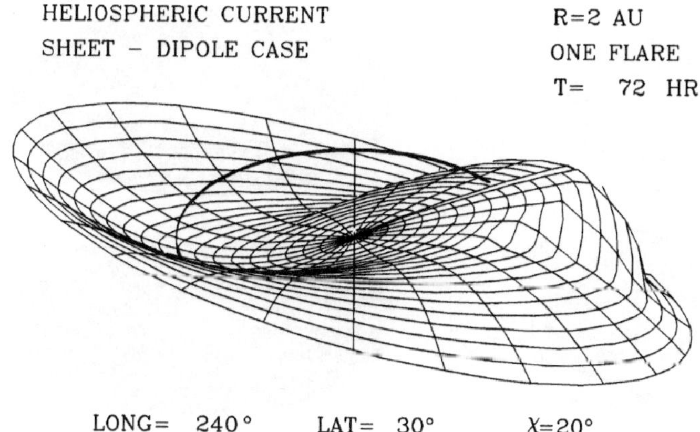

FIGURE 8.20. The current sheet deformed by a southern solar activity
Source: Akasofu, S.-I. and C.D. Fry

The situation may be complicated by the fact that the heliospheric current sheet is deformed by the solar ejecta. Figure 8.20 shows a simulated example of the deformed current sheet, which is produced by a southern solar activity. If the upward moving current sheet passes by the earth, the IMF orientation changes from the toward sector to the away sector, or vice versa, so that the By component will change the sign abruptly.

8.10. Optical Observations of Interplanetary Disturbances

In predicting a geomagnetic storm after a specific solar event, it is desirable to detect the advancing shocks and the flux ropes midway between the Sun and the Earth. A space probe at the midpoint is ideal, but is practically unavailable.

Interplanetary disturbances associated with CMEs are often called "ICME." CMEs are optically observed near the sun, not magnetically observed. What are generally called "ICMEs" observed in interplanetary space are magnetically observed objects. Thus, in addition to the possible geometrical incompatibility (CME being a thin shell or flux rope), we are dealing with the two different types of observations, namely magnetic and optical, so that we should be cautious in identifying magnetic flux ropes with CMEs without further study. The only exception in this regard is an optical observation of a halo and other structures in interplanetary space, which was observed by the SMEI satellite (Jackson et al., 2004). The observed halo structure agrees fairly well with the simulated shock wave structure that was constructed by Sun et al. (2005), which was based on the HAF modeling. Figure 8.21 shows an example of such a comparison (Jackson et al., 2006). This observation seems to suggest that the

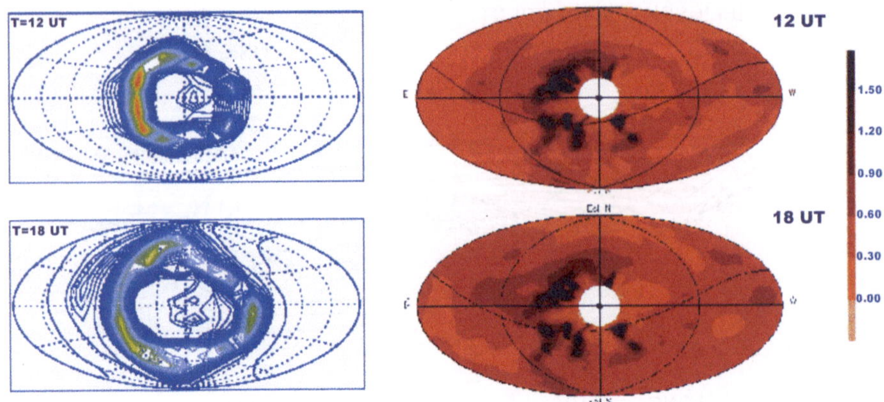

FIGURE 8.21. The observed halo structure by the SMEI satellite (right) and the corresponding HAF modeling results (left) in the sky map on May 29, 2003. In constructing the skymap, see Section 8.11.
Source: Jackson, B.V., A. Buffington, P.R. Hick, and Y. Yu, Earth-Sun System Exploration: Energy Transfer, January 16–20, 2006, Kona, Hawaii

optical phenomenon observed by the SMEI satellite was an interplanetary shock wave in this particular case.

8.11. Interplanetary Scintillation

The optical observation of the kind described in Section 8.10 is limited. For this reason, it is necessary to find other methods. One of them is to use interplanetary scintillation (IPS). In order to demonstrate this method, a study of an event in September 1978 was undertaken (Akasofu and Lee, 1989, 1990). We constructed successive 3-D surfaces of the shock wave and projected them onto the *sky map*, a map of the sky centered at the direction of the Sun (Figure 8.22). The available IPS observation during the event showed an intense IPS area in the upper left of the sky, in agreement with the projection (Hewish et al., 1985). As far as I am aware, this was the first comparison of the observed IPS and the model results. Such an observation can assure us that we chose reasonably well the necessary parameters of the solar event on the source surface before the arrival of the shock wave. Takao Saito and I found that comets between the Sun and the Earth are also useful in calibrating the simulation results, because comet tails tend to show some disturbance when the shock waves collide with them. The importance of the integration effort of the four disciplines in space weather research is to lead us to such new projects. The use of IPS observations and comets is a good example.

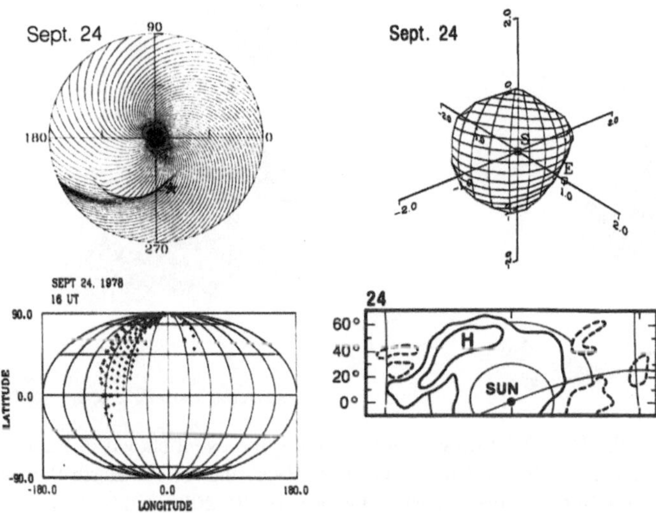

FIGURE 8.22. The 3-D shock front and its projection on the sky map at 16UT on September 24, 1978. The center of the map is the direction toward the sun. The shock front is located in the northwestern part of the sky. The observed IPS map for the September 1978 event is shown in the lower left; a high IPS region is seen in the northwestern part of the sky map on September 24.

Source: Akasofu, S.-I. and L.-H. Lee, *Planet Space Sci.*, **38** 575, 1990

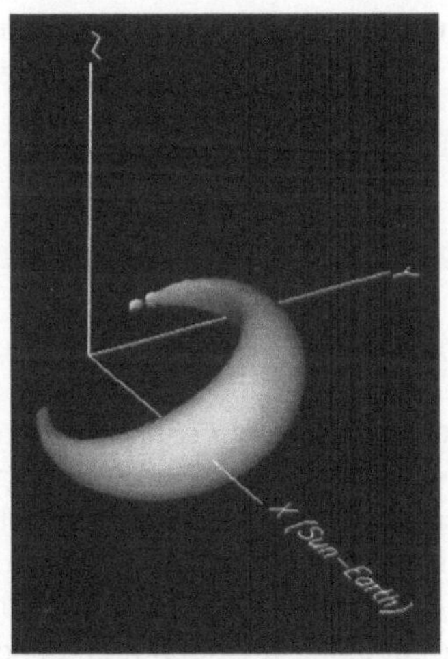

FIGURE 8.23a. The flux rope structure inferred from the IPS observations. (Three-dimensional iso-Δ Ne contour for Δ Ne0 = 4.5.) This plot is produced from the calculation with the best-fit model for the Bastille Day event. The x-axis corresponds to the Sun–Earth direction, and the xy plane is the ecliptic plane. The origin corresponds to the location of the Sun, and the length of a line in each axis corresponds to 1 AU.

Source: Tokumaru, M., M. Kojima, K. Fujiki, M. Yamashita, and A. Yokobe, *J. Geophys. Res.*, 108, SSH1, 2003

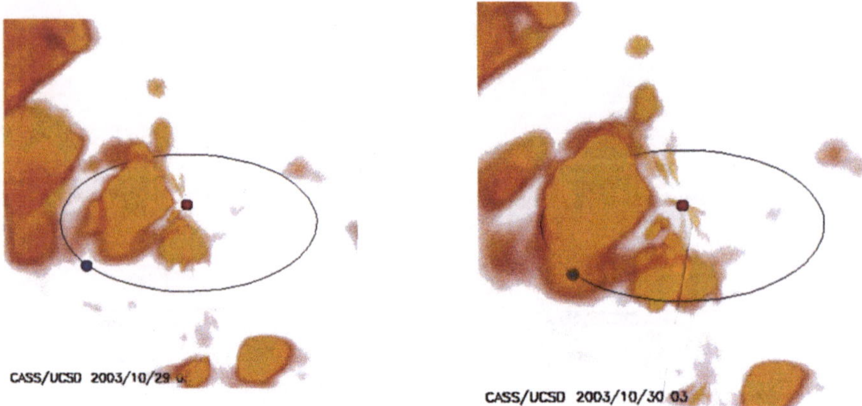

FIGURE 8.23b. The SMEI observation for the same event shown in Figure 8.23a. *Source*: Jackson, B.V., A. Buffington, P.R. Hick, and Y. Yu, Earth-Sun System Exploration: Energy Transfer, January 16-20, 2006, Kona, Hawaii

Tokumaru et al. (2003, 2005) detected both a flux rope-like structure and also a shell-like structure by analyzing IPS results. Their flux rope case is reproduced here as Figure 8.23a. Fortunately, the same event was observed by SMEI, as illustrated in Figure 8.23b. There is a reasonable agreement of the geometry, in spite of the fact that both were based on entirely different methods of observation. Such results show that IPS observations are one of the important ways to detect the advancing interplanetary shockwaves and/or solar ejecta in the midway to the Earth. We hope that a network of IPS observations will be set up to observe IPS signals for 24 hours.

8.12. Polarity of the Source Surface Field

Bothmer and Rust (1997), Crooker (2000), Lyatsky et al. (2003), Zhao and Hoeksema (1998, 2001), and many others found that there is a sunspot cycle variation of the leading field component of the loop (for example, the predominate leading field was directed northward during Sunspot Cycle 21).

However, when one deals with the largest-scale field of the sun, one must consider changes of polarity of the source surface field (on a sphere of radius of 3.5 solar radii), not the sunspot cycle. This is because the polarity of the source surface field changes at about the maximum epoch of the solar cycle; see Section 5.1. Therefore, Takao Saito and I re-examined the finding of Bothmer and Rust (1997) and found that we can improve Bothmer and Rust's statistics by noting the above fact. Although the number of examined cases is small, our finding is that the polarity of the leading field is the same as that of the IMF (thus, directed oppositely to the magnetic moment of the general field). Our results are shown in Figure 8.24.

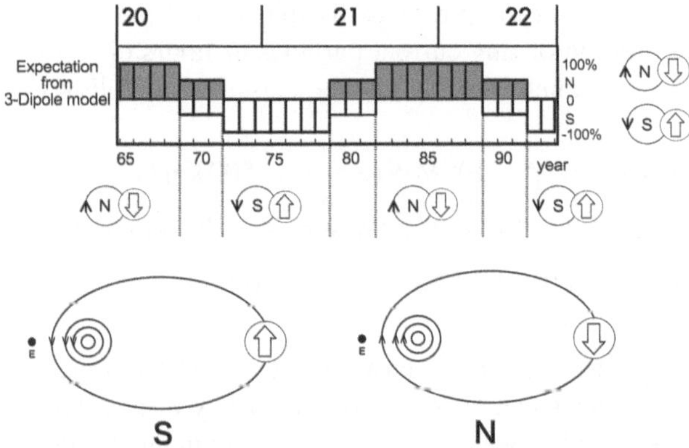

FIGURE 8.24. Expected polarity of the leading component of the loop during Sunspot Cycles 21, 22, and 23.
Source: Saito, Takao and S.-I. Akasofu,

It may be inferred that when the polarity of the interplanetary field and the leading component of the flux rope are anti-parallel, it is likely that magnetic reconnection on the front of the flux ropes can erode magnetic fields of the flux ropes; the flux ropes can survive when the polarity of the leading component is the same as that of the interplanetary magnetic field. The reconnection process between the flux ropes and the IMF was studied by Schmidt and Cargill (2003) and others. It appears that the source surface field is also important for the polarity of the ejecta from sigmoids (Pevtsov et al., 2001). A rough estimate of the magnetic flux of flux ropes is comparable with the interplanetary magnetic flux swept by the flux ropes.

8.13. Characterizing Geomagnetic Storms

The prediction of a geomagnetic storm requires the prediction of the Dst and AE indices as a function of time. For this purpose, note that electric power is given by (Section 1.9):

$$\varepsilon\ (\text{Megawatts}) = 20\,V\,(\text{km/sec})\ B^2(\text{nT})\sin^4(\theta/2) \tag{1}$$

where θ denotes the IMF polar angle. Knowing that Dst is proportional to the total kinetic energy of the ring current particles (Dessler and Parker, 1959), the calculated Dst can be obtained by:

$$\frac{\mathrm{d}Dst}{\mathrm{d}t} = \alpha\varepsilon - \frac{Dst}{\tau_R} \tag{2}$$

It is assumed that 70% of the power is dissipated in the ring current, so that α is 0.7; τ_R is the lifetime of ring current particles (7 hours or less). In terms of ε, the intensity of geomagnetic storms may be classified as follows ($\theta = 180°$):

$$\text{weak storms } \varepsilon \sim 0.25 \times 10^6\,\text{MW}(\text{e.g. } V = 500\,\text{km/sec}, B = 5\,\text{nT})$$

$$\text{moderate storms } \varepsilon \sim 1.4 \times 10^6\,\text{MW}(\text{e.g. } V = 700\,\text{km/sec}, B = 10\,\text{nT})$$

$$\text{very intense } \varepsilon > 8.0 \times 10^6\,\text{MW}(\text{e.g. } V = 1000\,\text{km/sec}, B = 20\,\text{nT})$$

The first important test of adopting ε (Megawatts = MW) in predicting geomagnetic storms is whether ε can characterize the variety of geomagnetic storms, or more specifically, whether we can infer the two geomagnetic indices AE(t) and Dst(t) as a function of time from $\varepsilon(t)$.

There is so far no theoretical study that relates ε to the AE index. This is because the magnetosphere responds to an increased ε in two ways, the directly driven component and the unloading component (Section 1.10). The directly driven component correlates fairly well with ε, but the unloading component does not and thus cannot be predicted at this time. The AE index includes both components. Therefore, the empirical relationship between ε and AE has to be established.

$$\text{AE(nT)} = -300(\log \varepsilon)^2 + 11700 \log \varepsilon - 113200 \tag{3}$$

Figure 8.25 shows, from the top, ε, calculated Dst, calculated AE and the observed AE for the March 1973 storm.

It can be seen that the Dst variations computed based on ε can reproduce fairly well both the observed characteristics of the storm and its time variations. However, it is obvious that the empirical relationship between ε and AE should be improved. The size of the auroral oval can be predicted in a similar empirical way.

It is emphasized that we cannot succeed in predicting geomagnetic storms until we can find a way to predict $\varepsilon(t)$, the IMF Bz component and/or the polar angle $\theta(t)$. Thus, the prediction of the quantities B, θ, and V after a specific solar event is most crucial in predicting geomagnetic storms.

8.14. Predicting Ionospheric Effects

Pat Reiff and her colleagues found that the ε parameter has a good correlation with the polar cap potential: $\Phi_{pc} = (0.93\,\varepsilon - 3.19)^{1/2}$. This potential drives a flow of ionospheric plasma from the dayside hemisphere into the polar cap region. Sergei Maurits and Brenton Watkins (1996) demonstrated that it is possible to forecast the electron density distribution of the F region in the polar cap on the basis of observed solar wind and the IMF (Figure 8.26).

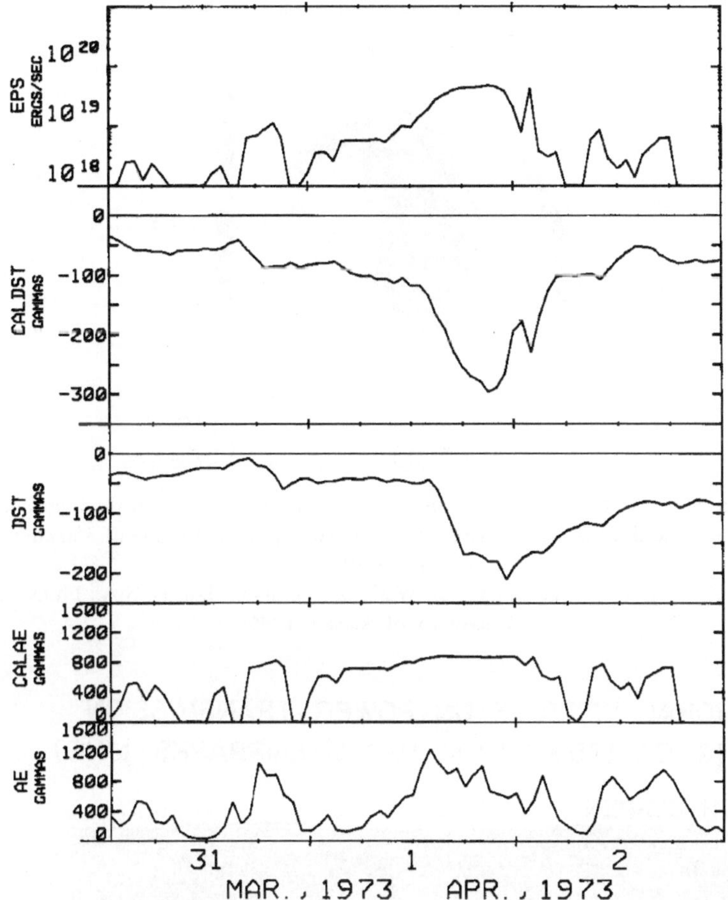

FIGURE 8.25. From the top, ε computed on the basis of the solar wind observations, the calculated Dst (CAL Dst), observed Dst, calculated AE (CAL AE), and observed AE and for the March 1973 storm.
Source: Akasofu, S.-I., *Space Sci. Rev.*, **28**, 121, 1981

8.15. Effects on Power Transmission Lines and Oil/Gas Pipelines

Although it has been emphasized by the space science community that solar events could cause serious problems for power transmission lines, there have so far been only a few studies to examine how geomagnetic storms can actually affect them. Bob Merritt, John Aspnes and I (1979, 1982) demonstrated that changing magnetic fields produced by the auroral electrojet induce electric currents in the neutral line of a three-phase transmission line and that these extra currents are converted into pulse signals in the circuit breaker system. It can be seen in Figure 8.27 that large impulsive changes of the Earth's current tend to

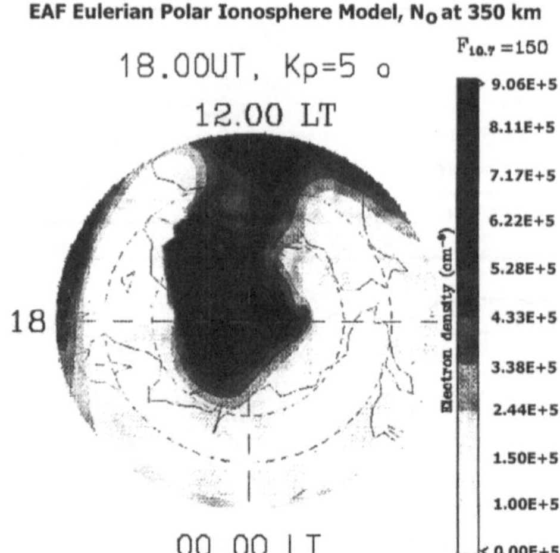

FIGURE 8.26. An example of the simulation of the $(\boldsymbol{E}\times\boldsymbol{B})$ flow of the ionospheric plasma from the dayside hemisphere into the polar cap at an altitude of 350 km (18 UT, Kp = 50).
Source: Courtesy of S. Maurits and B. Watkins, See also, Maurtis, S.A., Ph.D. Thesis, University of Alaska, 1996

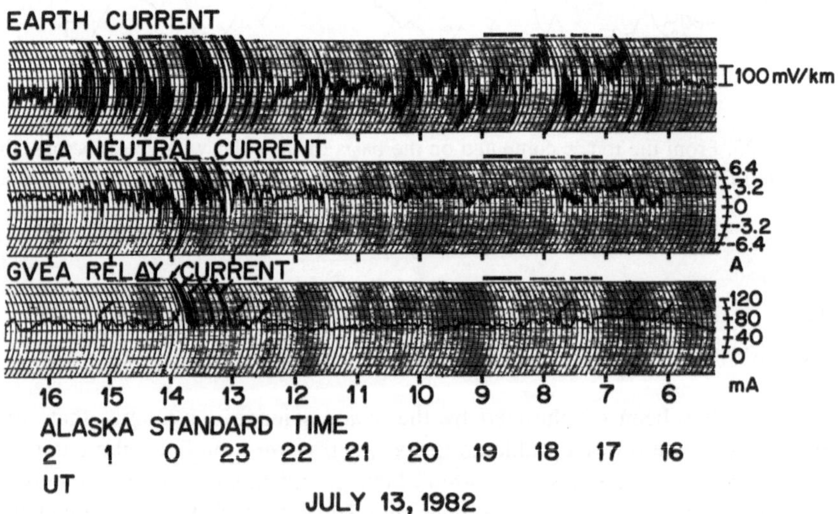

FIGURE 8.27. From the top, the Earth current recorded at the Geophysical Institute, University of Alaska, electric currents induced in the Golden Valley Electric Association (GVEA) power transmission line and the relay current in the GVEA substation on July 19, 1982.
Source: Akasofu, S.-I. and J.D. Aspnes, *Nature*, **295**, 136, 1982

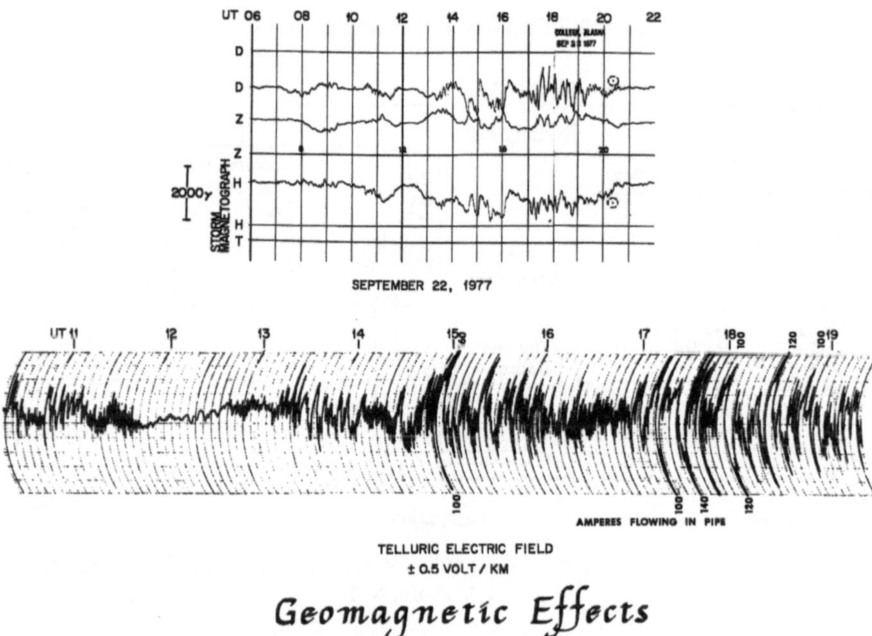

FIGURE 8.28. Upper: College magnetic record on September 22, 1977. Lower: Electric current in the Trans-Alaska Pipeline on the same day.
Source: Wescott, E., W. Sackinger, and S.-I. Akasofu (unpublished report)

produce pulses in the relay system. However, a simple mathematical relationship between them could not be found. It is generally very difficult to obtain such data from a power company. It was fortunate that many of the engineers at the local power company (Golden Valley Electric Association) were University of Alaska graduates. They were Bob Merritt's former students and were very helpful for this project.

During the construction of the Trans-Alaska Oil Pipeline, I wrote to the president of the pipeline company, saying that the aurora can induce strong currents in the pipe. He responded by saying that his corrosion engineers knew what to do about the problem. However, later they asked us to monitor the induced current during the construction, Gene Wescott, Bill Sackinger, Bob Merritt, and I found a rather simple formula; the current I in the pipe is given by $I = V/\Omega$, where V is the induced voltage (~ 1 volt/m) for moderate auroral activity (the total length of the pipe (~ 1000 km) and Ω the total resistance of the pipe. Thus, $I \sim 100$ amperes for $V = 1000$ volts and $\Omega = 10$ ohms (Figure 8.28). Currents leaking from the pipe to the ground cause corrosion of the pipe. Because of our finding, they are now monitoring the corrosion of the Trans-Alaska Oil Pipeline.

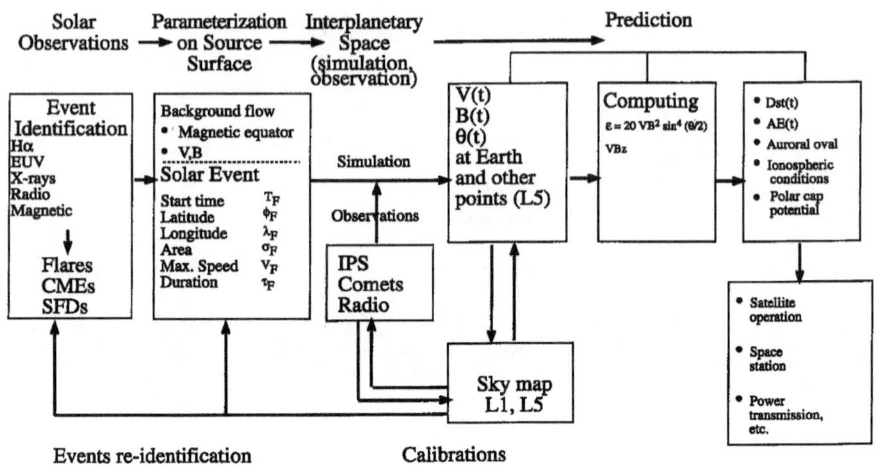

FIGURE 8.29. The geomagnetic storm prediction scheme in a block diagram form. *Source*: Akasofu, S.-I.

8.16. Geomagnetic Storm Prediction Scheme

As a summary, Figure 8.29 shows the geomagnetic storm prediction scheme as a summary in this chapter in a block diagram form. The figure is self-explanatory. It is satisfying to see that many solar physicists and magnetospheric physicists have started to work together in order to set up a space weather scheme, by learning each others' discipline. However, much more concerted efforts are needed for the success in the prediction of geomagnetic storms. In particular, we should focus our effort in predicting IMF $\theta(t)$; this requires the knowledge of the interplanetary magnetic structure associated with the magnetic flux ropes and loops

The needed efforts in succeeding in the prediction of magnetospheric storms may be summarized as follows:

1. Modeling of the background solar wind flow; see Chapter 6.
2. Parameterizing solar activity in the photosphere/chromosphere/corona and also on the source surface.
3. Finding what part of the solar atmosphere is ejected out during solar activity; Chapter 7 and Chapter 8.
4. Finding the magnetic field structure of filaments, coronal mass ejections (CMEs), and magnetic flux ropes; see Chapter 7.
5. Modeling the advance of the flux ropes and shock waves in interplanetary space by finding if the flux ropes are the so-called "driver gas" for the shock wave; see Chapter 8.

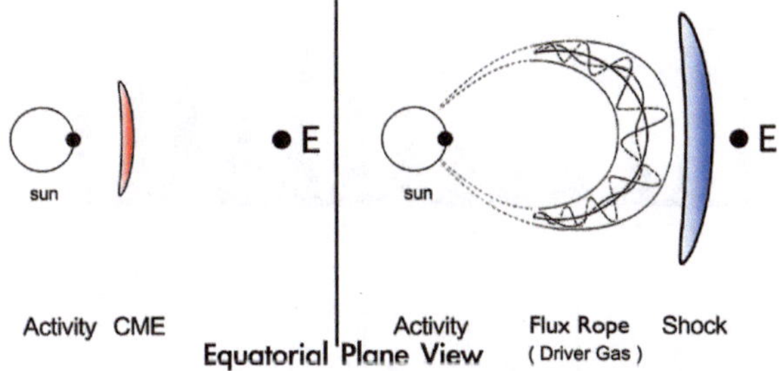

FIGURE 8.30. The relationship between solar activity, CME, magnetic flux rope (driver gas), and a shock wave. Near the sun, we optically observe an expanding thin shell like structure that is called the halo CME, while we observe magnetic flux ropes and shock waves by satellites and space probes in interplanetary space.
Source: Akasofu, S.-I.

6. Estimating the velocity, density, and IMF of the solar wind at the Earth and inferring the solar wind-magnetosphere dynamo power $\varepsilon(t)$; see Chapter 1.
7. Predicting the development of geomagnetic storms as a function of Dst(t) and AE(t) based on ε (t); see Chapter 1 and Chapter 8.
8. Predicting the size of the auroral oval; see Chapter 2.
9. Predicting the development of ionospheric storms; see Chapter 8.

At present, there are two sets of observations (Figure 8.30) in this regard. First, near the sun, we observe transient solar activities and CMEs, which launch solar disturbances into interplanetary space. Second, in interplanetary space, we observe "magnetic flux ropes" and shock waves. In addition to the major problem of establishing the relationship between three transient solar activities and CMEs, we have to learn how CMEs are related to the interplanetary observations by satellites. Magnetic flux ropes are thought to be magnetically connected to the photosphere, while flare phenomena are interpreted as results of magnetic reconnection, suggesting the ejection of plasmoids, which are detached magnetically from the photosphere.

Some of the scientific problems to be solved in the above prediction effort are as follows:

(1) Near the Sun

(a) What is the relationship between solar activities (filament disappearances/and sigmoid eruptions, trans-equatorial loops) and CMEs?
(b) Are magnetic helical structures of the filaments the source of magnetic flux ropes?
(c) Are CMEs magnetic flux ropes or shock waves?
(d) Can we infer the IMF Bz component from the helical structure in the filaments, the magnetic field distribution around sigmoids, and trans-equatorial loops?

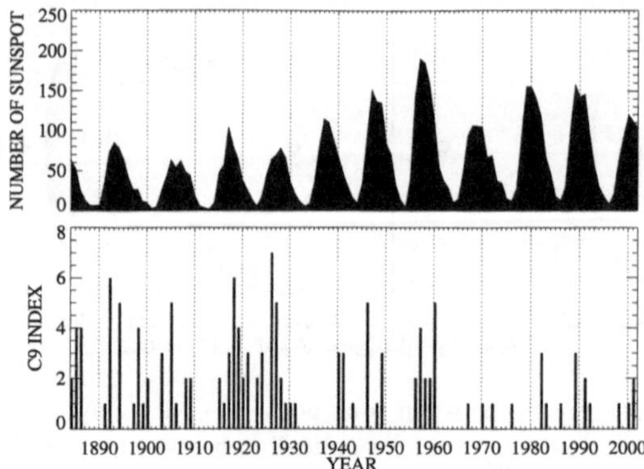

FIGURE 8.31. The relationship between the 11-year cycle variation of the sunspot number and the occurrence frequency of C9 storms.
Source: Akasofu, S.-I.

(2) Interplanetary Space

(a) What is the relationship between CMEs and various interplanetary disturbances?
(b) Are magnetic flux ropes magnetically connected to the sun or are plasmoids magnetically isolated from the sun?
(c) Can magnetic clouds be identified as the so-called "driver gas," which generate interplanetary shock waves?

(3) Magnetosphere

(a) Can we predict the development of individual geomagnetic storms, in terms of geomagnetic indices Dst(t) and AE(t) on the basis of $\varepsilon(t)$?

There are many other interesting statistical results between the solar and terrestrial relationship study. For example, in Chapter 6, it was shown that a higher sunspot number does not lead to more intense geomagnetic storms. It is shown here that the sunspot number and the number of C9 storms (the most intense storms) are shown in Figure 8.31. It is interesting to note that there were many intense storms before 1930, when the peak sunspot numbers in each cycle were rather low.

After 1930, the peak sunspot numbers increased, while the occurrence of intense storms decreased. The implication of such a statistical result is uncertain, but may have an important implication on the Sun or may simply be an indication that sunspots tend to suppress the solar wind as discussed in Chapter 6. It should also be mentioned that the occurrence of C9 storms is not related to any specific epoch of the sunspot cycles except the minimum period.

9
Beyond the Inner Heliosphere: The Magnetic Field Structure of the Outer Heliosphere: A Three-Dimensional Model

Establishing the three-dimensional structure of the heliosphere is an important problem in space physics. Some postulate that the heliosphere has a magnetosphere-like structure as it moves through interstellar gas. Further, we are aware that the magnetic field structure in the heliosphere deviates considerably from Parker spiral. One of the purposes of this chapter is to suggest that a crude model of the magnetic field structure can be made by considering the distribution of electric currents in the heliosphere, as suggested by Hannes Alfvén (1977). Figure 9.1 shows the distribution of the heliospheric current.

One can, indeed, consider that the familiar Parker spiral arises from two currents, both the radical (J_r) and azimuthal (J_φ) components, flowing perpendicular to the spiral magnetic field lines (Akasofu et al., 1980). We first consider the case when the solar main dipole is directed southward (the north polar region being the south pole). The unipolar induction generates the radially outward-flowing current from both polar regions (Alfvén, 1950, 1977). After reaching the pole of the heliosphere, the current flows along the outer surface (assumed to be spherical) of the heliosphere to the equator and then flows radially (J_r) toward the Sun.

The sector boundary can be taken as a thin current sheet where the current flows azimuthally on the equatorial plane eastward when the solar main dipole is directed southward; the IMF has the sunward component above and the anti-sunward component below the current sheet. In fact, the boundary is often called the heliospheric current sheet. Thus, the basic heliospheric magnetic field is composed of the magnetic fields produced by these currents, together with the intrinsic solar dipolar field. When the solar dipole reverses its polarity (because of the 11-year cycle variations), the direction of the currents reverses as well. Note that the current continuity and $\nabla \cdot \boldsymbol{B} = 0$ conditions are automatically satisfied in this model.

The solar system is also embedded in the interstellar magnetic field of the Orion arm. Therefore, the heliospheric magnetic field interacts with it.

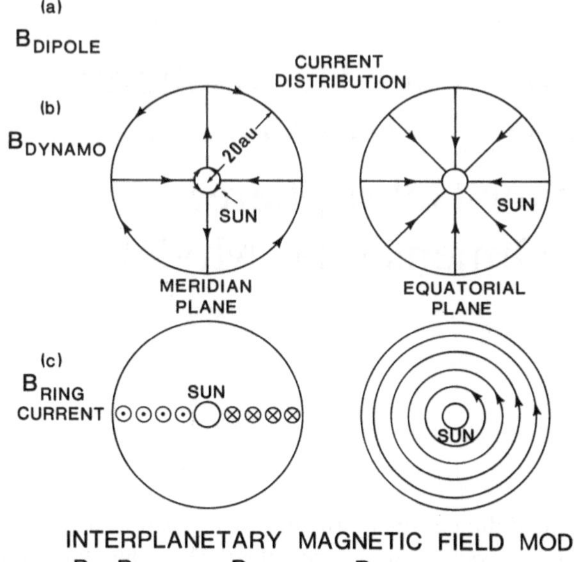

FIGURE 9.1. Heliospheric current system.
Source: Akasofu, S.-I. and D.N. Covey, *Planet Space Sci.,* **29**, 313, 1981

The magnitude of the field is inferred to be about 0.22 nT. The orientation of the field with respect to the heliosphere is not known. In modeling the heliospheric magnetic field, it is assumed that the orientation is vertical to the solar equatorial plane and directed southward, but only a small fraction of it (~5%) is allowed to penetrate into the heliosphere. Note that only a small fraction of the interplanetary magnetic field can penetrate into the magnetosphere.

Figures 9.2a and 9.2b show an example of modeling of such a situation. As far as I am aware, Figures 9.2a and 9.2b are the first to show the three-dimensional structure of the Parker spiral, although it is crude. So far, all I have seen are many hand drawings. Ed Smith and his colleagues (1995, 1997) reported, based on their Ulysses observations, that the spiral structure is present at high latitude in the heliosphere. The field tends to have a spiral structure, although it tends to be more radial than what Parker's model predicts. This may be because the solar wind speed is known to be higher in higher latitudes, producing an underwound condition. However, this particular model cannot handle such a situation. Depending on the orientation of the solar dipolar field, some of the high-latitude hemispheric magnetic field lines may be connected to the interstellar magnetic field lines (Figure 9.2b).

As described in Section 6.1, a magnetic field line originating from the source surface can be traced by following particles leaving a particular point on the rotating source surface. The resulting interplanetary magnetic field structure,

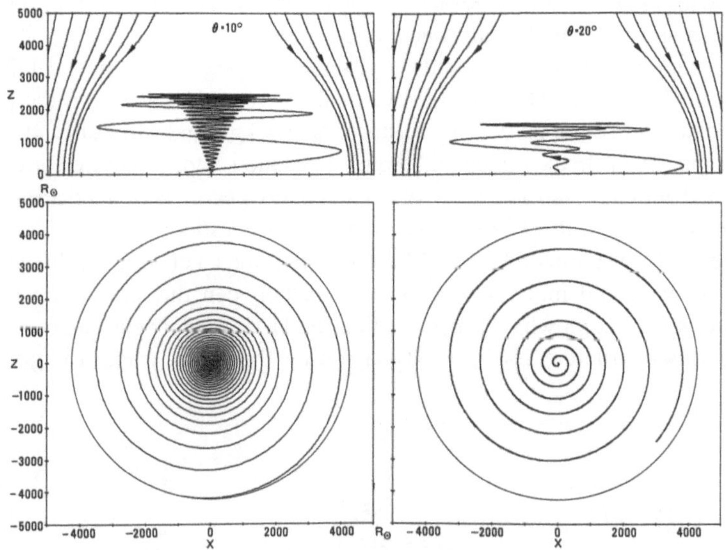

FIGURE 9.2a. Heliospheric current structure.
Source: Akasofu, S.-I. and D.N. Covey, *Planet Space Sci.*, **29**, 313, 1981

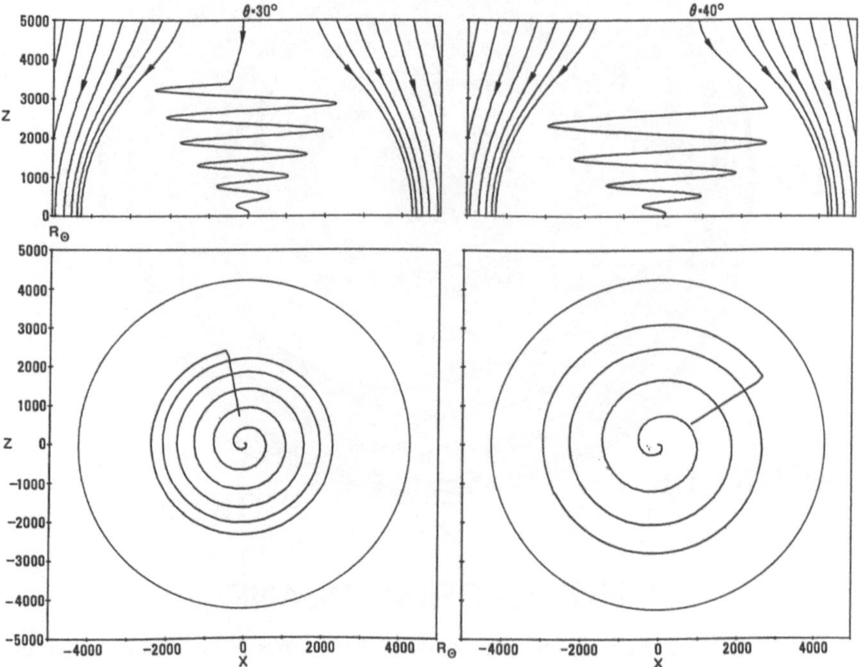

FIGURE 9.2b. Heliospheric current structure.
Source: Akasofu, S.-I. and D.N. Covey, *Planet Space Sci.*, **29**, 313, 1981

together with the corotating interaction region produced by the formation of the shock wave structure, is the familiar Parker spiral.

As an example, simulation results of a series of the June–August 1982 event is shown in Figure 9.3. There were seven successive major flares in the period.

Solar disturbances considerably distort the background structure. It can be seen that the seven shock waves and the background sector structure form a very complicated magnetic structure in the outer heliosphere. Based on Pioneer 10 and 11 missions, Smith (1990) noted that the observations are consistent with such models. Both Pioneer 10 and 11 were in the outer heliosphere during these periods, so we could make some comparison between the simulation and the observations. It is somewhat surprising that the HAF model reproduced well the observed shock structure for up to 10 days after the shock wave was generated

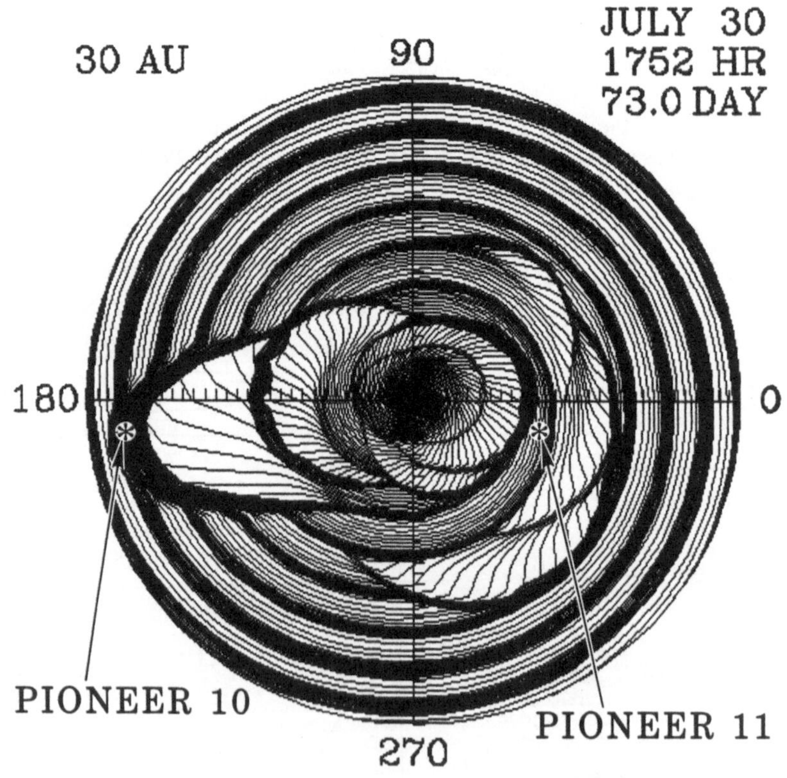

FIGURE 9.3. Heliospheric magnetic field 73 days after a series of major solar activities in June–August 1982.
Source: Akasofu, S.-I., W. Fillius, W. Sun, C. Fry, and M.J. Dryer, *J. Geophys. Res.*, **90**, 8193, 1985

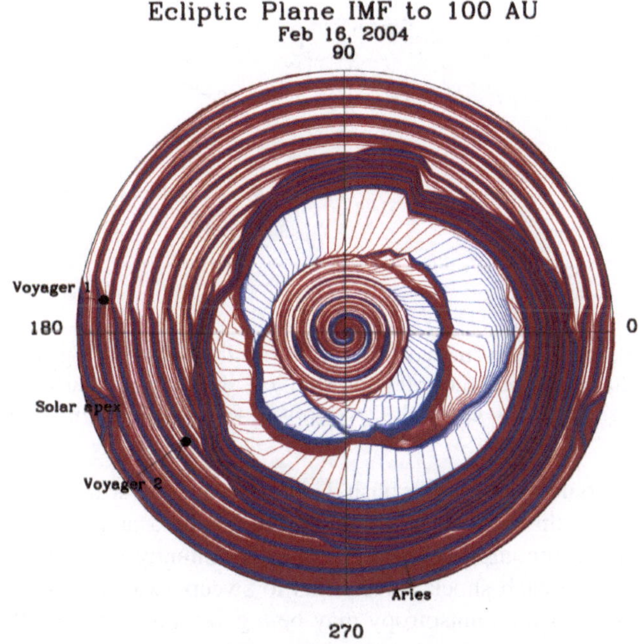

FIGURE 9.4. The IMF structure to a distance of 100 AU, perhaps the limit of the solar system. Note the location of Voyager 1 and Voyager 2: Sun (2005).
Source: Courtesy of W. Sun, 2005

near the Sun. As far as I am aware, this is the first semi-quantitative simulation of the disturbed condition of interplanetary space within 20 AU that is caused by a successive solar activity. Figure 9.4 shows a tentative IMF structure in the entire heliosphere, namely to a distance of 100 AU, on February 2004. Note that several shockwaves form a field structure that surrounds the inner heliosphere.

The shock structures are expected to have considerable effects on the propagation of galactic cosmic rays from interstellar space to the center of the heliosphere (Van Allen, 1996). Indeed, a significant decrease of cosmic ray flux was observed at the location of both space probes, as well as on the Earth (cf. Akasofu et al., 1985). This phenomenon is called the *Forbush decrease*, honoring Scott Forbush, who discovered it. Figure 9.5 shows an example of a large Forbush decrease caused by solar activities on May 23–24, 1967.

It is also well known that the cosmic ray intensity has a clear anti-correlation with the 11-year cycle of solar activity (Figure 9.6). This phenomenon was considered in the past in terms of the degree of diffusion of cosmic rays from interstellar space; it was thought that turbulence in the heliosphere during sunspot maximum years tends to prevent the diffusion of galactic cosmic rays toward the inner heliosphere. Another possibility is that the 11-year cycle variations of cosmic ray intensity result from such successive sweeping effects on cosmic ray particles by the shock waves. Indeed, it can be seen from Figures 9.3 and 9.4 that successive shock waves form a barrier surrounding the inner heliosphere.

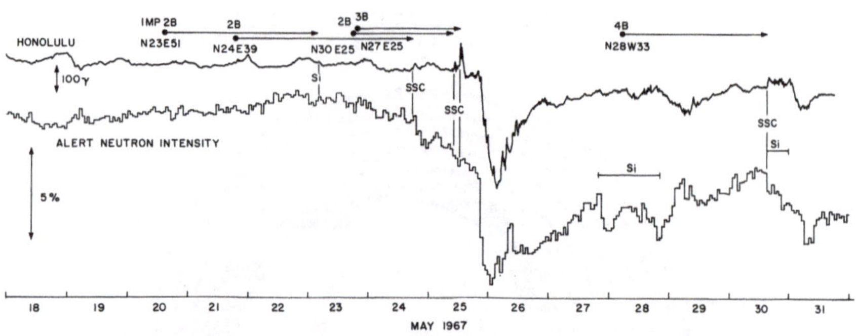

FIGURE 9.5. Honolulu magnetic record and Alert neutron intensity, together with solar activity during the period of May 18–31, 1967, showing a Forbush decrease. *Source*: Yosjhida, S., N. Ogita, S.-I. Akasofu, *J. Geophys. Res.*, **76**, 7801, 1971

It is interesting to note that the Forbush decrease is not uniform around the Earth. Figure 9.7 shows this non-uniformity. Sekiko Yoshida and I worked on this particular feature, although the space physics community has not shown much interest in it. Since each shock wave tends to sweep away cosmic rays from a limited direction, such an anisotropy may be a good indication of effects of the shock waves.

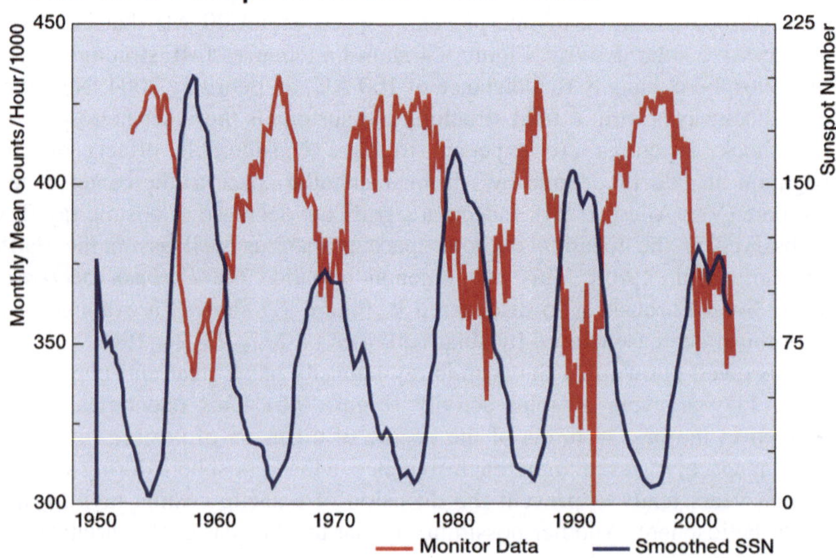

FIGURE 9.6. The anti-correlation of the sunspot number and the cosmic ray intensity. *Source*: SPATIUM, International Space Institute, 2003

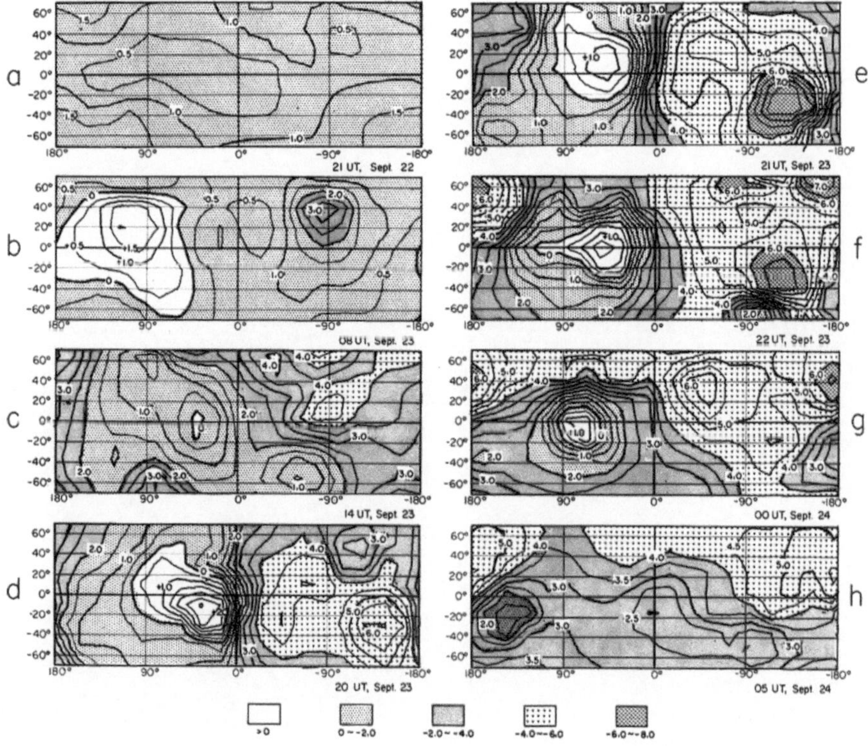

FIGURE 9.7. The distribution of cosmic ray intensity variations. *Source*: Yoshita, S., N. Ogita, S.-I. Akasofu, and L.J. Gleeson, *J. Geophys. Res.*, **78**, 6409, 1973

The magnetic field structure of the outer heliosphere needs much more modeling effort than in the past, because space probe observations are very limited. It is our hope that our effort is of use in providing some idea about the geometry of the heliospheric magnetic field structure during quiet and disturbed periods.

Epilogue

Paradigms and Paradigm Changes

If history teaches us merely a simple chronological description of past events, it is of little interest to us. Thus, the history of science is of little interest if it tells us only the standard stories, such as who discovered what, when and so forth. It was T.S. Kuhn who told us what to learn from the history of science, by introducing the concept of paradigm in his book *The Structure of Scientific Revolutions*.

Kuhn pointed out that in the history of science, there are periods during which there is a high degree of agreement, both on theoretical assumptions and on the problems to be solved within the framework provided by those assumptions. The resulting coherent tradition of scientific research is called a paradigm. Scientists whose research is based on shared paradigms are committed to the same rules (including established viewpoints) for scientific practice. That commitment and the apparent consensus it produces are prerequisites. The members of the community are tasked to solve puzzles defined by the paradigm. Like exercises and examples in our scientific textbooks, which are a product of paradigms, the solution is assured. In fact, the term paradigm originally is said to be related to an example in a textbook. Our discipline of solar-terrestrial physics is no exception in establishing many paradigms; we have witnessed a number of examples of such situations in this book.

Paradigm	Pioneer	New Paradigm/(initiated by)
Solitary Particles	Birkeland	Diamagnetic Plasma Flow (Chapman)
Closed Magnetosphere	Chapman	Open Magnetosphere (Dungey)
2-D Current System	Chapman	3-D Current System (Alfvén)
Auroral Zone	Loomis	Auroral Oval (Feldstein)
Fixed Auroral Pattern	Fuller/Heppner	Auroral Substorm (Akasofu)
MHD (E_{II}= 0)	Alfvén	"Thawing" (Alfvén) ($E_{II} \neq 0$)
Magnetospheric Convection	Axford/Hines	Surviving
Magnetic Reconnection	Hoyle/Parker	Surviving
Magnetic Tube (Sunspot)	Babcock	Not Available
Solar Wind (Thermal)	Parker	Not available

Almost all scientists inevitably spend most of their time in this puzzle-solving work. Scientists articulate, verify, elaborate, and consolidate those theories that the paradigm supplies, resolve some of the minor details (residual ambiguities), and attempt to reconcile anomalies. Scientists are generally practitioners engaged in such mop-up work. Even the brilliant mathematicians Euler, Lagrange, Laplace, and Gauss spent their lives elaborating Newton's paradigm, so that there is nothing to be ashamed of or embarrassed about. We do it all the time. However, there is a danger that bright young scientists tend to be attracted to the puzzle-solving problems and often become almost fanatic supporters of a particular paradigm. Furthermore, because a representation of reality is easier to grasp than reality itself, researchers tend to confuse the two and take concepts for reality. They forget that as their model becomes mathematically rigorous, it becomes increasingly detached from the real world (truth). Is a sunspot really a static, cylindrical magnetic tube of force emerging from beneath the photosphere?

There is often one popular model (paradigm) in each field that is the creation of one scientist. When his model becomes very popular, there is a tendency for a large number of scientists to swarm around it and attempt to improve it. The accepted truth about reality is nothing but a consensus of contemporary experts. Meanwhile, all other models are often forgotten.

Eventually, however, there may be a growing number of unsolved puzzles and anomalies for a particular paradigm. As a result, the scientific community's confidence in the paradigm is eroded. This crisis of confidence means that the agreement, which constitutes the sharing of the paradigm, begins to dissolve.

Kuhn observes, however, that even when confronted by severe and prolonged anomalies, scientists do not, in general, respond to the resulting crisis. They think that anomalies are just a few rotten apples and their accidental presence in a barrel does not discredit the other apples. Scientists will push the rules of their paradigm harder to find where and how far they can be made to work. Although they may begin to lose faith, they do not renounce the paradigm that led them into the crisis. They will devise numerous articulations and *ad hoc* modifications of their theory in order to eliminate any apparent conflict. Because of different assumptions used by different researchers in explaining the unsolved anomalies, the initial agreements are lost, causing controversies within a particular paradigm.

The transition from a paradigm in crisis to a new one, a *scientific revolution*, is inaugurated by this growing sense that the existing paradigm has ceased to function adequately. This occurs when scientists carefully examine the barrel and find more rotten apples. Kuhn further observes that scientists do not reject paradigms simply because they are confronted with anomalies or counter instances. Once it has achieved the status of paradigm, a scientific theory is declared invalid only if an alternative candidate is available to take its place. New facts alone do not destroy an outlived theory. As a result, the longer a powerful paradigm survives, the more damage it inflicts. When a powerful paradigm dies, there may be a long period of vacuum because no one can think of anything else. Since a paradigm is fated to die (otherwise, there is no progress) eventually, a long-lasting paradigm actually retards the progress in its field. Max Planck was quoted as saying:

A scientific truth does not triumph by convincing its opponents and making them see the light, but rather because its opponents eventually die and a new generation grows up that is familiar with it.

The emergence of a new paradigm candidate is most often far from a cumulative process, one achieved by an articulation or extension of the old paradigm. It is more often the creation of an imaginative mind spurred by an epiphany by a single scientist rather than a result of logical thinking; he suggests that what was thought to be rotten apples are oranges. The act of creation is intuitive, irrational, illogical, and, above all, unscientific (as A. Koestler puts it in his book *The Act of Creation*). Suffice it to say, there will be no inspiration and no breakthrough if one logically elaborates his thinking on the basis of an old, unworkable paradigm.

However, one must be careful about the sudden flash of inspiration. The sudden flash would not arise unless a researcher has had many months or years of struggle and agony in solving his/her problem. A hint of the flash should be available to everyone on an equal opportunity basis, but only one particular person can grasp the hint, because of his/her struggle in the past. A falling apple gave Newton a hint for universal gravity, but even a child is aware that an apple can fall from an apple tree. It is unfortunate that many stories of discovery are distorted, emphasizing only the sudden-flash aspect.

The unscientific nature of creating a new paradigm will most often bring a battle over its acceptance. I suggest that readers learn how the concept of plate tectonics became accepted after A. Wegener proposed the concept of continental drift; those who held the paradigm of geosyncline did not accept him until after World War II, when strong support began to emerge. Sir W.L. Bragg learned about Wegener's idea from Sydney Chapman. After learning more details from Wegener, he gave Wegener's paper to the Philosophical Society. Bragg (1967) mentioned:

The local geologists were furious; words cannot describe their utter scorn of anything so ridiculous as this theory, which has now proved so abundantly to be right.

Since an old paradigm must be fully developed near the end stage, mathematical rigor can be used as a powerful arm against such a creative act (which is often full of errors by its nature) to give an impression that imprecision is a defect of the new idea. Actually, such imprecision is almost a prerequisite for a pioneering paper.

As new paradigm candidates begin to emerge, scientists tend to respond in a way similar to members of any other community. Here, Koestler observed:

Like other establishments, they are consciously or unconsciously bent on preserving the status quo, partly because unorthodox innovations are a threat to their authority of the paradigm, but also because of the deeper fear that their laboriously erected intellectual edifice might collapse under the impact.

A scientific establishment is highly conservative and will attempt to preserve the power of its ruling group against any rebels. Thus, a pioneer often must stand-alone and be independent-minded on the fringe of the scientific establishment, and perhaps be a rebel (W.B.I. Beveridge in his *The Art of Scientific Investigation*).

During this turbulent period, called the preparadigm period, scientists get involved in passionate controversies. Eventually, one paradigm candidate gains the status of a new paradigm because it is more successful than its competitors. The chosen paradigm will be said to be beautiful, artistically creative, imaginative, inspirational, novel, and elegant after having been treated as a crackpot idea by those clinging to the old paradigm.

The emergence of a new paradigm does not necessarily mean, however, new progress in that particular field. As Koestler put it:

> Progress by definition never goes wrong. Evolution constantly does, and so does the evolution of ideas, including those of exact science.

In his address as retiring president of the American Association for the Advancement of Science, S.P. Langley (1889) noted that the progress of science is not like the march of an army towards truth, but

> ... not wholly unlike a pack of hounds, which in the long-run perhaps catches its game, but where, nevertheless, when at fault, each individual goes his own way, by scent not by sight, some running back and some forward: where the louder-voiced bring many to follow them nearly as often in the wrong path as in a right one: where the entire pack even have been known to move off bodily on a false scent...

All these confusions are left out of the textbooks by their authors, who are mostly compilers. Thus, students learn only that science progresses monotonically, asymptotically approaching truth. One of the mentor's tasks is to tell his/her students how science actually progresses. The mentors need to teach them to swim not only in a pool, but also in a swift river or ocean.

Suppose that a group of scientists agree that they are going to solve a Garfield jigsaw puzzle – we may call this the "Garfield paradigm." A jigsaw puzzle has many rules. First of all, you have to think of Garfield only, not other jigsaw puzzles, and you cannot use scissors. An important point here is that, like examples in a textbook, the jigsaw puzzle is supposed to be solvable if you do it right. All scientists working on the Garfield puzzle believe so, at least at the beginning.

Unfortunately, the frontier of science is not like a simple jigsaw puzzle for many reasons such as, for example, the limit of accuracy of observations; the pieces do not necessarily match together well; many pieces are still missing. Most scientists have not learned how to deal with such a jigsaw puzzle.

Suppose one piece does not seem to fit at all in the Garfield puzzle. In this situation, many scientists throw the piece away perhaps saying that it came from another puzzle. Some scientists in the Garfield paradigm think that they are not capable of solving the puzzle or they are not working hard enough. They reject, often violently, someone who suggests that the puzzle is not a Garfield puzzle, but another one.

One scientist will finally find that all the pieces, including the odd piece, match better together by supposing that the puzzle is actually the "Snoopy puzzle," not the Garfield puzzle. After some confusion, everyone begins to believe that he is solving the Snoopy puzzle.

It is important to note that scientific research consists of three steps. In the first step, both observations and analyses of a particular phenomenon should be conducted. In the second step, researchers are supposed to synthesize new and earlier observations and then formulate a new interpretation and scheme. The new scheme of a particular phenomenon must be proven quantitatively in the third step.

First Step	Second Step	Third Step
Observation, Analysis, Interpretation of a particular phenomenon	Synthesis by choosing a set of observations (new and earlier) and formulating a new theory or scheme	Quantitative examination of the new theory conceived in the second step
Observed Facts X, A, K, ?, α, Ω, β, P, τ, M, D, W, Y, C, π, O, ?, B	Prevailing Paradigm K → Ω →C, →Y→ M →Phenomenon New Paradigms (1) A→ β → P → α → O →Phenomenon (2) M, →Y→C→ Ω → K →Phenomenon (3) α → X →? → π → X →Phenomenon	

The chart above illustrates the three steps. In the second step, individual researchers choose a set of observed facts from a large number of observations and propose their own sequence in explaining the cause–effect relationship for a particular observed phenomenon. The chosen set may differ considerably depending on researchers, and may also differ from the set that corresponds to the prevailing paradigm. In some cases, a successful new paradigm can often include one or more old paradigms with a new and higher order of interpretation. In another case, the chosen set may be identical to the set of the prevailing paradigm, but the sequencing may be reversed; see Case (2) in the Second Step. A situation similar to this happened during the development of theories of stellar evolution. A red giant was a newly born star (younger than the Sun) in the old paradigm, but is an old star (older than the Sun) in the latest paradigm.

It is important to know the definition of creation in science in this context, because the second step is an act of creation in science. Creation in science consists of perceiving a new thought pattern on the basis of *already available data and theories*. That is to say, creation in science is synthesis by combining, relating, and integrating data that are often seemingly unrelated to each other. Thus, the second step is indeed this act of creation.

Perhaps Charles Darwin, a 19th century British scientist, is the most creative scientist in history. He had an amazing synthesizing power. He found that there are three kinds of coral reefs – the first one is, the reef surrounding the shore of an island. The second one is an atoll surrounding an island in the center, and the third one is simply a circular atoll. This observation was good enough. However, he went further. He inferred that these different types are related. By hypothesizing increasing of sea level, he suggested that the different types of reefs represent different states of the atoll formation. This example shows how great his synthesizing power was. In fact, his hypothesis has recently been confirmed by drilling at the center of an atoll. In another example, he observed:

> What can be more curious than the fact that the hand of a man, formed for grasping, that of a mole for digging, the leg of a horse, the paddle of the porpoise, and the wing of the bat should all be constructed on the same pattern?

Charles Darwin was one of the first scientists to conceive the concept of the evolution of life and developed the hypothesis. We are surrounded by many living creatures: single-cell organisms, starfish, squids, horses, dogs, cats, crabs, scorpions, mushrooms, flower-bearing plants, ferns, algae, etc. They do not seem to be related at all. With his great synthesizing power, however, Darwin inferred that all these creatures have evolved from a single-cell organism and have branched out into different species.

Now, returning back to the three steps, there are very few researchers who can engage in all three steps satisfactorily. Most of them can engage in one of the three steps. There is nothing wrong with that; most researchers are supposed to be good at one of the steps. The problem is that they tend to spend all their effort on only one of the steps and do not even realize that they are doing so. An excellent researcher who is good at the first step may be intentionally or unintentionally considering the second and third steps and thus can design a crucial observation to advance his/her discipline. An excellent researcher in the first or third step may suggest a new scheme in the second step or a new observation to discover an undiscovered element. Indeed, there are many set elements that have not yet been discovered; see the ? sign in the table. An excellent observation specialist may be able to identify one of them by carefully designing the observation on the basis of an intentional or unintentional consideration in the second step; see Case (3) in the Second Step. Such a case may be called a *discovery*.

As described in detail earlier, the second step is often misunderstood to be *unscientific*. In order to comprehend this statement, let us take the case of the extinction of dinosaurs. There are a large number of observed facts on this particular phenomenon. It is impossible to propose a theory that can explain all the observed facts. One has to choose a set of observed facts that are considered to be essential. Some researchers are confident that the cause of the extinction is internal (e.g., climate change), while some others believe it is external (e.g., an impact by asteroids) and set up a scheme to explain the cause of the extinction in either case. Actually, there are many observations to support both. Thus, in the history of this particular discipline, the paradigm shifts between the two (namely, internal or external causes) from time to time. The set depends greatly on individual researchers, so that *this choice is a very subjective process* and becomes also controversial.

It is for this reason that many researchers are uncomfortable with the second step and tend to avoid it by saying that it is not a scientific process. Thus, considerable courage is needed to pursue the second step. This is also the step to confront an established paradigm, if the new interpretation is radically different. After all, it is important to realize that in science the first and the third steps are designated to serve the second step. The second step is the one that leads to a great breakthrough.

Unfortunately, some of those who are mainly interested in the first step, or are not comfortable with pursuing the second step, keep making only their observations by saying that observations are most crucial. Yet, some others keep a new (important) observation under the rug, because it does not conform to the prevailing paradigm. Some of those who are mainly interested in the third step keep improving a particular theory that is a synthesis by someone else, by criticizing that the second step is not science. It is in such a way that some waste considerable amounts of supercomputer time, but believe they are doing good science.

There is no question that the third step distinguishes science from science fiction, so that it is a vital scientific process. Unfortunately, however, many of those who are mainly interested in the third step tend to insist that the third step is the only important aspect of science. If the second step is not correct, the third step will not be very useful (just an exercise), although the effort may not be totally wasted. There is another serious problem in the third step. For example, with proper initial and boundary conditions, one can solve, say, a set of MHD equations and show that one can reproduce an observed phenomenon. However, such a process does not provide any physical insight into the observed phenomenon. One has simply demonstrated that a particular phenomenon can be described by a set of MHD equations, so that the physical processes associated with the phenomenon have not been further elucidated in any concrete way.

Even if a new paradigm can ultimately become established, human nature dictates that the resulting revolution will soon turn into a new orthodoxy, with its unavoidable symptoms of one-sidedness, specialization, loss of contact with other provinces of knowledge, and ultimate estrangement from reality (Koestler). Since the new paradigms can never reach truth, a new crisis will eventually arise again, leading to a new revolution, namely a new synthesis, and the cycle starts all over again. History repeats itself. Scientific communities are no different from other establishments (except that scientific knowledge is cumulative, unlike the political one). After all, human beings called scientists created science, and so the history of science is nothing but a human drama. Nevertheless, unlike politics, perhaps, science makes progress.

Synthesizing

Since the Second Edition emphasizes synthesis, let me make one statement here.

I am neither a theorist nor experimental observer, although I deal with both theories and observations at times. When people press me about what I am, I have to say that I am a sort of "synthesizer". In fact, looking back over my publications listing, the ones most often quoted are those in which I tried to integrate and/or synthesize various aspects of a natural phenomenon under a simple concept, rather than the topical papers.

Before 1964, polar upper atmospheric phenomena, such as the aurora, magnetic disturbances, ionospheric disturbances, and many other features were studied almost independently by a great variety of instruments. Even auroral phenomena alone were observed almost independently by several spectroscopic instruments, all-sky cameras, radars, riometers, ionospheric sounders, magnetometers, VLF detectors, etc.

On the basis of a study of all-sky camera records, I thought that I could provide at least the time-frame of reference in studying all these observations. My 1964 paper on auroral substorms was designed for this particular purpose; it has become one of the most quoted papers (I was named one of the 1000 most-cited contemporary scientists in 1981) and is still being quoted these days (even 45 years after its publication) in studies on magnetospheric phenomena. This fact indicates that synthesis, the second step in scientific methodology mentioned earlier, is very important. I am suggesting in this book that we should integrate and synthesize various observations in solar, interplanetary, and magnetospheric physics.

Synthesizing natural phenomena is a very complex process. This is because it is not possible to conduct an experiment on a natural phenomenon by confining it in a chamber, in which one can maintain all the variables constant except one.

In studying a natural phenomenon, some researchers challenge by trying to grasp the whole phenomenon involved all at once, and obviously become discouraged or get into a maze. Some other researchers try to classify all possible processes and give up; even if they succeed in classifying them, it is not possible for them to obtain any hint of the chain of processes involved.

On the other hand, if one observes a natural phenomenon, one finds that many tend to repeat themselves; some of the examples we learned in this book are auroral substorms and solar flares. Further, when one carefully observes a repeating natural phenomenon, they are not identical. I used to warn my colleagues by saying "no two substorms are alike". Some researchers try to classify them, but I am afraid that such a methodology leads to a maze. One way to study it is to find a set of common features in it. However, some researchers become too eager to include as many features as possible in the common features. This is a mistake as explained below.

The most important thing in this process is to *keep the number of common features to a minimum* and try to establish the morphology of the phenomenon on the basis of those selected features. If your colleagues find the set of selected features in their own observations, they will trust and support the morphology. If you include some features that are not always present, so that your colleagues cannot find them in their observations, your morphology will be less credible to them. Without the strong support from your colleagues, your morphology paper will get lost. Once the basic morphology is acceptable to your colleagues, they will consider that other features can be understood to be auxiliary features without doubting it. If the selection is appropriate, the set will serve as the basis for studying the physical processes involved.

I presented, as a new example, an anti-parallel flow of photospheric gas under a magnetic archade. This case seems to have all the essential ingredients for solar flares (Section 7.5). The magnetic flux tube case may be a complex combination of such simple situations. It is puzzling why solar physicists try to choose solar flares in a very complex situation. It is my belief that a natural phenomenon consists of a few basic processes. It is important to try to find the simplest one among them.

At this stage of writing this book, we are still far from selecting a minimum number of common features in space weather research. I hope that this book will help young researchers to accomplish this task in the future. After all this, if I have to conclude this book by choosing one word, I will choose the word open-mindedness. Each of us scientists wishes to make some small progress in our discipline for a better understanding of Nature. This is all we can hope for. Open-mindedness is the only way to make one small step eventually in the right direction. This is what I have learned in my research life, although it is so difficult to be open-minded.

Sources of Figures

R.C. Carrington's sketch of a sunspot group.	Carrington, R.C., *Mon. Not. Roy. Astronom. Soc.*, **20**, 1860
Lord Kelvin	Terr. Magn. Geoelect.
E.W. Maunder	Courtesy of R.H. Eather
Figure I	Akasofu, S.-I., *Magnetospheric Substorms*, ed. by J.R. Kan, et al., AGU Monograph, 64, p. 3, Washington D.C. 1991
Figure II	Akasofu, S.-I. and S. Chapman, *Solar-Terrestrial Physics*, p. 542, Oxford University Press, Oxford, 1972
K. Birkeland	Courtesy of University of Oslo
Sydney Chapman	Courtesy of Geophysical Institute
Hannes Alfvén	Courtesy of H. Alfvén
James Van Allen, Carl McIlwain, and George Ludwig	Courtesy of J.A. Van Allen
Dungey's open magnetosphere	Dungey, J.W., *Phys. Rev. Lett.*, **6**, 47, 1961
The spiral structure of the interplanetary magnetic field lines	Hakamada, K. and S.-I. Akasofu, *Space Sci. Rev.*, **31**, 3, 1982
Heliospheric current sheet	Akasofu, S.-I. and C.D. Fry, *J. Geophys. Res.*, **91**, 13,679, 1986
Solar activity and coronal mass ejection	SOHO/(Instrument) consortium. SOHO is a project of international cooperation between ESA and NASA
Propagating shockwave during May 1977	Saito, Takao, W. Sun, C.S. Deehr, and S.-I. Akasofu, *JGR*, Accepted, 2006
Interplanetary disturbances	Akasofu, S.-I., K. Hakamada, and C.D. Fry, *Planet. Space. Sci.*, **31**, 1435, 1983

Plates 1–4	Photographed by Jan Curtis
Plate 5	Photographed by J. Yokota

Chapter 1

Figure 1.1a	Courtesy of University of Oslo
Figure 1.1b	Courtesy of N. Fukushima
Figure 1.2a	Courtesy of University of Oslo
Figure 1.2b	McCracken, K.G., U.R. Rao, and M.A. Shea, *M.I.T. Tech. Rep.*, **77**, 2, 1962
Figure 1.3a, b	Chapman, S. and J. Bartels, *Geomagnetism*, Oxford University Press, 1940
Figure 1.3c, d	Chapman, S., and V.C.A. Ferraro, *Terr. Magn.*, **36**, 77, 1931
Figure 1.4	Roederer, M., see also S.-I. Akasofu, and D.N. Covey, *Planet. Space Sci.*, **28**, 757, 1980
Figure 1.5	Akasofu, S.-I.
Figure 1.6	Störmer, C., *The Polar Aurora,* Oxford University Press, Oxford, 1955 (quoted by Van Allen, J.A., in Proceedings of the General Assembly, Berkeley, 1961, p. 99, Academic Press, London, 1962)
Figure 1.7	Akasofu, S.-I., J.C. Cain, and S. Chapman, *J. Geophys. Res.*, **66**, 4013, 1961
Figure 1.8a	Akasofu, S.-I. and S. Chapman, *J. Geophys. Res.*, **68**, 125, 1963
Figure 1.8b	Akasofu, S.-I. and S. Chapman, *Solar-Terrestrial Physics*, Oxford University Press, Chapter 8, 1972
Figure 1.9	Akasofu, S.-I., Space Sci. Rev., **28**, 121, 1981
Figure 1.10	Akasofu, S.-I., E.W. Hones, Jr., S.J. Bame, J.R. Asbridge, and Lui, A.T.Y., *J. Geophys. Res.*, **78**, 7257, 1973
Figure 1.11	Akasofu, S.-I., *Planet. Space Sci.*, **12**, 801, 1964 (see also S.-I. Akasofu and S. Chapman, *Solar-Terrestrial Physics*, Chapter 8, Oxford University Press, Oxford, 1972)
Figure 1.12	Akasofu, S.-I., *Physics of Magnetospheric Substorms*, Chapter 5, D. Reidel Pub. Co., Holland, 1977
Figure 1.13a,b	Akasofu, S.-I., *EOS*, **66**, 465, 1985
Figure 1.13c	Akasofu, S-I., *EOS*, **70**, 529, 1989
Figure 1.13d	Akasofu, S.-I., *Dynamics of the Magnetosphere*, p. 447, D. Reidel Pub. Co., Dordrecht, Holland, 1980

Figure 1.14	Akasofu, S.-I., *Space Sci. Rev.*, **28**, 121, 1981
Figure 1.15	Sun, W., W.-Y. Xu, and S.-I. Akasofu, *J. Geophys. Res.*, **103**, 11,695, 1998
Figure 1.16	Vampola, A.L., *J. Geophys. Res.*, **76**, 36, 1971
Figure 1.17	Akasofu, S.-I., W.C. Lin, and J.A. Van Allen, *J. Geophys. Res.*, **68**, 5327, 1963
Chapman and Akasofu	Akasofu, S.-I., Geophysical Institute, University of Alaska, 1961
Alfvén and Akasofu	Akasofu, S.-I., Geophysical Institute, University of Alaska, 1974
Chapman and Akasofu	Akasofu, S.-I., Geophysical Institute, University of Alaska, 1964
Alfvén and Akasofu	Akasofu, S.-I., Geophysical Institute, University of Alaska, 1974

Chapter 2

Figure 2.1	Courtesy of Yale University
Figure 2.2	Loomis, E., *Amer. J. Sci. and Arts*, **30**, 89, 1860
Figure 2.3	Akasofu, S.-I., *Aurora and Airglow*, ed. by B.M. McCormack, p. 267, Reinhold Pub. Co., New York, 1967
Figure 2.4	Carlheim-Gyllensköld, N., *Observations faites an Cap Thordsen*, Spitzberg, vol. II, Stockholm, 1882
Figure 2.5	Courtesy of University Tromsø
Figure 2.6a,b	Courtesy of Geophysical Institute, University of Alaska
Figure 2.7a	Akasofu, S.-I., *Space Sci. Rev.,* **16**, 617, 1974
Figure 2.7b, 2.8a	Akasofu, S.-I, 1968
Figure 2.8b	Snyder, A.L., Ph.D. Thesis, University of Alaska, 1972
Figure 2.8c	Snyder, A.L. and S.-I. Akasofu, *J. Geophys. Res.*, **77**, 3419, 1972
Figure 2.9a	Courtesy of G. Gassmann
Figure 2.9b	Buchau, J., J.A. Whalen, and S.-I. Akasofu, *J. Geophys. Res.*, **75**, 7147, 1970
Figure 2.9c	Akasofu, S.-I., *Department of Physics and Astronomy Report*, 1967
Figure 2.10	Kamide, Y. and S.-I. Akasofu, *J. Geophys. Res.*, **81**, 3999, 1976.
Figure 2.11	Iijima, T. and T.A. Potemra, *J. Geophys. Res.* **83**, 599, 1978
Figure 2.12a,b	Courtesy of C. Anger

Figure 2.13	Courtesy of University of Alaska
Figure 2.14, 2.15	Akasofu, S.-I.
Figure 2.16a	Akasofu, S.-I., *Space Sci. Rev,* **16**, 617, 1974
Figure 2.16b	Akasofu, S.-I., *J. Geophys. Res.*, **68**, 1667, 1963
Figure 2.17	Akasofu, S.-I., *Planet Space Sci.,* **12**, 273, 1964
Figure 2.18	Ames Research Center, NASA (see also Akasofu, S.-I., *Planet. Space Sci.,* 16, 365, 1968)
Figure 2.19	Courtesy of Geophysical Institute, University of Alaska
Figure 2.20	Anger, C.D., A.T.Y. Lui, and S.-I. Akasofu, *J. Geophys. Res.*, **78**, 3020, 1973
Figure 2.21	Akasofu, S.-I., *Planet. Space Sci.*, **24**, 1349, 1975
Figure 2.22a	Craven, J.D., Y. Kamide, L.A. Frank, S.-I. Akasofu, and M. Sugiura, Magnetospheric Currents, ed. by T.A. Potemra, *Geophysical Monograph*, 28, AGU, Washington D.C., 1983
Figure 2.22b	Courtesy of G. Parks
Figure 2.23	Akasofu, S.-I., *Space Sci. Rev.*, **19**, 169, 1976
Figure 2.24a,b	Akasofu, S.-I. and S. Chapman,*Planet. Space Sci.*, **24**, 785, 1962
Figure 2.25	Wagner, J.S., R.D. Sydora, T. Tajima, T. Hallinan, and S.-I. Akasofu, *J. Geophys. Res.*, **88**, 8013, 1983
Figure 2.26	Kimball, C. Chaston, J. McFadden, G. Delory, M. Temerin, and C.W. Carlson, *Geophys. Res. Lett.*, **25**, 2073, 1998
Figure 2.27	Voots, G.R., D.A. Gurnett, and S.-I. Akasofu, *J. Geophys. Res.*, **82**, 2259, 1977
Figure 2.28	Akasofu, S.-I., Upper Atmosphere Research in Antarctica, Antarctic Research Series, vol. 29, ed. by L.J. Lanzerotti and C.G. Park, *AGU*, Washington, D.C., 1978
Figure 2.29	Bhardwaj, A. and G.R. Gladstone, *Rev. Geophys.* 38, 295, 2000, Hubble Space Telescope Project, together with D. Michell, Earth-Sun System Exploration: Energy Transfer, January 16-20, 2006, Kona, Hawaii
Figure 2.30	Courtesy of D. Mitchell, 2006
The aurora and the comet's tail	Geophysical Institute, University of Alaska Fairbanks
Cheng Meng and S.-I. Akasofu	Geophysical Institute, University of Alaska Fairbanks, 1968
Charlie Kennel and Vytenis Vasyliunas	Kiruna Geophysical Observatory

IUGG General Assembly meeting (1975) attendees	Akasofu, S.-I.
AGU Chapman Conference Progrm Cover	Akasofu, S.-I.
Los Alamos Conference	Akasofu, S.-I.
People's Republic of China lecture	Akasofu, S.-I.
Japanese Academy Award	Akasofu, S.-I.

Chapter 3

Figure 3.1, 3.2	Chapman, S., *Terr. Magn.*, **40**, 349, 1935
Figure 3.3a	Axford, W.I. and C.O. Hines, *Can., J. Phy.*, **39**, 1433, 1961
Figure 3.3b	Courtesy of R.A. Greenwald, 2001
Figure 3.4	Kawasaki, K. and S.-I. Akasofu, *Planet. Space Sci.*, **19**, 543, 1971
Figure 3.5	Akasofu, S.-I. and C.-I. Meng, *J. Geophys. Res.*, **74**, 293, 1969
Figure 3.6a,b	Akasofu, S.-I. and C.-I. Meng, *J. Atm. Terr. Phys.*, **29**, 965, 1015, 1967
Figure 3.7	Akasofu, S.-I. and C.-I. Meng, *J. Atm. Terr. Phys.*, **30**, 227, 1968
Figure 3.8	Akasofu, S.-I.
Figures 3.9a–d	Akasofu, S.-I. and Y. Kamide, *J. Geophys. Res.*, **103**, 14,939, 1998
Figure 3.10	Ahn, B.-H., Y. Kamide, and S.-I. Akasofu, *J. Geophys. Res.*, **89**, 1613, 1984
Figure 3.11	Akasofu, S.-I. W. sun and B.-H. Ahn, 2000
Figure 3.12	Ahn, B.-H., S.-I. Akasofu, and Y. Kamide, *J. Geophys. Res.*, **88**, 6275, 1983

Chapter 4

Figure 4.1	Boström, R., *J. Geophys. Res.*, **69**, 4983, 1964
Figure 4.2	Akasofu, S.-I., *Geoophys. J. Int.*, **109**, 191, 1992
Figure 4.3	Akasofu, S.-I.
Figure 4.4a	Waters, C.L., B.J. Anderson, and K. Liou, *Geophys. Res. Lett.*, **28**, 2165, 2001
Figure 4.4b	Akasofu, S.-I.

Figure 4.5	Akasofu, S.-I., *Solar Terr. Phys.*, Part III, E.R. Dyer and J.R. Roederer, ed., D. Reidel Publishing Co., 1972
Figure 4.6	Brekke, A., J.R. Doupnik, and P.M. Banks, *J. Geophys. Res.*, **79**, 3773, 1974
Figure 4.7	Akasofu, S.-I., *J. Geophys. Res.*, **108**, 8006, 2003
Figure 4.8a	Sun, W., W.-Y., Xu, and S.-I. Akasofu, *J. Geophys. Res.*, **103**, 11695, 1998
Figure 4.8b	Sun, W., W.-Y. Xu, and S.-I. Akasofu, Geospace Mass and Energy Flow, *Geophysical Monograph*, 104, *AGU*, 1998
Figure 4.8c	Sun, W., W.-Y. Xu, and S.-I. Akasofu, *J. Geophys. Res.*, **103**, 11695, 1998
Figure 4.9	Akasofu, S.-I., *J. Geophys. Res.*, **108**, 8006, 2003
Figure 4.10a	Akasofu, S.-I. and W. Sun, *EOS*, 81-361, 2000
Figure 4.10b	Akasofu, S.-I.
Figure 4.11	DeForrest, S.E. and C.E. McIlwain, *J. Geophys. Res.*, **76**, 3587, 1971
Figure 4.12	Caan, M.N., R.L. McPherron, and C.T. Russell, *Geophys. Res.*, **80**, 191, 1975
Figure 4.13a	Akasofu, S.-I., *Nature*, **284**, 248, 1980
Figure 4.13b	Akasofu, S.-I., C.-I. Meng, and K. Makita, *Planet. Space Sci.*, **40**, 1513, 1992
Figure 4.14	Frank, L.A., J.B. Sigwarth, and W.R. Patterson, *Substorms-4*, ed. by Kokubun S. and Kamide, Y., Terra Scientific Pub. Co., Tokyo, 1998
Figure 4.15	Akasofu, S.-I.
Figure 4.16	Courtesy of L. Frank
Figure 4.17	Akasofu, S.-I.
Figure 4.18	Lyons, L.R., R.L. McPherron, and E. Zesta, *J. Geophys. Res.*, **106**, 349, 2001
Figure 4.19	Akasofu, S.-I., *Space Sci. Rev.*, **113**, 1, 2004
Figure 4.20	Akasofu, S.-I.
Figure 4.21	Akasofu, S.-I., *Space Sci. Rev.*, **113**, 1, 2004
Figure 4.22	Akasofu, S.-I.
Figure 4.23a	Akasofu, S.-I., *Ann. Geophys.*, **26**, 443, 1970
Figure 4.23b	Geomagnetic Data Center, Kyoto University
Figure 4.24	Fok, M.C., Earth-Sun System Exploration: Energy Transfer, January 16-20, 2006, Kona, Hawaii
Figure 4.25	Nosé, M., S. Taguchi, K. Hosokawa, S.P. Christon, R.W. McEntire, T.E. Moore, and M.R. Collier, *J. Geophys, Res.*, **110**, A09524, 2005
Figure 4.26	Mitchell, D.G., Earth-Sun System Exploration: Energy Transfer, January 16-20, 2006, Kona, Hawaii

Figure 4.27	Akasofu, S.-I. and S. Chapman, *Planet. Space Sci.*, **12**, 607, 1964
Figure 4.28	Akasofu, S.-I., *EOS*, **73**, 209, 1992
Figure 4.29	Akasofu, S.-I., *Scientific American*, p. 90, May 1989
Katherine and Sydney Chapman, and the Akasofu's	Akasofu, S.-I.
AGU Chapman Conference	Geophysical Institute, University of Alaska
International Conference on Substorms (ICS-2)	Geophysical Institute, University of Alaska
International Conference on Substorms (ICS-2)	Akasofu, S.-I.
Chapman Medal	Akasofu, S.-I.
Fleming Medal	Akasofu, S.-I.
Opening of the Van Allen Hall	University of Alaska
All remaining graphics in Chapter 4	Akasofu, S.-I.

Chapter 5

Figure 5.1	Courtesy of K. Hakamada, 2005
Figure 5.2	Akasofu, S.-I.
Figure 5.3	Courtesy of K. Hakamada
Figure 5.4	Saito, Takao, Y. Kozuka, T. Oki, and S.-I. Akasofu, *J. Geophys. Res.*, **96**, 3807, 1991
Figure 5.5, 5.6	Saito, Takao, T. Oki, S.-I. Akasofu, and C. Olmsted, *J. Geophys. Res.*, **94**, 5453, 1989
Figure 5.7, 5.8a	Akasofu, S.-I., L.-H. Lee, and T. Saito, *Planet. Space Sci.*, **39**, 1259, 1991
Figure 5.8b	Van den Henvel, E.P.J., *Science*, **312**, 539, 2006
Figure 5.9	Akasofu, S.-I. and Saito, Takao, *Planet. Space Sci.*, 38, 1203, 1990
Figure 5.10a,b	Saito, Takao, S.-I. Akasofu, Y. Kozuka, T. Takahashi, and S. Numazawa, *J. Geophys., Res.*, **98**, 5639, 1993
Figure 5.11a,b	Akasofu, S.-I. and C.D. Fry, *J. Geophys. Res.*, **91**, 13,679, 1986

Chapter 6

Chapter 6 - Section 6.1	All of the figures in Section 6.1 can be found in: Akasofu, S.-I. and C.F. Fry, *Planet Space Sci.*, **34**, 77, 1986 Akasofu, S.-I., Space Weather, *Geophysical Monograph*, 125, AGU Workshop, 2001 Hakamada, K. and S.-I. Akasofu, *Space Sci. Rev.*, **31**, 3, 1982
Chapter 6 - Sections 6.2 through 6.9	All of the figures in these sections and be found in: Akasofu, S.-I., H. Watanabe, and Takao Saito, *Space Sci. Rev.*, **120**, 27, 2005 Akasofu, S.-I. and L.-H. Lee, *Planet Space Sci.*, **37**, 73, 1989 and **38**, 575, 1990 Akasofu, S.-I., K. Hakamada, and C. Fry, *Planet Space Sci.*, **31**, 1435, 1983

Chapter 7

Figure 7.1	Courtesy of Royal Swedish Academy of Sciences
Figure 7.2	Akasofu, S.-I., *Planet. Space Sci.*, **32**, 1469, 1984
Figure 7.3a	Courtesy of Big Bear Solar Observatory
Figure 7.3b	Courtesy of Kitt Peak Solar Observatory
Figure 7.3c	Courtesy of K. Hakamada
Figure 7.3d	Courtesy of Kitt Peak Solar Observatory
Figure 7.4	Courtesy of Big Bear Solar Observatory
Figure 7.5a	Akasofu, S.-I., *Planet. Space Sci.*, **33**, 275, 1985
Figure 7.5b	Left: Akasofu, S.-I., *Planet. Space Sci.*, **33**, 275, 1985; Right: M.K. Georgulis, *Earth Transfer*, January 16-20, 2006, Kona, Hawaii
Figure 7.6, 7.7	Akasofu, S.-I., *Planet. Space Sci.*, **32**, 1469, 1984
Figures 7.8a	Akasofu, S.-I., *Planet. Space Sci.*, **33**, 275, 1985
Figure 7.8b	http://sohowww.nascom.nasa.gov/
Figure 7.9	Solar flare photo: Big Bear Solar Observatory; Aurora photograph: Kanazawa Astronomical Society
Figure 7.10	Akasofu, S.-I.
Figure 7.11	Wang, H., M.W. Ewell, Jr., and H. Zirin, *Astroph. J.*, **424**, 436, 1994
Figure 7.12a	Courtesy of C.S. Choe and L.C. Lee, 1995
Figure 7.12b	Kusano, K., Joint International Workshop on Space Weather, April 4-6, 2005, Tokyo

Figure 8.1	NASA/ESA SOHO Project
Figure 8.2a	Kyoto University Solar Observatory

Chapter 8

Figure 8.2b	Shibata, K. and M. Oyama, *Photographic Atlas of the Sun* (in Japanese), Shokabo, 2004
Figure 8.2c	Yurchyshyn, V., Earth-Sun System Exploration: Energy Transfer, January 16-20, 2006, Kona, Hawaii
Figure 8.3	Bothmer, V. and R. Schwenn, *Space Sci. Rev.*, **70**, 215, 1994
Figure 8.4a	Moore, R.L., A.C. Sterling, and G.C. Marshall, Space Flight Center Report, 2005, M.K. Georgoulis, Huntsville, Alabama
Figure 8.4b	Qiu, J., H. Wang, C.Z. Cheng, and D.E. Gary, *Ap. J.*, **604**, 900, 2004
Figure 8.5	Shibata, K., and M. Oyama, *Photographic Atlas of the Sun* (in Japanese), Shokabo, 2004
Figure 8.6a,b and 8.7a,b	Yurchyshyn, V., H. Wang, P.R. Goode, and Y. Dang, *Ap.J.*, **563**, 381, 2001
Figure 8.8a,b	Saito, Takao
Figure 8.9	NASA/ESA SOHO Project
Figure 8.10a	Chen, J. and J. Krall, *J. Geophys. Res.*, **108**, A11, 1410, 2003
Figure 8.10b	Riley, P., J.A. Linker, Z. Mikis, and D. Odstricil, *IEEE Trans. Plasma Sci.*, **32**, 1415, 2004
Figure 8.11	Webb, D.F., R.P. Lepping, L.F. Burlaga, S.E. DeForrest, D.E. Larson, S.F. Martin, S.P. Plunkett, and D.M. Rust, *J. Geophys. Res.*, **105**, 27251, 2000
Figure 8.12	Smith Z., S. Watari, M. Dryer, P.K. Manohara, and P.S. McIntosh, *Solar Physics*, **171**, 177, 1997
Figure 8.13	Lepping, R.P., J.A. Jones, and L.F. Burlaga, *J. Geophys. Res.*, **95**, 11957, 1990
Figure 8.14	Akasofu, S.-I.
Figure 8.15a,b	Akasofu, S.-I. and C.F. Fry, *Planet Space Sci.*, **34**, 77, 1986
Figure 8.15c	Akasofu, S.-I., *EOS*, **77**, 225, 1996
Figure 8.16a	Courtesy of D. Rust, Applied Physics Laboratory, John Hopkins University, 2001
Figure 8.16b	Left: Big Bear Observatory; Middle: SOHO image, NASA/ESA SOHO project; Right: Yohkoh project
Figure 8.17a,b	Sun, W., M. Dryer, C.D. Fry, C.S. Deehr, Z. Smith, S.-I. Akasofu, M.D. Kartalev, and K.G. Grigorov, *Geophys. Res. Lett.*, **29**, No. 8, 2002

Figure 8.17c	Courtesy of Geomagnetism Data Center, Kyoto University
Figure 8.18a	NASA/ESA SOHO project
Figure 8.18b	Courtesy of R.B. Decker, Applied Physics Laboratory, John Hopkins University
Figure 8.19a	Saito, Takao
Figure 8.19b	Saito, Takao & W. Sun
Figure 8.19c, d, e, f, g, and h	Saito, Takao, W. Sun, C.S. Deehr, and S.-I. Akasofu
Figure 8.20	Akasofu, S.-I. and C.D. Fry
Figure 8.21	Jackson, B.V., A. Buffington, P.R. Hick, and Y. Yu, Earth-Sun System Exploration: Energy Transfer, January 16-20, 2006, Kona, Hawaii
Figure 8.22	Akasofu, S.-I. and L.-H. Lee, *Planet Space Sci.,* **38** 575, 1990
Figure 8.23a	Tokmaru, M., M. Kojima, K. Fujiki, M. Yamashita, and A. Yokobe, *J. Geophys. Res.*, **108**, SSH1, 2003
Figure 8.23b	Jackson, B.V., A. Buffington, P.R. Hick, and Y. Yu, Earth-Sun System Exploration: Energy Transfer, January 16-20, 2006, Kona, Hawaii
Figure 8.24	Saito, Takao and S.-I. Akasofu
Figure 8.25	Akasofu, S.-I., *Space Sci. Rev.*, **28**, 121, 1981
Figure 8.26	Courtesy of S. Maurits and B. Watkins, See also, Maurtis, S.A., Ph.D. Thesis, University of Alaska, 1996
Figure 8.27	Akasofu, S.-I. and J.D. Aspnes, *Nature*, **295**, 136, 1982
Figure 8.28	Wescott, E., W. Sackinger, and S.-I. Akasofu (unpublished report)
Figure 8.29, 8.30, 8.31	Akasofu, S.-I.

Chapter 9

Figures 9.1, 9.2a,b	Akasofu, S.-I. and D.N. Covey, *Planet Space Sci.,* **29**, 313, 1981
Figure 9.3	Akasofu, S.-I., W. Fillius, W. Sun, C. Fry, and M.J. Dryer, *J. Geophys. Res.*, **90**, 8193,1985
Figure 9.4	Courtesy of W. Sun, 2005
Figure 9.5	Yosjhida, S., N. Ogita, S.-I. Akasofu, *J. Geophys. Res.*, **76**, 7801, 1971
Figure 9.6	SPATIUM, International Space Institute, 2003
Figure 9.7	Yoshita, S., N. Ogita, S.-I. Akasofu, and L.J. Gleeson, *J. Geophys. Res.*, **78**, 6409, 1973

Further Reading

Akasofu, S.-I., B. Fogle and B. Haurwitz (Ed.), *Sydney Chapman*, Eighty, University Corporation for Atmospheric Research, Boulder, Colorado, 1968.
Akasofu, S.-I., Evolution of Ideas in Solar-terrestrial Physics, *Geophys. J. R. Astro. Soc.*, 74, 257, 1983.
Akasofu, S.-I., *Polar and Magnetospheric Substorms*, D. Reidel Publishing Co., Dordrecht-Holland, 1968.
Akasofu, S.-I., *Physics of Magnetospheric Substorms*, D. Reidel Publishing Co., Dordrecht-Holland, 1977.
Akasofu, S.-I., S. Chapman, and A.B. Meinel, *The Aurora, Handbuch der Physik*, Vol. XLIX/1, Springer-Verlag, Berlin, 1966.
Akasofu, S.-I. and S. Chapman, *Solar-Terrestrial Physics*, Oxford Univ. Press., 1972.
Alfvén H., *Cosmical Electrodynamics*, Oxford University Press, 1950.
Alfvén H., *Cosmic Plasma*, D. Reidel Publishing Co., Dordrecht-Holland, 1981.
American Geophysical Union, Pioneers of Space Physics 2, Reprinted from *Journal of Geophysical Research*, 101, 5, 1996.
Armstrong, E.B. and A. Dalgarno, *The Airglow and the Aurorae*, Pergmon Press, London, 1956.
Aschwanden, M.J., *Particle Acceleration and Kinematics in Solar Flares*, Dordrecht, Boston, Kluwer Academic Publishers, 227p., 2002.
Aschwanden, M.J., *Physics of the Solar Corona: An Introduction*, Dordrecht, Boston, Springer, xxi, 842p., 2004.
Banks, P.M. and G. Kockarts, *Aeronomy*, Academic Press, New York, 1973.
Benz, A.O., *Plasma Astrophysics: Kinetic Processes in Solar and Stellar Coronae*, 2nd ed., Dordrecht, Boston, Kluwer Academic Publishers, xvii, 301p., 2002.
Beveridge, W.I.B., *Seed of Discovery*, Norton, New York, 1980.
Beveridge, W.I.B., *The Art of Scientific Investigation*, Vintage Press, Random House, New York, 1957.
Biermann, L., Kometenschureife und Solare Korpuskularstrahlung, *Zeitschriffefur Astrophysik*, 29, 274, 1951.
Birkeland, K., *On the Cause of Magnetic Storms and the Origin of Terrestrial Magnetism*, Norwegian Aurora Polaris Expedition, 1902–1903.
Bryant, D., *Electron Acceleration in the Aurora and Beyond*, Institute of Physics Publishing, Bristol, 1999.
Burgers, D.J.M. and L.H.C. van de Halst (Ed.), *Gas Dynamics of Cosmic Clouds*, North Hallard Publishing Co., Amsterdam, 1955.
Burlaga, L.F., *Interplanetary Magnetohydrodynamics*, Oxford University Press, New York, 1995.
Carlowicz, M.J. and R.E. Lopez, *Storms from the Sun: The Emergency Science of Space Weather*, Washington D.C., Joseph Henry Press, ix, 234p., 2002.

Chamberlain, J.W., *Physics of the Aurora and Airglow*, Academic Press, New York, 1961.

Chapman, S. and J. Bartels, *Geomagnetism*, I, II, Oxford University Press, 1940.

Collier, M.R., T.E. Moore, K. Ogilvie, D. Chornay, J.W. Keller, S. Boardsen, J. Burch, B. El Marji, M.-C. Fok, S.A. Fuselier, A.G. Ghielmetti, B.L. Giles, D.C. Hamilton, B.L. Peko, J.M. Quinn, E.C. Roelot, T.M. Stephen, G.R. Wilson, and P. Wurz, Observations of Neutral Atoms from the Solar Wind, *J. of Geophys. Res.*, 106, A11, pp. 24, 893–24, 906, November 1, 2001.

Cook, A.H. and T.F. Gaskell (Ed.), *The Earth in Space*, Royal Astronomical Society, London, 1968.

Cowling, T.G., *Magneto-hydrodynamics*, Interscience Publishers, Inc., London, 1957.

Dungey, J.W., *Cosmic Electrodynamics*, Cambridge University Press, 1958.

Dyer, E.R. (Ed.), *Solar-Terrestrial Physics/1970*, D. Reidel Publishing Co., Dordrecht-Holland, 1972.

Eather, R.H., *Majestic Lights*, American Geophysical Union, Washington, D.C., 1980.

Egeland, A. and J. Holtet (Ed.), *The Birkeland Symposium on Aurora and Magnetic Storms*, September 18–22, 1967, Centre National de la Recherche Scientifique.

Engvold, O., J.W. Harvey, C.J. Schrijver and N.E. Hurlburt (Eds), *The Physics of the Solar Corona and Transition Region: Proceedings of the Monterey Workshop, held in Monterey, California, August 1999*, Dordrecht, Boston, Kluwer Academic Publishers, 2000–2001.

Foukal, P.V., *Solar Astrophysics*, 2nd rev. ed., Weinheim, Wiley-VCH, xiv, 466p., 2004.

Fuller, V.R. and E.R. Bramhall, *Auroral Research at the University of Alaska*, 1930–1934, University Alaska Publication, 1937.

Gillmor, C.S. and J.R. Spriter (Ed.), *Discovery of the Magnetosphere*, American Geophysical Union, Washington, D.C., 1997.

Golub, L. and J.M. Pasachoff, *The Solar Corona*, Cambridge University Press, 1997.

Gubbins, D., Geomagnetic polarity reversals; A connection with secular variation and core-mantle interaction?, *Rev. Geophys.*, 32, 61, 1994.

Hoyle, F., *Some Recent Researches in Solar Physics*, Cambridge University Press, 1949.

Humboldt, A.v. *Cosmos*: *Physical Description of the Universe*, Vol. 1, London, 1871.

Kallenrode, M.-B., *Space Physics: An Introduction to Plasmas and Particles in the Heliosphere and Magnetosphere*, 2nd updated ed., Berlin, New York, Springer, xiv, 482p., 2004.

Keiper, G.P. (Ed.), *The Sun*, University Chicago Press, 1953.

Kellner, L., *Alexander von Humboldt*, Oxford University Press, Oxford, 1963.

Lord Kelvin, President Address, *Proceedings of the Royal Society of London*, 52, 299, 1892/93.

Kennel, C.F., *Convection and Substorms*, Oxford University Press, 1995.

Koestler, A., *The Act of Creation*, Picadov, 1964.

Kohl, J.L. and S.R. Cramner (Ed.), *Coronal Holes and Solar Wind Acceleration*, Kluwer Academic Publishers, Dordrecht-Holland, 1999.

Kuhn, T.S., *The Structure of Scientific Revolutions*, 2nd ed., University Chicago Press, 1962.

Lang, K.R., *The Sun from Space*, Springer, New York, 2000.

Langley, S.P., *Amer. J. Sci.*, 3, 37, 1, 1809, see Barr, E.S., *Amer. J. Phys.*, 28, 42, 1960.

LeGalley, D.P. and A. Rosen (Ed.), *Space Physics*, John Wiley and Sons, Inc., New York, 1964.

Lindermann, F.A., Note on the Theory of Magnetic Storms, *Phil. Mag.*, 38, 1919.

Lyu, L.-H. (Ed.), *Space Weather Study Using Multipoint Techniques*, Oxford, Elsevier, xiv, 360p., 2002.
Mackin, K.J., Jr. and M. Neugebauer (Ed.), *The Solar Wind*, Permagon Press, 1966.
Massey, H.S.W. and R.L.F. Boyd, *The Upper Atmosphere*, Hutchinson, London, 1958.
Maunder, E.W., *Mon. Not. R. Astr. Soc.*, 65, 2, 1905.
McCormac, B.M., *Aurora and Airglow*, Reinhold Publishing Co., New York, 1967.
Menzel, D.H., *Our Sun*, The Blakiston Co., Philadelphia, 1949.
Merrill, R.T. and P.L. McFadden, Geomagnetic Polarity Transitions, *Rev. Geophys.*, 37, 201, 1999.
Newton, H.W., *The Face of the Sun*, Penguin Books, 1958.
Odishaw, H. (Ed.), *Research in Geophysics*, Vol. 1 and 2, The MIT Press, Cambridge, MA, 1964.
Parker, E.N., *Interplanetary Dynamical Process*, Interscience Publishers, John Wiley and Sons, New York, 1963.
Parker, E.N., C.F. Kennel, and L.J. Lanzerotti, *Solar System Plasma Physics*, Nutt-Holland Publishing Co., Amsterdam, 1979.
Piddington, J.H., *Cosmic Electrodynamics,* Wiley-Interscience Publishers, New York, 1969.
Rees, M.H., *Physics and Chemistry of the Upper Atmosphere*, Cambridge University Press, Cambridge, 1989.
Russell, C.T., E.R. Priest and L.C. Lee (Ed.), *Physics of Magnetic Flux Ropes*, Washington, D.C., American Geophysical Union, 685p., 1989.
Sharma, S.A., *Disturbances in Geospace: The Storm-Substorm Relationship*, Y. Kamide and G.S. Lakhina, (Eds), Washington, D.C., American Geophysical Union, viii, 268p., 2003.
Song, P., H.J. Singer, and G.L. Siscoe (Eds), *Space Weather*, Washington, D.C., American Geophysical Union, 440p., 2001.
Spitzer, L., Jr., *Physics of Fully Ionized Gases*, Interscience Publishers Inc., London, 1956.
Stern, D.P., A brief history of magnetospheric physics during the space age, *Rev. Geophysics*, 34, 1, 1996.
Stewart, C. and J.R. Spreiter (Ed.), *Discovery of the Magnetosphere*, American Geophys. Union, Washington, D.C., 1997.
Störmer, C., *The Polar Aurora*, Oxford University Press, 1955.
Svestka, Z., *Solar Flares*, D. Reidel Publishing Co., Dordrecht-Holland, 1976.
Thomson, J.J. and G.P. Thompson, *Conduction of Electricity Through Gases*, General Publications, Tronto, 1903.
Zirin, H., *Astrophysics of the Sun*, Cambridge University Press, 1988.

Name Index

Index